Samuel Sullivan Cox

Search for Winter Sunbeams in the Riviera, Corsica, Algiers, and

Spain

Samuel Sullivan Cox

Search for Winter Sunbeams in the Riviera, Corsica, Algiers, and Spain

ISBN/EAN: 9783337240844

Printed in Europe, USA, Canada, Australia, Japan

Cover: Foto ©berggeist007 / pixelio.de

More available books at **www.hansebooks.com**

SEARCH

FOR

WINTER SUNBEAMS.

SEARCH

FOR

WINTER SUNBEAMS

IN

THE RIVIERA, CORSICA, ALGIERS, AND SPAIN.

BY

SAMUEL S. COX,

AUTHOR OF "THE BUCKEYE ABROAD;" "EIGHT YEARS IN CONGRESS," ETC.

> " By what means to shun
> The inclement seasons, rain, ice, hail, and snow,
> Which now the sky, with various face begins
> To show us * * * while the winds
> Blow moist and keen, shattering the graceful locks
> Of these fair spreading trees; which bid us seek
> Some better shroud, some better warmth."
>   PARADISE REGAINED.

.

NEW YORK:

D. APPLETON & COMPANY,

1, 3, & 5 BOND STREET.

1880.

TO MY CONSTITUENTS

SIXTH CONGRESSIONAL DISTRICT OF THE CITY OF NEW YORK.

To you, I have the honour and pleasure of dedicating these "Sunbeams" of travel. They were made bright by your confidence, and cheerful by your indulgence ; without which I could not have pursued them, into far and almost untrodden paths—in " search" of the health so needed, and I trust, secured—for the duty which you have devolved upon me.

WESTMINSTER PALACE HOTEL,
    LONDON, *Sept.* 1, 1869.

CONTENTS.

PRELIMINARY CHAPTER.

EXPLANATION OF THE TITLE—SUNBEAMS.

CHAPTER I.

ATTRACTIONS OF THE RIVIERA, NICE, CANNES, AND HYÈRES.

CHAPTER II.

THE RIVIERA.—MENTONE.

CHAPTER III.

MONACO : ITS SCENERY, HISTORY, POLITY, PRINCE, MYTHS AND HELLS.

CHAPTER IV.

CORSICA AFAR AND NEAR.

CHAPTER V.

AMID THE MOUNTAINS OF CORSICA—A STRANGE, WONDERFUL LAND.

CHAPTER VI.

CORSICAN HISTORY—MOUNTAINS AND COAST.

CHAPTER VII.

THE CLAIMS OF CORSICA AS A HEALTH RESORT, ETC.

CHAPTER VIII.

ADVENT INTO AFRICA.

CHAPTER IX.

ALGIERS.

CHAPTER X.

AMONG THE KABYLES.

CHAPTER XI.

BLIDAH AND MILIANAH—THE ARABS.

CHAPTER XII.

PLAIN OF SHELLIF.—TENIET-EL-HAAD.—CEDARS.—DESERT.

CHAPTER XIII.

CONFLICT OF CIVILISATIONS.—FAREWELL TO AFRICA.

CHAPTER XVIII.

SUMMER IN SPAIN—SCENES, SOUNDS, AND SENTIMENTS.

CHAPTER XIX.

TOLEDO.

CHAPTER XX.

ALCOLEA, CORDOVA, AND ARANJUEZ.

CHAPTER XXI.

MADRID—THE CORTES—JUBILEE OF THE CONSTITUTION.

CHAPTER XXII.

MORE OF MADRID—THE ESCURIAL.

CHAPTER XXIII.

SARAGOSSA—THE MAID—THE BORDER—OUT OF SPAIN.

SEARCH

WINTER SUNBEAMS.

PRELIMINARY CHAPTER.

EXPLANATION OF THE TITLE—SUNBEAMS.

יהי אור ויהי אור—"Be light!" "And light was."—GEN. i. 3.

HEATHEN teacher, Longinus, found in this extract the sublimest expression. What the world would have been 'without a Sun,' the poet Campbell brings home to us by the tenderest of illustrations. There is a Light which never was on sea or land. It consecrates the poet's dream. It is not this illumination, whose beams I would presume to search for; neither is it the *lumen siccum* of the philosopher, nor the inner radiance of the spiritual life; for these may be found in the untravelled tranquillity of scholastic leisure and domestic repose. My pursuit was for that Sunbeam, which makes the tree grow and the waters flow—which makes the earth warm, and the air sweet—which has, in its prismatic rays, colours to paint the beautiful flower and the pallid cheek. The Sunbeams I craved are those which flow from that source whence "the Stars repairing in their golden urns draw light." I sought not the radiation which brings glowing heat, so much as that which yields temperate warmth, and with warmth, relief, and with that, health. Hence, my search was for Winter Sunbeams. Although the search extended into the Summer, it ended—so far as my experience went—upon a Swiss mountain peak, where, 11,000 feet above the sea, the Sunbeams were discreetly mitigated by the aspect, and tempered with the temperature of a hundred snow-clad mountains.

It would impair the cheerful spirit of these pages, were I to dwell on the physical infirmity which compelled me to seek a gentle clime; and yet, perhaps, the only utility—and I may add the primary object—of this volume is, to point out a path to health, which is such, because ' irradiate with beams Divine.'

What a beam of light can accomplish—has it not been the theme of many a girlish composition, as well as of scientific disquisition? The regeneration of the Earth with every Auroral advent, is it not a perpetual hymn of praise to the Creator of light, and an ever-during and radiant rainbow-covenant of His love? What a beam can do, even after its untired tour of millions of miles, and through uncounted periods of duration—how it gilds the starry vault, and builds the heavenly structure—how before it the clouds about our star dissolve and the blackness of night is dispelled; and how chaos becomes cosmos, is but the type of what can be wrought upon the human frame and spirit by its sanitary and cheering influence.

It is impossible sufficiently to aggrandize the wonderful functions of Light. All the 'powers' of the Earth emanate from the Sun. It gives us coal, food, wind, and water. Efforts have been made to calculate some of these mechanical powers, but they are incalculable. Engineers may 'estimate' the work of the Sun in evaporation—A billion of tons a day, over one hundred and fifty millions of square miles, may, as they tell us, be lifted two thousand feet high every day, equal to a 90,000,000,000 horse power. Even this measures but the infinitesimal part of the Sun's power, as the earth receives only 1–2,000,000,000th of the heat of the Sun. How this wonderful power may be made available—how these horses may be harnessed to the chariot of the Sun, is less interesting to mortals, than how its sanitary qualities may be utilized.

What medicinal efficacy the Light has; how it works in the unseen chambers of the brain and body; how it plays from the optic to every other nerve ; what relation it bears to the elements of our physical and spiritual nature—may, doubtless, be determined by the same law under whose administration it gives growth to the tree, glory to the grass, and splendour to the flower. But what that law is, is one of the mysterious arcana of knowledge.

We know that light is life-generating and health-sustaining; that without it, man becomes blighted, even as the parched grass of the field. Take away the light, and both serenity of mind and strength of body are gone. The very tissues of the body degenerate in utter darkness. Take away the light, and the body becomes blanched, etiolated, and wasted. Go to the colliery or the dungeon, and you go to the limbo of ghosts; not ruddy, healthy men and women. I need not refer to the catalogue of diseases belonging to darkness. 'Weeping and wailing,' even in this world, is an inheritance of many of our race before thrust into future 'outer darkness.' Aside from heat, light has its beneficent function. Experiments have shown that life itself will not be generated from the egg so soon in the dark as in the light. Dr. Hammond once tried an experiment on the tadpole. That lively little batrachian was kept in its inchoate

condition for 125 days, by confinement in a lightless
vessel; but fifteen days only were required for its
magnificent maturity out of its wriggling nonage,
under light!

We know how light affects the skin and its hue.
This is illustrated under various conditions and latitudes; in the nightless Arctic regions, where if the
sun goes down half the year, an aurora, or a snow hill,
or ice mountain, keeps it up; and thus, by reflected
light, colours the Esquimaux into olive and brown;
or on the African coasts, where the intensity of the
perpendicular rays secretes in the cuticle a dark pig-
ment of wonderful gloss and glory! How it affects
the stature, the blood, the hair, the liver, the whole
body in fact, inwardly and outwardly; how it makes
men muscular and healthy among mountains, and tiny
and feverish among marshes; how it makes one man
or race sanguine, and another nervo-bilious in tempera-
ment; how it affects conditions, customs, and life,—
puberty, marriage, sterility, and longevity; how the
Light, or its absence, makes men savage or civilized,
passive or active, doltish or intelligent, stationary or
progressive; how in one latitude, and under certain
conditions, these effects may follow; while under
another latitude, and under other conditions, other
effects may ensue,—these are discussions which Hippocrates, Newton, Mead, Jackson, Lardner, Brewster,
Lindley, Balfour, Draper, Hammond, Page, Herschel,
Arago, Florence Nightingale, Sanson Alphonse, and
Dr. Forbes Winslow have considered in elaborate
treatises. It is enough for me to refer to them, in
elucidation of my statements.

In regard to the effects upon the human mind and
health, induction has left us its répertoire of facts;
and has led us to generalize about the very colours
of the spectrum. Red and yellow stimulate the brain;

blue depresses, by exhausting the vital energy; scarlet makes bulls, and some men, madly pugnacious. It has been likened to the sound of a trumpet. Green, violet, and all the sweet tints which Nature paints so often, and which suit the eye, even as the dark cave suits the eyeless fish—these soothe and caress, and have in their tints the elements of cheer and health. Predominating in the radiance, these elements make the life blood of Nature, organic and inorganic. With these elements predominant, come ventilated and lighted houses, fewer inhabited cellars, less corruption and uncleanness, and consequently less disease and death. Not artificial light, which has its offensive effluvia, but the sweet sunbeams,—this is the light which should replace the dark corners of the earth, as the precursor of the full meridian of that better day coming. The utility of light has its application in the construction of hospitals, nurseries, bed-rooms, houses, streets, cities; and, as we shall find out, in the arrangement of gardens, orchards, and forests! Physicians have given us the record of diseases generated on the shady side of a building or street, and of cases cured by removal out of a dark into a bright room or locality.

'Obscurity hath many a sacred use,'

as Bailey sings in his 'Festus'; but its medical uses are not so apparent, valuable, or sacred. Wounds heal more quickly, delirium departs sooner, and convalescence comes more rapidly when the system is under solar radiation.

There are exceptions to these deductions. The eye need not be blinded by excess of light. If you must go to Egypt and its hot sands, or to Algiers and its white streets and houses, or to Russia with its white snows, or to parts of France with its white chalk, or to

India with its perpetual glare,—you may expect an over-excitement of the retina. Use dark glasses, or carry a yellow umbrella, or do as the parasol ant of Trinidad does, carry a green leaf over your head, if you can find no turban or other cover, and thus save eye and health from the injurious red and yellow rays. Thus, in travelling you may reverse the gloomy picture which the blind Milton so pathetically makes out of his darkness; for then, with the seasons there will return day, and the sweet approach of eve and morn; the sight of vernal bloom, and summer's rose; the flocks and herds and human face divine; not cloud instead, nor ever-during dark; nor from the cheerful ways of men cut off; nor the book of knowledge presented with an universal blank! Then, you are not only prepared to be restored in health, but you are actually restored. Then you can enjoy all these effects of light in their æsthetic relations to the outer world. Indeed, but for these relations constantly recurring in my journeying, there would have been little for me to note. Wherever I found the volume of Nature open, there I found sunbeams to illumine and beautify. Whatever was seen in or around the sky, be it star or mountain, bird or tree; whatever was painted in water, or written on rocks, had been glorified from this golden source.

Light, like Sound, the learned tell us, comes to us in waves. Professor Maury has said, with philosophic truth and poetic beauty, that the organs of the human ear are so ordered that they cannot comprehend colour, any more than the eye can see sound; yet, that we may hear over again the song of the morning stars; for Light has its gamut of music! The high notes vibrate with the violet of the spectrum, and the red extremity sounds the bass; and though the ear may not catch the song that the

rose, lily, and violet sing, it may—for aught we know
—be to the humming-bird, the butterfly, and the
bee, more enchanting than that which 'Prospero's
Ariel' sung to the shipwrecked mariners.

Ah! there is more meaning here for you, my shipwrecked brother, than meets the ear. The song
which Light sings to your wearied spirit, may be
the tone or tonic which will stimulate your flagging,
suffering life! The gamut of Sunbeams, is it not a
medical prescription, not in dead Latin, nor measured
by Arabic signs, but in living letters of gold?

This idea of light may be considered fanciful. It
is so, to a small extent, but not so much so to the
scientific man. Whether he take the theory of Newton or of Huygens—whether he regard light as minute
particles, projected with inconceivable velocity from
the sun, or as the undulations of an elastic ether—
he will find in it many analogies to the phenomena of
Sound. Light is so rarified as to offer no obstruction
to the sun and stars in their movements; but its
vibrations are so substantial, that by them our eyes
are struck, our nerves moved, and the sensation
of light produced, just as the vibrations of the air
make sound for the ear. The frequency of the pulsa-
tions or vibrations of the air determines the note ; so
the frequency of the appulse of light impinging upon
the eye determines (as it is held) the colour of the light!

It may be most interesting, indeed, to search for
the sunbeams which make music for the birds, butterflies, and bees, and which these interpret to us ; but
when health is in question, the hard facts of meteoro-
logy and optics become of paramount consequence.
The cool, but pleasant air, interfused with lustre,
the absence of damp and chilliness, not only are
conditions which create for the human economy
an appetite, and thus improve digestion and nutri-

tion; but they also sing a song of joyous health to the diseased mind and dyspeptic soul.

If a lunatic be one—*qui gaudet lucidis intervallis,* according to a legal dictum—what is he who rejoices through all the winter days in sunlight without intervals—for *lucidity* is light. Blackstone calls him a lunatic who sometimes enjoys his senses, and sometimes not, and that frequently depending on the change of the moon. What, then, is he, who, hiding in dark houses, or worse, in dark offices, or worse still, in mines of the earth, but rarely 'enjoys his senses;' and whose mind is affected by the peculiar and polarized light from an old volcanic, crazy reflector like the moon. It is the sun which makes the joy, broken by no intervals. It is the sun, with its power to sustain, which, not less potential than its power to create, gives us that intelligent repose which is one of the conditions of health. It is difficult even to use the languages of men, in order to express the full enjoyment of all the senses, without metaphors drawn from the light. Hence the Bible is full of imagery about sun, and moon, and the perfect day. The very acme of all joys—the joys of heaven—is expressed in the words: 'And there shall be no night there.'

To the same law, which the infant of my vignette obeys, when he endeavours to catch the beams about his head, the matured man yields, when for his defective body he requires fresh vigour and elevated vitality. However unsentimental and humiliating it may be, he 'must be as a little child,' and seek sunshine, even across oceans and zones. Even Dr. Fahrenheit must be consulted. The pores of the skin—as an excretory organ, and as a purifier of the blood—do not perform their functions in the cold damp winters of the north. Hence, sore throat, influenza, and bronchitis prevail; and when neglected, are aggravated; and aggravated, the blood purification, in warm weather, is thrown on the lungs and air passages. They having too much to do, the burdened blood is poisoned. Hence, inflammations and fevers, and finally the tragic destroyer, consumption, closes the scene. Consumption has come to be regarded, therefore, as a disease of debility. It attacks those whose vitality is deficient hereditarily, and those who are injured by excesses, either of vice or of work. Here enters the sunshine, and with its dry, bracing radiance—under proper dietetic, and medicinal rules—restores vitality.

I do not desire to cumber these pages with essays about phthisis and membranes. I am happily required only to write about the enjoyments which create, and which are the proof of renovated health.

In recounting these enjoyments, I but add fresh eulogy to the sunbeams. The associations which Art, Nature, Time, and History have inwoven with natural scenery, and which bestow that rational delight, which an all-bountiful Maker intended as one of the means of health; did more than give enjoyment, because they were sought under appropriate conditions of latitude and longitude, under the protection of mountain walls and hygienic provisions, without which all search for sanitary sunshine, in winter or summer, is in vain. Survey our star, from China to Peru, and you will find no lovelier land or sweeter sun,—none more opulent and fruitful, as well in vegetable glories, as in all that heroism and romance have illustrated; and what is better, none more salubrious or health restoring, —than the sunlit and sea-kissed shores of the Mediterranean!

Selecting a circuit of travel from this—the most favoured part of our globe—the writer found physical scenery and phenomena of rarest attraction. He found a people, picturesque, composite, and interesting,—the

2

result of systems unique, yet diverse, and of blood and
daring the most heroic and adventurous; interesting,
composite, and picturesque, because compounded of all
the virtues and vices of the pre-historic and historic
nations. Phœnician, Hebrew, and Egyptian; Greek,
Roman, Goth, and Moor; Frank, Spanish, German,
and English—have here done what Human nature,
under manifold and various conditions, can do, for
Poetry, Art, Science, War, Commerce, Government,
and Liberty! Italian, Spanish, Greek, Jew, Turk,
German, English, and French—the people who claim
Homer and Virgil, Dante, Shakespeare, and Milton;
who honour Joshua, Charlemagne, Abderrahaman,
Bonaparte, and Wellington; who erect monuments
to Mahmoud, Henry IV., Charles V., Lorenzo the
Magnificent, and Gonsalva; who reverence Columbus,
de Gama, Angelo, Murillo, Gutenberg, and Newton,
and how many more, whose names stand pre-eminent
in the history of our race, have made these shores
resplendent with genius and action. Surely, an
American, seeking moderate excitement, aloof from
the moil and toil of active affairs at home, could not
have chosen a better theatre in which to recall and
observe, under the sunshine of the latitude—the associations which most adorn our race.

My circle began at the Riviera under the Alps; it
includes Corsica; thence enters into Africa, and passes
through Spain and Southern France, until, again in
the Alps of Italy, it ends, with a view so eminent, that
it seems to comprehend the whole sweep of nearly a
year's tour of travel.

When I reached the point, where the current of
ordinary travel became visible, I dropped the pen.
But Corsica, Algiers, and Spain—about which the
body of the volume speaks—are not hackneyed
themes, or trodden ground. Corsica, indeed, is almost

a *terra incognita.* It is the connecting link between the two continents of Europe and Africa. It is in the centre of the basin of the Western Mediterranean. Its mountains are midway between the Alps and the Atlas. They have all the fruitful vigour of Atlas, with the rugged grandeur of the Alps, and the vegetable growth of each. Volney gives Corsica three zones. Up to 1800 feet, the climates of Italy and Spain are found, with the date palm of Elche and the *chamaerops humeris* of Algiers, the oranges of Nice and Blidah, and the lemons of Malaga and Mentone, — the oleander, cistus, lentiscus, and myrtle, which make its macchie, like the shrubs of the Riviera and Algiers. Above 1800 feet, and thence to 6000 feet, France is reproduced, with its vines, olives, chesnuts; and its forests of ilex, ordinary oak, pine, and beech. These last indicate the third zone—the climate of Norway; and reach above 6000 feet. The larch which is indigenous to Corsica, and especially the ilex, are gigantic trees, and their forests are covered with snow half the year.

Hence in Corsica you have an epitome of my whole circle of travel, and of the continents of Europe and Africa. Before visiting the island, I presume to print two chapters about the Riviera; partly with the view to show why I started from so shining a spot, in search of other and distant sunbeams.

CHAPTER I.

THE RIVIERA—NICE, CANNES, AND HYERES—THEIR ATTRACTIONS.

T H E beginning of the year 1869 found me in the Riviera. Where that sun-favoured land is, may be best seen by glancing at a map of Italy or France. It is a mountain amphitheatre. A little correction of the natural irregularities would make it a semi-circle; with Hyères at the western, and Leghorn at the eastern end; and Genoa sitting upon the apex of its arch, while Corsica points her Cape Corso northwardly, like an index finger, directly to the 'superb' city.

There are, in fact, two Rivieras; 'Levante,' on the east, and 'Ponente,' on the west. The mountains and the sea join here to give their glories to the scenery; and the sun looks down upon them, during the winter, with smiling serenity. The mountains are rugged and bold, and the sea blue and bright. The picture is that of Beauty reposing in the arms of Strength. Here all that we imagine of Italy, as the loveliest of lands, finds its nearest approach to realization. The army of Alps from Switzerland, Savoy, and Dauphiny, which the traveller down the valley of the Rhone never ceases to observe and admire, marches down to the sea near Toulon, and its summits 'fall in,' like good soldiers; while a detachment, called the Maritime Alps, move by the left flank along the coast until they effect a junction with the Apennines. Whither they tend after that, whether under the sea to Corsica, Sardinia, and Sicily, or along the backbone of the Italian peninsula into Sicily, and thence into Africa, to join the serried ranks of the Atlas, I may hereafter inquire when I follow them thither. My purpose now is to mark only that mountain section, whose shields protect the shores of the western Riviera. It would be sufficient for this purpose to call attention to the segment of the semicircle which binds Hyères to St. Remo, and which includes a string of sunlit brilliants, Nice, Monaco, and Mentone being the chief gems.

The writer passed a winter along this part of the Riviera, principally at Nice and Mentone. A little rain now and then, sometimes a harsh wind, hardly any frost, snow, or ice, and nearly all the time sunshine —so bracing and elastic as to have in its beams the vitalising qualities before described, made the winter pleasant and memorable for health restoring. Although under medical direction, he had the happiness to have a physician who did not prescribe low diet and water, nor indoors and opiates, nor leeches and blisters. His therapeutics were oxygen, and his pharmacy sunbeams. He treated me, as he did the plants and flowers of his garden. The sickly plant which came under his nurture, like his patient, had no inheritance of ills. The innate constitutional vigour—like that of other constitutions, political and otherwise—had survived many infractions, and even 'amendments.' When parasites attack the tree, they hide their deadly work under the fair exterior of moss. To remove the parasite, you must scrape away the moss. The fungi will then die. By trenching, digging, draining, and loaming the roots, the organization of the tree, by its native vigour and the sunshine, will do the rest. So with the body. Give it pure atmosphere, aliment, and plenty of sunbeams to digest; let the eye drink and the lungs eat, and even if exposed to the winters of Labrador, under a tent of spruce and skins, the lungs will fill and expand into health, the voice become rounded in tone, and the body strung into vigour. Avoiding the languors of a heated latitude, and the chills of a frosty sky, with not so much exercise as to become weary, nor so much rest as to grow torpid, but about 60° of Fahrenheit by day, and 50° by night, and your restoration has begun. Thus avoiding extremes, and keeping to the Ovidian mean —*medio tutissimus ibis,*—the person, like the tree, may throw off disease, and be braced into a new birth and being. The peevish, troubled, dyspeptic patient, the disgust of himself and the horror of his friends, by inhaling the air and living in the shine of the Riviera, will find himself, before he knows it, on a donkey, climbing mountains, gathering the violet, hyacinth, and narcissus, wandering under old olive orchards, soaking in sunlight on the warm rocks of the shore, or lounging among the fishermen of the beach, or, in some other way, growing into a cheerful and contented, because healthy person.

It was on the evening of the 6th of January that we arrived at Nice. December had not been harsh, even in London, Paris, or New York. We had seen flowers gathered in London in December, and France, from Paris to the Mediterranean, was all 'beaming.' But there was something at once inspiring and restorative in the glad sunshine, when, on the morning of the 7th of January, from a balcony of the Hôtel Royale, in Nice, I looked down into an unwintry prospect of roses in bloom, palms in flower and fruit, and children laughing and playing in the garden walks under orange-trees, dressed in summer costume!

When in the afternoon we drove to Mount Cimies, under a loving sunlight, upon roads lined with trees,

which were laden with gold orange-orbs, and, under a
shadowy grove of ilexes, alighted at the door of the
Franciscan convent, it did not require the excitement
of the adjacent Roman ruins, nor the priestly vivacity
and courtesy, to allure us to forget all bodily ailments.
The Franciscan brethren did what they could to make
their cloisters luminous ; except this, that the ladies
were not allowed to go in. I wandered amidst the
flowers of the convent gardens ; then into the cemetery, where the 'fathers' of the past sleep,—admired
the marble figures, and joining the disfranchised sex,
who were conciliated with the bouquets I brought,—
visited the Bath of the Faeries, and returned to Nice,
after a satisfactory survey, from on high, of our promised land.

Then, we join the mixed crowd of people, of all
grades and ranks, who, in picturesque groups, gather
to hear the music on the plaza. Around the palms
there gather princesses and more questionable people,
who here resort to hear the band and see each other,
—*spectatum veniunt, veniunt spectentur ut ipsæ.* After
driving down the English promenade to witness the
little carriages, all buoyant with fashionable flounces,
and women hid in clouds of drapery, drawn by swift
and graceful Sicilian ponies, whose feet twinkle and
manes crinkle as they race along; after driving by the
Casino, wondering at the white, garish, ostentatious,
almost regal, edifices, and wondering still more at the
battalions of women in baskets, and on their knees by
the blue sea, purifying the clothes worn by these gay
throngs,—washing them in the very face of all this
meretricious display,—we retire to our rest, with a
somewhat nebulous idea of Nice and its winter guests
and gossipers.

Thus the days and weeks glide smoothly at Nice,
and nearly thus, they ebb away for a month. They

are varied by drives and walks. You may visit Smith's exquisite Folly, a wilderness of building and gardening, overhanging the rocky shore. It was erected by a rich Englishman, who became *involved*—in the laby-rinths of his own taste. You may walk to the moun-tains behind the city, whence views are commanded of all the region about these splendid shores. You may, with a select circle, make an occasional dash into Monaco, where the pet lambs of fashion gamb*le* on the green, and the mercenary wolves ('tigers' Ameri-cans call them) in softly-cushioned lairs lie in wait for them. You may go to church in Nice,—Greek, Catholic, Protestant, Episcopalian, or Presbyterian, and hear the President of the United States prayed for, in the company of Victoria and Napoleon. You may visit the Turkish baths, the cafés, and the club. You will not be at a loss for American society. On my first visit to a café I knew that two well-dressed young men who walked up to the counter were my countrymen, for they quarrelled as to which of them should pay for the Bourbon! You may attend the costume balls at the Cercle, got up by noble dames for charity. There you will see the aristocracy, who, like the swal-lows, seek the bright sunshine of the South for winter. Some friend will point out the black-eyed Corsican, who is Prefect of this Department; and his wife, arrayed in purple and fine laces. You will note, too, the celebrated Turkish statesman, Fuad Pacha, pale and dying, but plucky. You may see the celebrated Madame Ratazzi—one of the Bonapartes—with her distinguished husband the ex-Minister of Italy. How she shines in the centre of a group of gentlemen! She is in the costume of an Oriental queen. She wears an immense necklace of diamonds and emeralds; a black, pointed, velvet crown, on the points of which jewels glitter in profusion, gives grace to her reginal

mien.   Her neat Zouave jacket, and her long, golden
and white, scalloped train, with two sets of scarfs tied
behind with floating streamers—one for the skirt and
the other for the sleeves; these set off the beautiful
lady, whose name has been much used in the Journals,
and whom I mention, because she is the impersonation
of Nice life.   Her husband has had several duels on
her account.   He is an intellectual man.   He wears
glasses, has a small head, is slim in person; and follows
her about with a smiling ostentation of pride in
her, very interesting and unique.

The balls and other inventions of the enemy of
health and life were not the attractions which drew me
to Nice.   I came to bask in the Riviera sunshine.
Hence my place is in the fairy voiture.   Under crack
of whip, and with dash of pony, you may drive to the
Var, skirt the bay, cross the bridge over the waterless
river, turn up to the right, and, before you are aware,
observe and be amidst rising vistas of snow-clad mountains, mountains withdrawn and withdrawing into
others; within whose nooks little villages nestle and
climb, and up whose heights the terraced vine, olive,
and orange, and the elegant embowered villas reach
almost to the bleak, rocky summits.   Or, you may
leave the sea-shore of Nice, upon a cold, raw, or mistral
day, when dust prevails,—for some such days do come,
—and in the upper valleys, where the tender lemon
grows in the shelter, and where the mountains by their
barriers make the air soft and sweet, clamber amidst
the velvet swards of the terraces; or visit, if you will,
St. André and the magical grotto, where the incrus-
tations and medallions are made by the water trickling
over the moulds, by the same process which makes
the grotto so beautiful with stalactites.   Or, to fill up
another day under the sun with something new, you
may wander in the orange orchards of Madame Cla-

risse, where the little honied mandarins grow, and
where the contadine gives you the anomaly of a sweet
lemon !

If an American man-of-war should happen to lie in
the harbour of Villa Franca, and the feeling to see
the flag rages, you may satisfy the national craving by
an hour's drive, and, at the same time, witness the old,
narrow streets, and picturesque people of this odd
village; and from the road thither, look down on
Smith's Folly by the sea; on the villa of the songstress
Cruvelli, who is caged by this beautiful shore; or,
rising higher than the main road, and leaving the
carriage, ascend Mount St. Alban, gather flowers by
the way, see the forts and soldiers, and in one glance
gather a bouquet of beautiful views from the bluest
of seas, the gayest of cities, the brightest of skies,
the greenest of olive forests, and the purest of snow
mountains !

If time hangs heavy on your hands, day by day
you may thread the fine winding roads leading from
Nice, among the mountains; make a visit to the
'obscure vale,' where the torrent has torn a gorge
through the pudding stone, and the sunbeams are
few in its bed, between the narrow walls; where
boulders and flowers, children and oranges, are in profusion, to add to your diversion. Upon our trip to
the 'obscure vale', we had ten children in our train,
carrying boards for *pontons* in those parts of the vale,
where the water is. We bought on our return, for
one sou, a cluster of large oranges, ten on the branch !
You can, if you choose, pass your time under the
gnarled and aged olives with a book; or wander
through groves, where lemon, almond, and orange trees
by the thousand press their fruit down to your very
lip; or, if you would, you can go to the factory, where
the geranium, heliotrope, orange, rose, and violet, are

weighed by the ton, thrown into vats, mingled with horrible hog's lard, to come out the rich scents, pomades, oils, and essences of my lady's toilette; or ramble up to the château in the heart of the city, under the tall juniper trees, and see the play of light and colour upon the translucent sea, or the flash of breakers white as snow against the rocks, or watch the steamers go out upon its blue bosom, faintly staining the sky with a streaming pennon of smoke. These employments may serve to make the time dance away in the sunbeams of the Riviera. To live at Nice, and not to be in the air, is like going to a feast with sealed lips. The reason why Nice is so full of people; why one half of the houses are hôtels and 'pensions' is, that it has more than its share of that

> " Great source of day—for ever pouring wide,
> From world to world, the vital ocean round."

On an afternoon, upon the promenades, you will find hundreds of invalids in wheel chairs drawn by servants, or in their own pony phaetons. A large proportion are of the aged. In this summer-in-winter they realize the idea of Hippocrates: "Old men are double their age in winter and younger in summer." Such a company of the sick, however, does not tend to make the promenade cheerful. We need the bright, ruddy, chirruping children, playing about with their nurses, to give redoubled lustre to the solar radiance; and these we have every where searching for sunbeams.

There were, indeed, days, or rather mornings and evenings, and certainly nights, when the twisted old olive roots, lighted by the pine-cones and aglow with *their* radiance in the fire-place, were necessary to give a cheer to our room. We had some exceptional days. The Sun hid under the piles of mountain clouds, the mistral came, and once with a sprinkle of snow and a dash of rain.   Yet I think Nice is better protected from the winds than Cannes, and Mentone is milder, and less subject to the mistral, than either.

I spent a day at Cannes, at the villa of Mr. Eustis of Louisiana.   The air was all balm, the sky all beauty; and the villa of my friend, and especially those of the Duke of Vallambrosa, and Mr. Woolfield adjacent, are perpetually enamelled in my memory.   Flowers and fruits, shrubs and trees, made summer here, as in the villas and gardens around Nice.   These delights, however, were enhanced by enjoying them with the grandson of Lafayette, Mr. Eustis' guest, who has all the affection for Americans which his name evokes from us.   A middle-aged man, of brilliant intellect, and social disposition, he remains a plain republican of moderate views, and refuses on principle all preferment or office from an imperial source.

The view from the villa of my friend over the sea, including the Estrelle mountain range, and St. Marguerite isle—famed as the prison, where the 'Iron Mask' and Abd-el-Kader were confined — renders this spot as attractive as it is possible to be made by sun, sky, and land in harmonious combination.   I have seen, in no place, grounds so beautifully situated, disposed and ornamented, as those of Cannes; especially those already mentioned, but Vallambrosa excels.   The Duke has, around his mediæval château, amongst his exotics, Australian plants, and tropical trees, including the aloe of Algeria, the orange of Portugal, the palm of Egypt, the magnolia of America, the jujube of Spain, and those especially whose leaf and flower are perennially green.   His castle is on a rock.   A flag flies from a tower.   A chapel is attached.   Running waters mingle their music with that of the birds in his pine groves.   Roses, oleanders, camelias, myrtle, oranges, lemons, all that hath fragrance,

colour, and grace for the winter, as for the summer,
here flourish. A visit to Lord Brougham's neighbouring villa completed my delight at Cannes. Workmen were fixing fountains and repairing pipes; digging
about the trees, and arranging the turf, under the supervision of a nervous, timid-looking, elderly man—in
slippers and gown—who is the brother of the late
Lord Brougham, inheriting all his property and some
of his oddity, but nothing of that *vivida vis animi*
which makes the name of Brougham immortal!

Lord Brougham was the precursor of this winter
exodus towards the South. He led the way to Cannes,
when for a song lands were sold, whose value now
makes men millionaires. Here he prolonged his life,
and made its burden easy. Here the sun warmed him
into intellectual vigour—long after his natural force
was abated, and when men thought that he lagged
superfluous, at least for his fame.

In writing of select spots where Sunbeams perform
that benignant office, I could not limit my observations to Cannes, or to Nice. These places are but
samples of the Western Riviera. Every part of it
illustrates what I would say. From Toulon to Nice,
and winding with path and road inland to Grasse and
Draguignan,—along the blue sea, and apart from it,
and into the garden of Provence,—on the southern
slopes of Les Maures and Les Estrelles, you will find
in the flora and in the sunbeams health-giving signs.
The aloe lifts its stately stem, the umbrella pine
spreads its broad branches, the pomegranate blushes,
the cactus, the cork tree, the magnolia, the heliotrope,
and the jessamine, the aromatic shrubs of the Corsican
macchie, and the roses of Cashmere, come forth to
testify to the favour of heaven toward this ancient home
of the troubadour. The mistral seldom ruffles this
paradise. Hence Hyères, St. Maxime, Antibes, and

Cannes, are already becoming adorned with Swiss
châlets, Italian villas, and English cottages; while
walks and gardens, hôtels and pensions, picturesque
roads and mountainous paths, are peopled all the
winter, with the brain-fagged, lung-diseased, throat-
sore, and body-broken pilgrims to these shrines of
health.

The air of the western Riviera is dry, the sky clear,
and almost cloudless; sometimes cool and rainy days
interrupt the spring-like weather, sometimes dusty days
come.   These occur, however, more and more rarely
as you go eastward toward Mentone.   There the
stimulus is not so great in the air; but the sedative
influences are by no means relaxing.   Volumes have
been written by eminent men to analyze and detect the
shades of difference between these resorts.   Thermo-
meters and barometers have made their record. The
wind is watched by disease-detectives, and what it
brings on its wings, either for hurt or healing—and
from what quarter—all is chronicled.   Each ill, from
neuralgia to consumption, from the gouty toe to
the distempered brain, has been subjected to the *vis
medicatrix* of the learned, as affected by these haunts
of health.   I have perused much of this literature; it
is sometimes conflicting as to minor details, but it is
generally accordant in the main facts.   These are:—
that all along this shore there are, *first*, many curious
vestiges of bygone times, from the castellated rock of
Hyères, to the Saracenic tower above the Pont of St.
Louis, on the Italian border, to draw the mind from
the body and its ills; *secondly*, that there is a luxuriant
vegetation, and an exotic flora in the open air, which
make the aspect of the scenery varied and pleasing;
and that the rare flowers of the conservatories and
gardens of the north are here found in wild profusion;
—at once a sign of a mild climate, and a provocation

to wander and gather ; *thirdly*, that not only do the mountains make scenery unparalleled, but they shelter from harsh winds the shore, and all that grows and lives upon it, beneath their auspicious screen ; and, *fourthly*, as the climax of my conclusions, that the air is so guarded by mountains from winds, chills and damps, and so tempered with sunbeams ; so mild and yet so bracing, so full and resinous of emanations from pine and fir, and redolent of violet, rose, jessamine, and cassia—that, from November to April, the despairing invalid may prolong his life ; the suffering, mayhap, lead a painless life ; the desponding receive genuine exhilaration, and the consumptive, with care, receive cure—ABSOLUTE CURE! All these climatic virtues spring from the potential sunbeam ; and the best theatre of its wonderful 'winter performance' is to be found on the Riviera ; and its most eligible point, in my opinion, is at Mentone !

CHAPTER II.

THE RIVIERA.—MENTONE.

"O! I seem to stand
Trembling where foot of mortal ne'er hath been,
Wrapped in the radiance of that sinless land
Which eye hath never seen."
Anonymous; attributed to Milton.

THE human body is a magnet. If one colour of the spectrum—the violet—can render a needle magnetic, what may not all the colours in unity do for the human magnet? Whether light be corpuscular, undulatory, or otherwise; whether it consist of small particles or tiny waves; whether it be vapour emitted from great solar conflagrations, flaming a thousand miles high;—one thing is true, its shower upon Mentone is grateful and golden. According to science, Light is elementary iron, sodium, magnesium, nickel, copper, zinc, hydrogen, and what not. According to the classic myth, the flood is metallic—golden; and, on its undulations, came Zeus (the Burner), to hold loving intercourse with mortals. I accept the chemistry and the poesy —both. Whether it be iron or gold, we can understand why, deprived of light, the vegetable and animal kingdom — including man — degenerates and dies. Second-hand moonbeams, and third-hand fire will not save. Having started on my search for Sunbeams, I hope to be allowed, like other philosophers, to make my facts and findings conform to my pre-conceptions.

What a flood was that, upon whose jewelled bosom we were borne into Mentone! How sweetly it flashed

over that vulgar and denser medium, the sea, as we
followed the Corniche road from Monaco. The iron
—if iron it be whose element is light—'entered the
soul.' It was medicine to the mind, and so easy to
take in the open air, sugared with the fragrance of
rose and jessamine, upon that lofty route along the
mountain sides! Here and there the light was checkered
by walls, over which the orange and lemon looked, and
in windings where the myrtle and laurel grew, and
where the greyish-green of olive groves gave its
sacred contrast to the scene. If that light was iron
to the blood, it was gold to the eye. Mentone is
expressly arranged under the sun, for the spectrum.
Not only is it guarded by the tall walls of the ranges,
some four thousand feet high, which—one above the
other, in tiers—throw their well-bossed bucklers about
it; but down to the margin of the sea, its two semi-
circular bays—between Cape St. Martin and Pont
St. Louis—glisten with unwonted beauty. Without
making comparisons with other places—and certainly
not with places on the Riviera, where all is so lovely—
one cannot see the geranium, heliotrope, camelia,
magnolia, verbena, and rose flowering into the air of
mid-winter,—and in Europe, too, and in a Canadian
latitude of forty-three degrees and more—without feeling what Science tells: that the sea itself is here
more tenderly tepid, the sky more serenely clear, the
atmosphere more elastically dry, and the solar radiance richer than elsewhere in the unappreciated
opulence of health. If not better as a climate, for
some purposes, than other places—such as Pau,
Madeira, Cuba, Algiers, Palermo, Florida, Egypt, and
Sonora—certainly the Creator, when He called the
light down upon this shore—' saw that it was GOOD !'
How much less of moisture there is here than at
Cannes, Nice, or other places; how much more suit-

able for human ills of throat, skin, lung, or nerve,
let the doctors settle.   I had only one thing to settle
—myself.   I did that in a 'pension' at Mentone,
about the last of January, 1869, under the care of
these solar beams and of Dr. Henry Bennet.   It is no
derogation from the merits of the latter to say that
he works very happily under the former.   In fact, he
could not work at all without them.   Pronounced
consumptive himself, he prolongs his life in this climate,
where he winters ; and where, upon the mountain side,
he cultivates the very rocks, giving to them the colours
of the prism, and to the terraces a beauty and fragrance,
from fruit and flower, which make them like a garden
of the Orient.   What is still better, he helps by his
example and advice many a worn-out patient, whose
organism is thereby renewed in vigour and prolonged
in days.

Properly speaking, the bay within whose shore Mentone lies, and where these marvels are wrought, is
between the point, not far from Nice, called L'Hospice
and palmy Bordighera on the east.   But it seems more
snugly esconced between Cape St. Martin and the
Murtula point, near the Border.   This division leaves
out of this chapter Monaco, and its magnificent
panorama of fort and palace, sea and mountain.
Monaco has a chapter for itself; but, having some
drawbacks, it must wait until I pay my respects to the
more moral Mentone.

Why I have preferred Mentone to other places of
the Riviera will appear as I ramble amidst its mountains, and repose under their shelter.   The rocky
ridge of the east end of the bay, where most I used to
saunter, makes not only a natural boundary for Italy
and France (at least France found it so), but its
height is a guard from eastern winds.   The lower
range of olive and orange hills behind the town, is

itself guarded by lofty amphitheatres. These shelter it from east, north-east, and north. Still loftier Alps, some of them 7000 feet, running inland far, shelter it still more from the north, east, and west. Thus is Mentone left to the influences of the sweet south.

Its temperature is, therefore, mild enough to grow the lemon, which is too tender a fruit for the frosts which sometimes visit Cannes and Nice. The shape of the shore—bending south-westerly—the mountains following inversely, makes the space between shore and mountain outlets for wide valleys. The south wind is not at once chilled by the snows of the mountains; while the Boreal blasts are tempered by passing over the intervening lower spaces. Hence its warm, yet bracing airs. Hence, not too much moisture—little rain and so much fine weather. Hence, again,—since I am here to be out of doors,—I am glad to miss the mistral of Provence and the siroccos of Africa. Consequently, I do not wonder that in twenty-seven years, the thermometer has descended here only thrice below the freezing point. Nor will the reader be surprised that there was scarcely a day of winter here, when we did not find sunbeams, under which to take walks, make excursions, and gather wild flowers. This, too, even though it has been an exceptionally cold winter in southern Europe; for at Rome, Naples, Florence, and Genoa, the snow and the frost have made those charming cities almost inhospitable.

It is not my purpose in these preliminary chapters, to reproduce personal experiences. What is to be seen in the wooded valley of Gorbio; what delights are to be found in climbing up to red-rocked, romantic Roccobruna; what varied attractions Boirigo and Cabrolles present; and what flowers you may gather, and what sketches draw, in the vale

of Chataigniers, even up to the rugged summit of
St. Agnes, crowned with a feudal castle; what drives
upon the broad causeway, shaded with plane trees
in the Carei, leading you upward and onward toward
Turin into a little Switzerland; what green terraces
you may count rising from the vales of this route—
terraces rising by the hundred to the mountain-tops
crowned with pines and snow; into what gorges you
may wind upward to Castiglione, whose cascades and
rocks afford many a study for the artist and stanza
for the poet; what an Arcadian spot you enter on
the road to the jagged end of Cape St. Martin, whose
extreme point is a wild ten-acre lot of ravelled, ragged,
and black limestone, against and over which the sea
washes, and at times tempestuously roars; what sweet
little violet ravines and eglantine nooks, solemn plateaux
of olive shades, and wayside nests for orange and
lemon; what highways westwardly to Monaco and
eastwardly to St. Remo; what enchanting walks
amidst the boulders, caves, rocks, and sands of the
shore; these, if pictured by a masterly pencil, would
interest. They are but the incidents of luscious days
of convalescence, passed under the sun, to etch
which even is not my forte or my purpose. I speak
of them only as the decorations of that antechamber
of the temple of the sun, into which I am privileged
to guide the worshipper. The result from this kind
of life has been to make new blood, and to give
normal vitality. The principle which cures the un-
strung larynx and cicatrizes the wounded lung, has
here had the open air for its adjuvant, and the rarified, sweet light for its handmaiden; and its highest
benefaction consists in the restoration of those hitherto
believed to be *incurable!* What other adjuncts are
required, let others discuss. What doubts are to be
solved in the minds of the incredulous, let them be

solved by studying the new and progressive ideas which medical science has evolved! I do not write of causes, but effects.

It is no new thing—solar heat—for the prolongation of human life. It is an old medical maxim: *Levato sole levatur morbus.* Pliny, speaking of his uncle, says: *Post cibum aestate si quid otii, jacebat in sole.* I can well understand why in the eastern and southern land, the *solario* was built on the house-top, where the diseased were placed and the healthy took what was called their solar-air-bath. My baths of sun and air were taken upon the terraced mountain sides in Dr. Bennet's garden. I present in these pages a lithograph, copied from a water-colour drawing by a lady, for which I am under obligation to Dr. Bennet. It is from his volume of 'Winter Climates in the South of Europe.' In it, you will perceive but little of the garden; that I reserve for another sketch. The Saracenic tower of the foreground, which overlooks the panorama of Mentone to the west, is far above the Italian boundary line. That line is formed by the gorge of St. Louis,—a picture of which the reader will see, and over which the Corniche road leaps with an elegant bridge. The distant range of the sketch is that of the Estrelles, near Cannes. The view presented between the Estrelles and the gorge in the front, is of such varying loveliness in shape and colour, on land and sea, that it is impossible to keep the eye from its embrace. Nowhere does Nature unclasp a volume so full and splendid with illustration!

I had not been long in Mentone, before I was ordered up to this spot. I obeyed. The tower had, flying from its rounded top, an English flag. This was the sign that the Doctor was expected. Sometimes the Italian, sometimes the French, and sometimes the

star-spangled banner floats in the air—signs for the
initiated, that the garden is open to all, or that
the Doctor is busy or away, or something else.   It
has a curious utility, and has had a curious history,
—this old tower of the middle ages.   It was once
used as a 'look-out' to the sea, that the people of this,
then isolated and unfrequented coast, might be alarmed
in time, before the Turks and Moors swept in with
their corsairs, to capture, despoil, and destroy.

Let us walk up to the garden among the rocks.
We pass a Roman ruin—never mind! Seats for rest
line the path. We reach the gate. We read: '*Salvete
amici !*' That welcome is inscribed, as the hospitable
Pompeians used to inscribe it, on the portal.   Out of
these rocks—once bare, by skilful terracing, and in
three years' time, the Doctor has made with the aid
of water, which he leases from the proprietor of
Grimaldi village above, and out of the Sunbeams
which he has conscripted, and for which he has nothing to pay,—long avenues of vines, flowers, trees,
and scented shrubs, whose forms, colours, and odours
invade and capture the sense, as if from some unearthly, dreamy realm. Even his rude stone pillars,
which appear in the picture, are clad in the festoons
of Eden. The very stones seem floral. In a rustic
arbour a hammock is swinging; and a fountain is
dripping—whispering and dripping—over the fern
leaves of the grotto. It is a monotone of tranquil,
liquid lyrics. If you prefer a divan to a camp stool,
there it is! There is a table too, with a green tapis,
pen and ink, books and periodicals, and a lunch of
mandarin oranges, good white bread, and pure red
wine. Do you want more for comfort? This is the
salon de reception; and while swinging in the hammock, and enjoying its *dolce far niente* delights, far
up in the Elysian air, and warmed by the very rocks,

which have retained the geothermal heat of last summer for some purpose, either to man or flower, —I began to feel that Nature is exceedingly kind, the garden exceedingly paradisiacal, and the Doctor exceedingly clever! Who needs the Médoc of the lunch, or the novels and poems on the table or other adventitious aids, to enhance the enjoyment? Immersed in this warm bath of light and air, the eye will float out over range after range of mountains. It will catch the sparkle of the myriads of crisp waves of the many-hued sea beneath. Or it may watch the line of snowy breakers which swell and surge against Cape St. Martin; or it may pierce the misty distance, to see half-veiled the Isle of St. Marguerite, near Cannes. It may look below at the sea-gulls playing upon and above the waves, or at the soldiers of the two nations on the bridge of the boundary gorge, searching passengers for contraband; or very near, glance at the game of croquet, as it is played by the Doctor and a Russian princess in the little plateau of a lower terrace, near his conservatory, or it may observe the gardener, Antoine, petting his children—the flowers, and plucking violet, camelia, or rose, either for princess or caprice.

But most the eye rests upon the sea beneath. It is so easy to drift on its surface. Then you see so far into its clear depths. It is said that light cannot permeate further into the water than 700 feet. I have read that in the arctic regions shells are seen at 80 fathoms; and at the coral banks of the Antilles, it is said that the sea is as clear as the air; that coral and seaweed of every hue display all the tints of tropical gardens. Light has so much to do with these ' marine views' that physicists have made many experiments as to the photo-chemical and physiological qualities of the Beam on and under the ocean wave. Why the

waters below our garden are so lighted ; why the very
bottom can be seen, with its sand, weed, and rock,
determining the various colours of the surface, depends,
of course, upon its transparency.   Various media may
give varied colours to the ocean.   The Gulf of Guinea
is white ;   the sea is black around the Maldives ;
there is vermilion off California, owing to infusoria ;
the sea is very red near Algiers, owing to water plants ;
and the Persian Gulf is green.   Around the Poles the
transition is rapid from ultra-marine to olive-green.
Minute insects have much to do with all these prismatic
phases ; but in a sea not so deep, the bottom has most
concern with the hue.   Chalk gives green ; yellow sand
gives dark green ; a dark ground gives black or brown ;
and grey results from brown soil.   But in the sea
beneath the Doctor's garden all these colours, especially
the blue tints, enchant the gaze ; for they are as exquisite, changing, and evanescent as the colours of
the opal.   And these colours have their gamut—

"Hear I not—
The Æolian music of the sea-green plumes?"

Thus bathed in body and mind in these refreshing
elements, and so far from the hum and hurry of active
life—from which I had just emerged—is it strange
that I failed to remember that it was winter, and in
Europe, too, or that I was an heir to the ills of flesh?
Thus the winter wears away, amidst rock-land,
sun-land, and flower-land, but very like dream-land.
Of course, this experience is varied.   Even the song
of Philomel, oft repeated, loses its charm.   You may,
as I did, leave the hammock, and its reveries and
rest, and if sure of foot and head, clamber through
the olive and lemon terraces above the garden, into
the red, rough rocks.   You may follow the precarious goat-paths, below Grimaldi town, and investigate
the grottos where the shepherds hide with their flocks

from the sun's glare in summer. You may thus
obtain a more extended view; see the towering fort of
Ventimiglia, and the white specks which indicate the
villages of St. Remo and Bordighèra. You may, tired
of rocky eminence, wander below into the Corniche
road, and observe the vetture, *en route* for Genoa—
full of English and American tourists rushing by—in
sleepy content or disgust at these rocky defiles, little
knowing the paradises above them. Or, still below
the road you may follow the broken paths along the
sea, under frowning red rocks, and look in upon the
' bone caves,' wherein the skeletons of the prehistoric
men were found; or in rocky inlets and sheltered
nooks, play with the white surf, as it rolls up the
music of multitudinous pebbles, or smoothly washes
the white sand; or if you desire human society, you
may saunter among the washerwomen who line the
rivulet, which steals down into the sea, and gossip in
the worst patois with these useful dames. I remember them well. I shall never put on a particle of
pale linen without invoking a blessing on them. They
are a kindly company. When, venturing out one
day upon the rocks, little recking of their slipperi-
ness, and not observing a big wave, which was
approaching, and 'abounding in grace,' from one
mossy boulder to another, I fell in. Having emerged,
all dripping and green—did these angels of the tub
laugh? No! I did; but it was a sorry laugh.
Whether because accustomed to the washing process, or whether from inbred courtesy,—they arose
—each a Venus from the soapy foam—arose from
their baskets, and proffered sympathy and succour!
Bless ye! children of the sunny sheen and unsavoury
suds—bless ye! When afterwards I strolled that
way, was I not known? and their few sous per day,
were they diminished by reason of their gentleness

to the awkward stranger? I do not, however, recommend hydropathy for steady treatment at Mentone. Still, one *drop*—into the sea—is not much.
The sea scarcely feels it.

You may still more vary the life at Mentone by
two kinds of excursion. The pony-basket phaetons
of Nice are there; and what Nice has not, Mentone
has, the promenade on a donkey. This latter is a
most delightful recreation. If you are weary of the
garden, or the shore, or the mountain and olive
groves; if the hurdy-gurdy of the Tyrolese tramp,
or the song of the strolling Figaro, fail to interest;
if you can extract nothing from the Catalonian
peddler of shawls and knives; if the lemon girls, with
full baskets on their heads, trudging into town, knitting as they march, lose their picturesqueness; if
the old town, with strolling acrobats, its noises, mar-
kets, narrow streets, churches, cemetery, and castle,
fail to amuse or interest; if the drawing of the conscription, and its dancing, rollicking, singing saturnalia, do not arouse you out of ennui, then run
down to Nice for the Carnival, dash over the mountain behind Sicilian ponies to Monaco, and try your
little game on Monsieur Blanc and his Roulette, drive
over to Bordighèra, and ramble amidst the palm trees,
and make endless bouquets of wild flowers; or what
is best, endorse the Mentone donkey!

A donkey ride is the climax of Mentone diversion.
It will take all day, for owing to the change of
temperature after sunset, you must be home before
five. Send for the fair Judith, or her mother, and by
nine o'clock you will find donkeys at the door, caparisoned for the ascent. Choose your animal. They are
all sure-footed; but there is a choice. Get up an
imposing cavalcade. It looks romantic, thus to be
moving up the narrow, stony paths, which are trodden

only by the lemon girls, with their heads basket-laden,
or by peasants driving their goats and sheep. Visit
some of the towns which hang about the edges of the
ridges, or go to the top of the beautiful ridges themselves, overlooking the sea. There you may make tea
with the pine cones, or take lunch in the midst of
aromatic shrubs.

The Mentone donkey is as handsome for a donkey
as he is sagacious. Very rarely is he otherwise than
docile. The voice of the driver from behind suffices
to guide the animal. Our pet donkeys were 'Bijou'
and 'Grisette.' They answered to their names with
alacrity and serenity. Thus you may ascend to
Grimaldi, Castellare, Roccobruna, and St. Agnese,
rising from 800 to 2400 feet above the sea. There
is an agreeable sort of timidity in thus creeping upon
these safe little animals along the perilous, precipitous
points, though it is more agreeably safe to follow
the bridle or goat-tracks, under the shadow of the
olives, and among the flower and grassy terraces. It
is charming to go still higher, and play at hide-and-
seek with the sunbeams, which glance through the
resinous pine-trees. And it is still more buoyantly
delightful, when you attain your best elevation, to
gaze off into the blue-tinted air over sea and shore,
and drink in, with the pure oxygen, the full spec-
tacle!

The donkey is an archæologist. At least, he fur-
nishes the opportunity to investigate the ancient castles,
which were formerly built, away from the sea, in spots
almost inaccessible. The pirates of Barbary, in making
their raids for fair women and strong men, hardly
ventured so high for their slaves. The castle at Roccobruna especially is hard to ascend—even if there were
no foe to prevent. Under the guidance of the children,
who meet you with flowers, and are ready to direct

you for a pittance, you may sound the old haunts
and cisterns, and scour the nooks and prisons, from
turret to foundation. These make Roccobruna the
most interesting of the mountain towns. What adds
to this interest, is the fact, that the town has slid
away from the mountain. Its red rocks of lime
stone and pudding-stone have been torn by some
convulsion or caprice from their 'old fellows'; and
the town is curiously dropped about, and amidst
the chaotic *débris*. Many of the families live in and
between the fractured rocks.

We have visited the houses of these cheerful mountain people. They earn but little with their lemon, oil,
and wine; but that little is much to them. Go with
us into one of those twisted, contracted streets. It is
hardly wide enough for the pannier of the donkey.
There are many stories to the rude, stone, glassless
houses. Many families live in one house. The first
floor is for the donkey. His musical note advises
you of his proximity and confinement. You ascend
some dark stairways, and enter a neatly swept apartment with one window; it is adorned with pots of
flowers. The floor is of red square earthen tiles; you
see two small beds, a bureau, old chairs, some prints
of saints, and, at the head of the bed, a dry white
palm leaf, already blessed! A very little kitchen is near,
only five feet square, with its cooking furnace, and
windowless; huge piles of large, long loaves of bread
with a big bottle of wine are massed together in a
corner, both bought for a week in advance. All the
dirt is given every day to the scavenger who takes it
off to the terraces where it has its uses.

Going into one of those rooms in the old town of
Mentone, I find that it accommodates a man and
wife, and one child. The bread for the week costs
thirty-six cents. The woman of the house is one of

my washerwoman-acquaintances of the sea-side—her
name is Lisa. She earns, from sunrise to sunset, about
thirty cents for her work. Her husband works on
the railroad for the same wages. They have little or
no cooking. The bread is bought at a co-operative
bakery; and the wine is its only bibulous accompaniment. Said she:—'Sometimes we have soup. If my
husband comes home first, he lights the faggot, and
makes it; if I come home first, I do it. Once a week
we have a pound of fish or meat—generally reserving
this treat for Sundays; though when the railroad
presses, he works often on that day.' After com-
mending her frugality, she told us the secret of her
discontent—the skeleton in her little household. She
was labouring so hard to pay a debt of her first husband's last sickness; it was 200 francs. It has a year
to run, but the interest is in arrear. How to make
ends meet, pay the interest, and sink the principal, by
her small earnings, this is the Sphinx's problem of
Lisa's life. Chancellors of the Exchequer and Secretaries of the Treasury ponder similar problems about
larger sums, and fret less about them than Lisa. We
gave her our financial theory, and, what was better,
we helped her. It is so easy here, with so little, to
do so much. The flowers, wherewith every day our
rooms were adorned, betokened the gratitude of the
woman. She could not do enough for us. She acted
as our guide among the relics of the old town.
Meeting so many children, I remarked that they were
all girls. 'Ah! Monsieur,' she said, 'it is the air,
without doubt.' We asked her to show us the cemetery
at the top of the town. She consented. Bonnetless,
with a little handkerchief on her rich dark hair, her
little girl tripping ahead, we began the ascent amidst
the narrow and dark alleys. An old lady meets us.
'Where are you going, Lisa?'—'To the cemetery!'

—'Well,' sang out the crone, 'I won't go there till they carry me.' We reached the cemetery. I was almost sorry that we had asked Lisa to take us thither. She had not been there since her first husband's funeral; and though he was buried here, she had been too poor to mark his grave, even with a nameless wooden cross. Here was a fresh crucifixion of her love. Her child persisted in asking for her papa's grave. The tears came. She went within the chapel to hide her sorrow, where, doubtless, in fervent prayer —full of young memories—she found a prism of sunbeams through her tears! Around upon this rocky apex of the town, the wealthier families are buried. Their white marble tombs, in the shape of little chapels, are lighted with tapers, decorated with engravings, images, and paintings, and covered with bouquets of living flowers. I am not sure but this is as sweet a burial spot as I ever saw; for it is secluded from and above the world, and overlooks all the rare scenery of ravine and mountain, sky and sea!

I found the peasants of the Riviera, like Lisa, honest, simple, kind, and not ill-looking. They have not lost their original virtues by the increase of the foreign and invalid population, or by the enhancement of their lands, labour, and produce. This small Italian town, like the other towns of the Western Riviera, is becoming surrounded with suburban villas and pensions; but the primitive costumes and habits remain. It is seldom you see any one lazy, certainly not the women. How often we meet that old woman —a sample of the whole—who trudges along with her donkey, knitting; sometimes elevated between two casks of wine and the pine-branches, and sometimes afoot, but always knitting. When the railroad is completed from Monaco, and through to Genoa, all this undercliff of health stations, from Hyères to St.

Remo, will be filled, the winter through, with visitors, tourists, and health-seekers.   By that time the hanging garden of my friend, amidst the rocks of Grimaldi, above the Border, will have been enlarged, and still more beautified.   Already he has completed the purchase of manifold more piles of rock; and soon he will throw fairy bridges over the abysses, around which I have climbed with him,—erect more terraces, employ more hydraulics, and cut more paths on the mountain sides.   When I return to the scene, I shall no longer find the goat-caves, the limestone seats, and the elixir of the sunbeams, amidst their present ruggedness.   I shall find all changed into the elegances which a cultivated taste and a skilful gardener can evoke out of the very rocks of this favoured shore.

CHAPTER III.

' Aggeribus socer Alpinis, atque arce Monoeci
Descendens.'   VIRG. ÆN. vi. 830.

' Power from hell,
Though heavenly in pretension, fleeced thee well.'
                                                        COWPER.

T H E Prince—sovereign and independent—
who rules this microscopic realm of Monaco,
has a laureate. I have seen his verses, and
read his prose. His prose is poor poetry,
and his poetry is purely prosaic. He closes some
prose by saying :—' If Horace had known Monaco,
he would not so often have sung about the Tiber.'
I agree with the laureate. Another writer has written,
that to visit Nice and not see Monaco is like going to
Rome and not seeing the Pope. So I resolved to go.
In fact I have been there several times, and by several
routes. I have seen it by day and by night, under
light and shadow; but how shall I describe it?

I am at a loss whether to paint it under gas-light,
moon-light, star-light, or sun-light. As I am in search
of 'sunbeams,' a photograph would be best; although
a photograph might not give all the shadows which
Monaco, completely pictured—after Rembrandt—
morally and physically, deserves. I have not been
drawn hither by the enchanter, who flourishes his
râteau over the green *tapis* of the gaming-table, but
by the irresistible fascination of sky, sea, and land.

Last evening, when I arrived from Mentone, the
globes of gas were illuminated throughout the gaming
grounds of Mount Carlo, and in the palace grounds
across the bay. This morning I have wandered about
Carlo, in the fresh air and sunshine, and the enchantment is enhanced. But wandering soon grows irksome.
You want to repose and dream. If the reader will sit
on one of the green chairs which line the parterres of
the Casino (or gambling palace) fronting the sea, and
under the parasol which the palm, orient-like, lifts
over him, and follow my errant eye, he may see a
picture worth remembering.

Look out seaward first! Directly in front is my
favourite isle, Corsica, now under the deep veil of distance, which this morning lends to it no enchantment,
for the isle is not in sight. A white cloudlet hangs
over the spot where the island, I trust, still remains.
To the east and on the left, upon the horizon's rim,
is a dark, azure band, which, as it runs round to
the west, becomes fainter, till it fades into a glistening
line of silver. Between that rim and your eye is the
expanse of the sea. It is streaked with currents of
lighter hue; some running as straight as canals, and
some meandering like rivers, 'at their own sweet will.'
Scarcely a wave is on its bosom; only diamond
points, rippling in the sunshine. Six ships, sails set
and blanched, are seen; but without motion apparently.
They seem like 'painted ships upon a painted ocean.'
The steamer from Nice with its flag of smoke, which
itself takes the blue hue of the air and sea, moves
along, leaving behind it a bright line made by the
fretting of the tranquil element. Below us, on the
right, is the little harbour, above which, separated
from the Casino grounds of Mount Carlo where we
sit, is the fort, palace, and village of Monaco, the old
Phœnician harbour of Hercules, and now the little

Nation, dropped on a point of rock into the ethereal and marine blue. A fishing boat and a pleasure boat; these and no other signs of life are on its waters. From our point of view we may see down into the lucid depths of the sea. Its colours rival the flowers of the garden where we sit. But how shall I transfer into black type that which is so varied in hue and delightful to the sense? A writer has said that the sea upon this Riviera is a book of which one never tires. Charles Lamb was asked which play of Shakespeare he liked best. He said, the last one he read. So my last look at this wonderful water is the best. It is not alone the richness, but the fickleness of its hues which pleases. Its mutations vary with the hours, and run through the chromatic scale in two senses— music and colour; from lily-white to ebony, and from red and orange to green and blue. From its low murmur, like a sigh or sob, it rises into thunder, especially when the mistral of March moves along the shore. Even the local laureate, in his prose, rises into poetry in speaking of the music of the shore: ' *Toujours la nappe bleue de la mer semble toucher les cordes d'une lyre en baisant les points des rochers.*' It has been likened in colours to a dove's neck, bright green, dark purple, soft ultra-marine, the blue of a burnished steel blade, the glance of a diamond, and its foam to snow!

The prospect of the shore is, however, still more inviting. Far off to your left lies, under the feathery mist, Ventimiglia, with its old castles now held by Italian soldiers; and beyond it, is Bordighèra—home of palms. These villages look like specks of white upon the dark edge of the distant shore. Between them and Cape Martin is a beautiful bay, within which lies Mentone. The bay and town of Mentone form the figure 3, and divide the old from the new town.

Mentone, however, is hid from our present view by
Cape Martin. The mountains do not seem to me so
rugged and cragged seen from Monaco; only one
above me has this look. The rest are in curves. The
eye can easily float along their tops, and be gratified
while looking down from them upon the olive-trees
which fill and decorate the terraces, or into their
valleys, where, snug and green, the lemon orchards
perennially flourish—float until it comes against that
bold, irregular, jutting, fire-torn and water-worn, castle-
like mountain, behind which Turbia village lies unseen, called 'Dog's Head.' Wherefore called I know
not. Perhaps it is Canis called, because *Canis non.*
It looks more like a tortoise.

Stay! I remember now that *Tête du chien* is a corruption for *Tête du camp;* for here Cæsar established
the head-quarters of his legions after the conquest of
Gaul. *Huc usque Italia dehinc Gallia!* Here, too,
you may see, just peeping above the heights, from
our seat at the Casino, a Roman tower, in romantic
ruin. Across the bay, and immediately beneath the
'Dog's Head,' is the rocky promontory called Monaco.
Its walls, pierced for cannon, surround it. It is about
two miles in circumference; its fort overlooks the
sea. Its palace overlooks the valley which separates
it from the mountain above. The fort and palace
look puny beside the giant mountains above them.
From a sea view of Monaco it looks like an island
rock. This rock is about 460 feet high. Green vines
and shrubs cling to the almost precipitous sides of the
cliffs. On the other side, towards the west, there
is some ground cultivated with olives and figs. My
laureate says, in his volume, that sailors, as they sail,
can gather the big Barbary figs from the shore.
Doubted. You may perceive an open plaza above
on the rocky principality. It fronts the palace, within

whose walls art is represented in its utmost opulence. Venetian mosaics and frescoes, attributed to Caravaggio and Carlone, and precious marbles and rare paintings are to be found within these walls. It is said that the Prince is a great patron of the finest of the fine arts. In the gardens around the palace are seen umbrella-pines, tall cypresses, the lentiscus, gigantic aloes, tamarinds, roses, and other flowers of tropical luxuriance and quality.

If you would saunter over the rock, you may find some dozen brass field-pieces, unlimbered and lying on the ground. They bear French marks of the time of Charles X., with the Latin motto on them: "*Ultima ratio regum.*" Very little rationality they will display on behalf of his Serene Highness Charles III., Prince de la Roulette. If you would leave these precincts of insignificant royalty, you may dive down into the narrow streets of the old town, where donkeys only go, and where, when they are *loaded*, you may be sure they do more execution than the cannon in the plaza. The town of Monaco is like all the villages of this shore. It is made of stone; the houses can almost grasp hands across the street; they were built for fortresses as well as domiciles, and the people live in them pretty much as they did when Augustus Cæsar stopped here on his return from the wars.

Remember that the rock of Monaco, and the town proper—for a view of which, I refer to the frontis-piece of the volume—are across the harbour from us as we sit in the Casino grounds of Monte Carlo; but both places are within the little principality. I have not done with my glance along the mountain-range above, or the sea-shore below us. You will perceive a road winding along the sides of the moun-tains, as far up as the brown rocks of Roquebruna. The road, now and then, hides its line in clusters of

olives, now emerging from an orange grove, now
boldly turning about the edge of a precipice, and
then creeping into the shadows of a terraced mountain. This is the famous Corniche road, on which
it is sacrilege not to travel. I have been over it from
Nice to Bordighèra. Although it runs to Genoa, yet
the best views are from that portion which I have
traversed. Ruffini opened his 'Dr. Antonio'—that
exquisite gem of a novel—with a description of this
highway. Flanked by the Mediterranean, and the
Alps, and Apennines, and, covered with a sky that
seldom frowns, he has pictured it in words as an artist
would from his palette: "The industry of man has
done what it could if not to vie with, at least not to
disparage, Nature." Numerous towns and villages,
some gracefully seated on the shore, bathing their feet
in the silvery wave, some stretching up the mountain-
sides like a flock of sheep, or thrown picturesquely
astride a lofty ridge, with here and there a solitary
sanctuary perched high on a sea-washed cliff, or half
lost in a forest of verdure at the head of some glen;
marble palaces and painted villas emerging from sunny
vineyards, gaily flowering gardens, or groves of orange
and lemon trees; myriads of white casini with green
jalousies scattered all over hills once sterile, but now,
their scanty soil propped up by terrace shelving above
terrace, clothed to the top with olive trees—all and
every thing, in short, of man's handiwork, betokens
the activity and ingenuity of a tasteful and richly-
endowed race.

The road, in obedience to the capricious indentations of the coast, is irregular and serpent-like; at one
time on a level with the sea, it passes between hedges
of tamarisk, aloes, and oleander; at another it winds
up some steep mountain side, through dark pine
forests, rising to such a height, that the eye recoils

terrified from looking into the abyss below; here it
disappears into galleries cut in the living rock, there
it comes out upon a wide expanse of earth, sky,
and water; now it turns inland, with a seeming deter-
mination to force a passage across the mountain; and
anon shoots abruptly in an opposite direction, as if
bent upon rushing headlong into the sea. The variety
of prospect resulting from this continual shifting of the
point of view is as endless as that afforded by the ever-
changing combinations of a kaleidoscope. Could we
but give this sketch of Ruffini a little of the colouring
—real colouring of the country—what a picture we
should make of it! But we cannot. It is past the
power of words to paint the brilliant transparency of
this atmosphere, the tender azure of this sky, the
deep blue of this sea, or the soft gradations of tone
tinting these wavy mountains, as they lie one over
the other. Rarely will the traveller, searching for
summer sunbeams in the winter, find a heaven so
bright as that which makes a canopy, with the olive
and the orange, over the Corniche!

For a long time this route was only a donkey track.
There was formerly so little, and such unfrequent
communication between the points of the coast, that
the pirates of Barbary could and did easily pounce
on any village and ravage it, before succour could
come. Persons are living yet, who remember when
a convent of friars was carried off in a raid by these
marine robbers. Besides, the Italians were jealous of
the encroachment of the French, and for a long time
discouraged the making of a road. But in 1828,
owing to a heavy snow storm, the King of Savoy
could not get from Nice to Turin. He embarked on
the sea. It was so boisterous that he put back. Then,
the people turned out *en masse*, and made the road
for him. Napoleon I. improved it. Now you may

see going to and from Nice and Genoa, hundreds
of *vetture* every day. They are loaded with baggage
and filled with travellers. This is the best route to
Italy, except over the railway, by Mont Cenis. This
winter that route has not been open all the time, by
reason of the snow. Nor is it favoured by invalids,
and even less by tourists, who desire to see the
Riviera. Just before the holy days at Rome the
throng of carriages was immense. The railroad from
Marseilles is finished as far east as Nice ; in fact to
Monaco, and by next September it will be at Mentone.

Boom ! boom ! boom ! Right under our seat at
the Casino gardens, shaking the vases of geranium
which crown the balustrades, the explosion of rocks
resounds ! Look over ! Several hundred workmen
who are building the railroad walls between the Casino
and the sea, are standing stock still, watching for the
next explosion. The smoke rises and floats away ;
a boy upon a rock sounds a horn as a warning, and
boom ! boom ! boom ! the explosion again and louder !
Rocks of tremendous size fall far into the sea ; the
sound is echoed by other explosions beyond ; the
mountains catch up the sound, and the reverberations
tell of the work on the railroad which inflicts wounds,
apparent on the coast as far as Cape Martin—wounds
soon to be healed and hidden by the vegetable glories
of the climate.

The railroad has been begun since my laureate
wrote, or he would not have called 'Monaco a field
of repose for all the living. No one labours here. That
misery is unknown. The sun does all the work and
works for all.' But the sun does not make railroads, though it may, with the aid of irrigation,
stanch the wounds or hide the scars which labour
is making along these romantic mountains. It may,
too, bring balm by its beams to the tired and worn

throat and lungs of the invalid. In that way, the sun may prove a physician, more skilful than those accredited by collegiate authority.

If you choose, you may come to Monaco from Nice by the rail, or you may come by the little steamer, or by the Corniche road. You will not, however, see more than four miles of sunbeams out of the sixteen, by the railroad between Nice and Monaco. There are twelve miles of tunnelling through the hard limestone. You pass under beautiful and painted villas—under gardens which would shame the Hesperides, literally of golden fruit and perpetual flowers; you pass through the olive gardens of Cimies, an old Roman city, cross a viaduct, flash under Monte Albano, many hundred feet under the fort, then out again, into the sweet sunshine which flashes on the old towers, houses, and beautiful bay at Villa Franca; here you may see the fig, as you see it in the Orient, but it is a brief glance; another gorge, and you emerge on Beaulieu, under the shelter of a blue bay, whose eastern side is the rock of Monaco! You are nearly two hours on this trip through 'little Africa,' as this region is named—so named because of its tropical luxuriance of soil and African warmth of sun, even in mid-winter. Flowers and fruit, palms and cactuses, bright patches of the green gardens and blue sea alternate with the darkness of the tunnels; glimpses of daylight all splendid, and of night all Egyptian. Yet I prefer the trip on the Corniche road in the Nice corricolo, or in post-chaise. Thus you may see the landscape all the time; and what may you not see? Eza, a village of 600 souls upon the very apex of a mountain rock coped with a tower, where are the remains of a temple of Isis 2000 years old; Turbia, full of people just out of the Roman days, so primitive they seem, and ruins speaking of the early and eminent with the immensity, richness, and variety of the prospect.

You will understand, therefore, that Monaco is very accessible. Every day crowds of the noble and rich, sojourners of the Riviera, from Cannes, Nice, and Genoa, come to Monaco to gamble and to gaze ; to gamble and to lose; to gaze and to wonder; to study, if you will, the strange history of this anomalous little kingdom and its institutions. Before Nice was annexed to France, our laureate sang of Monaco and its incongruities in a poem I have seen, addressed, "à son altesse sérénissime Charles III." thus :—

> " A deux pas de la France, au bord de la mer bleue,
> Sirène qui caresse en frappant de sa queue
> Son rivage embaumé,
> Est un endroit charmant: rocher ville fantôme,
> Etat, principauté, république, royaume,
> Par les siècles formé."

But this picture was written before Italy ceded Nice and Mentone to France; before Roccobruna, like the rocks themselves, revolted from its old attachment, and became a part of the French empire. This picture was made before the present Prince of Monaco —independent and autonomous in his government— conceded to Monsieur François Blanc the right to establish the 'speculative science' of Rouge-et-noir in his dominions.

I believe M. Blanc is a German, though the name sounds French. He owns the gaming bank at Homburg, where he lives and fares sumptuously amidst grounds of imperial extent and beauty. He pays the Prince a rental of 100,000 dollars, or a half million francs per year for the concession. He pays all the taxes of the principality and for the gas-lights of the place. There is no octroi or customs duty. It is forbidden to the people of this Lilliputian realm to

play; and by arrangement with France, the business
people of Nice are also prohibited. Yet I think the
authorities of France do wink slyly at the invasion of
the agreement.

The people who are forbidden by the agreement
with M. Blanc from playing, console themselves with
this smart wit of a Frenchman : 'Sometimes Rouge
wins, and sometimes Noir, but always—Blanc.' To
the devotee of Trente et Quarante, this play upon
words needs no explication.

I said that the law was not observed. An artist at
Nice told me that Napoleon was rather remiss, or if
he was not, M. Blanc and the prince were, and that
Napoleon ought to capture Monaco. He said, shrug-
ging his shoulders: 'Napoleon took Sebastopol; but
could not take Monaco! Pourquoi non?' The prince
spends his winters in his palace here, when he does
not spend his time and money in his hôtel at Paris.
His right to Monaco is hereditary, as the heir of the
Grimaldis, whose ancestor Giballin Grimaldi, brother
of Charles Martel, fought the Saracens, and was rewarded for his prowess and success by being made,
by William I., Count of Arles, independent and sovereign. This sovereignty was recognised by the treaty
of the great powers at Vienna. Mentone and Roccobruna created a revolution in 1848, and drove out
the prince. They enjoyed independence till 1860.
Then the prince sold out his rights to these villages
to France for three millions of dollars; as much as
we gave for Louisiana! I think, notwithstanding
the ostensible independence of the prince, he may be
a sort of feudatory of Napoleon. Some fine day he
will wake up, 'annexed' to France. The people,
however, will not like it. Even Mentone people—
those who are subject to conscription—pine for the
old days when they were of Monaco. I was present

when the conscription was drawn in Mentone, and saw the terror of the young men when drawn, and their revelry, song, and saturnalia as they danced through the streets, when they were not drawn. I said to one, 'Wherefore so much hilarity? Your turn will come again.' 'Yes,' he replied; dolefully adding, 'Mentone was happy ten years ago. We then belonged to Monaco, Monaco never waged war and had no conscription.' This was true of late years; but in the olden time the Prince was the ally of Italian or French; and further back in the olden time, the Prince was always a warrior, and bound to Piedmont in feudal relations. Still it must be confessed, the little principality has survived many strange revolutions during its twelve hundred years of existence. We bow to it accordingly.

Some months after this paragraph was penned, the writer visited Spain and Italy. He never lost sight of the Grimaldis. They have enough of the mixed bloods—Mahomedan and Christian, Moor and Italian —to keep them from being de-vitalized. When revelling in the halls of the Alhambra, or rather clambering above them into the halls of the Generaliffe, he spied the genealogical tree of the Grimaldis. It was illustrated by the portraits of the tribe. But suppose they *are* of Moorish descent! what then? Nothing. Suppose they did turn against their race; what then? Much every way; and chiefly this: that the perfidy is a token of character for which all the 'hells' of Monaco, or under it, have no prison adequate in the supply of caloric and sulphur! I should, however, be ungrateful for the courtesies extended by the administrator of the Grimaldis, did I not do justice to that family.

I have said that the family were bound to Piedmont in feudal relations. While at Turin, in June, I sought

the Armoury. Stopping before the effigy of a giant,
clad, full length, in mail,—mail tinct with silver, and
rich in every joint and rivet of its harness. I ask,
'Who?' The guide says, 'Grimaldi of Monaco!'
'Not he of the present?' I said; 'This one is big
enough to hold in his breast the present Prince;
big enough to hold in his iron frame all the small
princes around. What is his name?' I inquire. 'I
do not know,' said the guide. The coat of mail is
marked with an 'F.' Perhaps he was named Felice,
or perhaps, Ferdinand! He was one of the 'giants
of those days,' when the Charles Martels wielded
swords equal in length to our muskets. His feet and
legs were decorated with white boots. His armour
was nailed with brass. His blade was Toledan. His
plumes were white and black, symbolic of his Moorish
and Christian blood! His vizor was down. He
seemed as inscrutably mysterious, as would the zero
in algebra, or on the roulette tapis, to a Kabyle. But
the very armour and the mien of this knight, gave
him the substance and spirit of endurance. Amidst
the jewelled swords of the Savoy dukes, the torn
battle-flags of 1831, 1848, and 1860, which Italy has
preserved to honour her army and its leaders, amidst
the effigies of mailed kings and soldiers of the past
and present, upon horse and on foot, surrounded
by men-at-arms and musqueteers, this form of the
Grimaldi stands proudly eminent.

It was pleasant to see, afterwards, the easy, modern,
elegant Prince Charles; and to know that time tames
the hot blood, and reduces the grandest forms. The
mailed hand of the middle age, which wielded the
battle-axe, is now gloved with Alexandre's best. The
head which was hid beneath the vizor is circled by
the glossiest of silken hats. The great white boots
above the knee are now of patent leather, and very

petites. Whereas the earlier Grimaldi glowered through and under his helmet, fierce as a Moor and brave as a Christian, and fought many a bloody fight, and many a bloodless tourney; we are saddened to learn that the present Grimaldi is almost blind; and whereas the progenitor was a warrior whose sword was of undoubted metal, whose 'coated scales of mail, o'er his tunic to his knees depend,' and who, by his spirit and puissance, won Monaco,—the present prince, a Christian gentleman, rents his gambling franchise for a roulette hell, at one hundred thousand dollars per annum!

It is, however, also pleasant to learn, that the present Prince, chief of a dynasty dating from A.D. 968, is only fifty years of age; that he may live many years yet to enjoy his ample rental and domain; and that he is both courteous and benevolent; that he lives for most of the year at his château of Marchais in France, never visiting Spain, or caring for his possessions there, although he is a grandee; and that his châteaux in France, Spain, and Monaco are ever open to visitors. It is quite interesting to read, as I have in a French Paper, that a company of gentlemen lately visited his château of Marchais, and finding a modest man in the grounds, slapped him familiarly on the back, saying, 'My brave man! Think you we can visit the château?' 'Certainly—Shall I show you in?' '*Allons!*' After the unrecognised Prince had shown them about, one of the company offered him,— *him*, the grim Grimaldi—a 'small *white* piece of money.' Did he accept? Aye, and with a smile; for saith the chronicle, is he not the only sovereign of the earth, all of whose subjects are gentlemen? Is it not a fact of history, that the Emperor Charles V. ennobled all the people of Monaco? Who will say hereafter that the people of Monaco, and by conse-

quence, those of Baden-Baden, Homburg, and Wies-
baden, are not all gentlemen—*ad unguem ?*

Again, to balance my description discreetly, let us
remember that these Grimaldis of Monaco have done
much in peace and war. They furnished four grand
admirals to France. There the Saracen was deve-
loped, not to say worse. They gave to Genoa eleven
doges, and one captain-general to Florence, and one
chief grandee to Spain. Many generals and diplo-
matists have they furnished, and once one of their
number was of the order of the Golden Fleece. This
last Grimaldi did not however, allow roulette.

I must drop a few grains of allowance on my description of the Grimaldis and of Monaco. Their
history is so obscure that one might be allowed to
romance about them. The simple fact is that Monaco
was one of those allodial domains which, by reason
of the favour of the strong in the feudal days,
escaped feudal confiscation. The reigning family
of the Grimaldis died out in the male line. Our
Prince is in such strong contrast with the big knight
in armour at Turin, because he represents the female
line. Antonio was the last of the warlike males. He
died in 1731. His daughter married a Frenchman,
Thorigny, who was a relative of the Talleyrands.
Hence, at the Congress of Vienna, the title, which
was disputed by the Genoese family of Grimaldis, was
preserved, by the influence of Talleyrand, to the present family. The king of Sardinia was at first suzerain over the Prince. In 1848, when Mentone and
Roccobruna revolted, Piedmont annexed them, and
placed her soldiers there. In 1854 the Prince endeavoured to regain Mentone, but failed. France, on the
annexation of Nice, bought the rights of the Prince,
and Napoleon is now Suzerain. Still the Prince
reserved the ancient privilege, and is called Sovereign.

Certainly, he has an independent princely rental from the Casino, although it is situated outside of his rock-founded village and fortification. M. Blanc can afford to keep the Prince in his realm. Last year, notwithstanding his expenditure in the beautifying of the Casino and its grounds, he cleared sixty thousand dollars. He is purchasing more land, and is to build another hôtel and gambling house. His franchises in Germany, under the Bismarck policy, expire in 1870, and, wiser than M. Benazet of Baden, he is preparing a permanent investment here.

But what of the Casino, and its gambling hell? I would prefer to wander about the enticing grounds, drink the inspiring air, watch the people come and go out of the cafés, shooting galleries, photographic shops, and carbineer offices, rather than go within. I should like to see the army of the Prince manœuvred upon the plaza. A humorous friend tells me, that this army once consisted of three—rank and file ; and when, on a great occasion, the order was given— 'by twos,' the third man, in despair, incontinently rushed into the sea! This is almost as tragic as the scene on an Irish schooner ; 'How many are you below?' sung out an Irish mate. 'Three of us, sor!' 'Then half of you come up!'

But if you will go within the Casino, there is much to see. There is much of decorative Art in the interior ; many fine frescoes, and much gilt on the walls. The music in the concert saloon is especially grand, and so are the bedizened servants of M. Blanc. Besides, you must doff your hat, and yield up cane and umbrella. You approach with reverence the presiding goddess of the place—Fortune. The green tables are surrounded by her votaries. The chink of gold and silver—none of less denomination now than five francs—resounds above the din and hum of con-

versation.   One half of the players are women, dressed
in all the peculiar attire and fashion 'of the period.'
The players are generally well dressed.   There are
many mysterious young women, and raffish, middle-
aged men—a strange mixture of lorettes, musicians,
actresses, counts, blacklegs, and old countesses.   The
predominant genius of the place, who might stand for
Fortune, is a young woman of touzled hair, natty
high-heeled shoes—heels in the middle of the sole, and
gorgeous dressing upon the back!   There does not
seem to be much eagerness for the play; I look in
vain for the fierce-eyed, anxious gambler of the stage
and novel.   A Russian Princess—who is reputed to
have lost several fortunes—monopolizes the first table
of Roulette.   She plays notes; and in a musical way,
too; for her voice—calling her numbers in French
—keeps time to the notes she drops.   Occasionally
she gathers, with her neatly-gloved hand, rouleaux of
gold and clusters of bills, and that, too, with as much
sang froid as one of her own native ice-bound
streams.   While she plays, all stop; for she takes
up quite a majority of the thirty-six numbers.
Around the other tables—of roulette and trente-et-
quarante—people with sharp eyes and sharpened
pencils note down the alternations of the game and
play at intervals.   They concoct in their minds—as
many a victim has done before—infallible rationalia,
or systems of successful play.   Such and such figures
and colours, 'they say,' have come up to-day, as
winning: '*Argal*, they will come again!'   They
watch the sequence of the figures, and suffer the consequence.   The croupiers rake in the money easily;
twirl their little ivory balls deftly; and deal their
cards gracefully.   There are some six tables.   You
will find a tiger under or about each; but his paws
are velvet, and his claws are unseen.   You can hardly

comprehend—unless you play at one of the tables long enough to be familiar with the monotonous French jargon—what the croupier says, as he calls on the betters to make their game, or announces the results : Here one sings : '*Noir gagne et couleur et rouge perd!*' That is Rouge et Noir ; and the jingle announces the receipt and dispensation of the money. Over there, at the Roulette : '*Le jeu est fait, rien ne va plus!*' and down goes the specie. The *mise* is placed ; it cannot be changed after the words are uttered. Then the chances are announced, '*Pair*' or '*Impair;*' '*Passe*' or '*Manque;*' '*Rouge et Noir;*' and, if fourteen comes up, '*Quatorze, rouge, pair, manque.*' Chink, chink, chink ! Adieux to many a fond Napoleon ; and the game goes on. Now and then this procedure is relieved by a vivacious quarrel. A woman of the *demi-monde* claims—or maybe she thieves—the winnings of an unsophisticated novice. He makes a feeble resistance ; she insists ; he retires, discomfited and blushing. Then a sophisticated *habitué* makes a quarrel with one of those accomplished women, or with the croupier, perhaps. It looks like a fight. But they never fight ; a quiet man—is it M. Blanc ?—drops in and sweetly settles it. 'Make your game, gentlemen!' On goes the game, till the midnight hour comes, and the company goes. A rush is made for the carriages, and the Nice cars, or the cloak room ; the lights are put out. Nothing but the pure sea 'beats the banks' at Monaco till next day at noon ; when water, shore, sky, sun, trees, flowers, and all the allurements of the natural scenery are forgotten again in the pursuit of fickle Fortune.

Many should be glad to forget the fortunate or unfortunate associations of Monaco in the pleasing myth which attributes its origin to Hercules, or in

4

recalling to memory the sweet story of the beginning of Christianity on its shore. Joseph Addison visited Monaco in 1701. His travels in Italy are scarcely ever read now-a-days; but having before me a chance edition—printed 'at Shakespeare's Head, over against Katherine Street in the Strand, MDCCXXVI'—I looked into it to see what the Spectator might say of Monaco. I regret to say that he says so little, and that little is but quotation from the classics. We have some 'rough passages,' not in the style, for that is all serene, but in his December voyage to St. Remo. Having observed several persons on the Riviera, 'with nothing but their shirts, and without complaining of the cold,' and noting the palm-trees, he sailed for Genoa; but adverse winds compelled him to lie by for two days. 'The captain thought the ship in so great danger, that he fell upon his knees and confessed himself to a Capuchin who was on board. But at last, taking advantage of a side wind, we were driven back as far as Monaco!' He then quotes Lucan's description of the harbour, not omitting the stormy part of it, but adds scarcely a descriptive word of his own. He translates Lucan into verse:—

> 'The winding rocks a spacious harbour frame,
> That from the great Alcides takes its name;
> Fenced to the West and to the North it lies;
> But when the wind in Southern quarters rise,
> Ships, from their anchors torn, become their sport,
> And sudden Tempests rage within the Port.'

But there is no association with Monaco so attractive as that which connects it—not with Hercules, or, as Virgil does in the verses heading this chapter, with Cæsar—but with a fair, good, sainted Christian woman. Such an association proves its nobility by a higher than imperial charter, compels our reverence, and does much to redeem it from its bad fame as an elegant gambling resort. It is the story of Dévote, a beautiful and spiritual virgin of Corsica, who clung to the religion of our Saviour through all the Pagan persecutions. Her purity of life, her self-abnegation, her firmness when urged on pain of death to deny her faith and bow before the deities of Rome, would make her sainted in any calendar. She was ordered to execution by the Roman magistrates of Corsica. In dying, she committed her soul to the Saviour, and in the article of death a dove was seen to ascend from her mouth to the sky. Some Christians took her body, bore it to a bark, and sailed with it toward Africa. She appeared to them, and persuaded them to take the body to that place whither the dove should fly from her mouth on the voyage. This proved to be Monaco, and they followed her direction; bore the body to Monaco, and buried it near where now is the Church of St. George; and there it remains. In view of the licensed hells of the principality, one would think it would hardly rest here in peace!

CHAPTER IV.

CORSICA AFAR AND NEAR.

'Is this the man of thousand thrones,
 Who strewed our earth with hostile bones?
 And can he thus survive?'
                    BYRON's 'Ode to Bonaparte.'

'CORSICA!' 'Corsica!' 'L'île de Corse! These and other like exclamations sprang to the lips of many visitors at Monaco, in the afternoon of February, 1869, while the writer was sojourning in that little realm. The occasion was the uprising of Corsica from the bluest of blue seas. To our vision the island was nigh, but in fact it was a hundred miles away. Our point of view was the far-famed, if not ill-famed, Casino.

Certainly, the hundred gamblers of both sexes—some of whom were princesses, and some of the *demi-monde*; some English millionaires, and some Russian nobles—were all intent on the *rouge et noir* until the cry went up, 'Corsica!' 'Corse!' from the windows of the Casino, where a company were gazing seaward. A rush, a lifting of the hands and eyelids, exclamations of delight; and soon Corsica is forgotten by the gambling throng. Not so by the writer. He resolved to enter upon, *pedis possessio*, the promised land.

I have said that the isle which we saw was over a hundred miles off. From the French coast it is only ninety; but the grand mountain range which we had seen was distant 120 miles. Its base was

concealed below the horizon. The sphericity of
the earth sinks about 3000 feet of Corsica from
the eye. Therefore, in order to obtain a view of the
Corsican coast, with its serrated ridge of mountains
on the extreme edge of the sea, you must 'rise
to the occasion' by an ascent either in a balloon,
or to the higher points of the Maritime Alps—at the
back of Nice, Mentone, Cannes, Monaco, or some
other part of the Riviera. From the top of the
Turbia or the Berceau, immediately behind Mentone,
and to which the imperturbable donkey makes his
adventure daily, the coast itself of Corsica may be
seen when the isle is visible. Below, on a fair day
and under favourable conditions, are seen the higher
inland peaks of Corsica, Monte Rotondo, Monte
D'Oro, Monte Cinto, and others, ranging from 8000
to 9000 feet high.

The newspapers of the Riviera were full of admiration for the splendid view we had enjoyed. The oldest
inhabitant had not seen Corsica so plainly. Some
even thought that they had seen through the island
into Monte Christo, Dumas's pet, and some into Elba;
but these were enthusiasts, who have second sight!
Generally Corsica does not honour the continent
with a view more than once a month; and that
too, most vividly, just after the sun rises. That
orb, of course, is at first behind the granite mountains of the island. Being on the move, he sur-
mounts them; and lo! they die out of our vision
as soon as he rises above them. Still, even when Corsica is not seen from the continent, one may discern
where it is. Clouds, by a law of their own, are
anchored over the splendid mountains. They are as
immoveable as if they were not vapour. There are
conditions when mountains 'melt like mist and solid
lands, like clouds they shape themselves and go.'

Here there is a reversal of this law of the poet and philosopher. The granite of eastern Corsica seems to have been frozen in fire, and unmelted by the ages. The clouds this evening as they are piled in grand, solemn, heavy, golden, orange and red masses, have all the moveless massiveness of a mountain's gravity. I do not know how it would impress others; but if a New York citizen, with a dash of poetry in him, from the top of Trinity steeple, or higher, could get a glimpse of the White or Alleghany Mountains, or an Englishman from Dover cliffs could discern Mount Rosa or Blanc, the effect could not be more mysterious or charming, especially if those mountains were tipped or streaked with silver snows! It is this charm or mystery that prompted me thitherward. But, being under medical care, I had to obtain medical consent. On consulting my physician, and afterwards our companion in travel, Dr. Henry Bennet, he at once gave consent. In fact, if you will read his book on 'Winter in Southern Europe,' you will see that Corsica, as a health station, or sanitarium, had already been observed by him on his own travels. What deductions he makes from his observations, the reader will perceive by the perusal of his volume. But of this hereafter. Enough that we had our doctor's cheerful advice. With it he gave us many letters to medical friends at Ajaccio, Bastia, &c., and, all alive to the trip, we started from Mentone for Monaco, where we took the cars in the afternoon for Nice. At seven in the evening we were aboard the steamer 'L'Industrie,' ready to plough toward that isle where Ulysses once found a harbour —where Seneca found his exile so irksome—where the European world found Napoleon, and—to descend from the great to the small—where a beautiful morning, in early March, found the writer.

Corsica is out of the route of European travel. It has not been written up or down. It is one of Bayard Taylor's 'Bye Ways' of travel. The very fragrance of the isle—the product of its uncultivated lands—has preserved it from the curiosity and vandalism of common travel. A few Americans, a few English, and a few French, on business, or for sporting, or health, have come to Corsica within a year or so; but their sojourn is confined to Ajaccio, or to a race across the island in a diligence to Bastia. As it takes twenty-four hours to cross the island, the tourist confines himself to gazing out of his coupé or sleeping through the romantic and variegated mountain and coast scenery.

I am afraid that I cannot adequately convey my impressions of this island. I must content myself with a few pictures of the land and people, not omitting some of the experiences of travel and some historic associations.

First, to get here. We found a neat and dashing steamer ready for us at Nice on Wednesday for its weekly visit to the island. The sun was going down over the range of the Estrelles, near Cannes, as we clambered up the gangway. Two ladies were along with me. Between us three, however, we commanded four languages—English, French, German, and Italian. The Italian was altogether the most useful. The people on the Corsican mountains leap to hear it from a stranger. It is indispensable in Corsica. The Corse tongue is very near Tuscan. It is mixed Arabic and Italian. It is a patois, but an Italian can comprehend it. A strange-looking gentleman, with broad-brimmed hat, knee breeches, and long dark robe, accompanied us in our boat from the shore to the steamer. Who was he? He talked French, looked like a Greek priest, and acted like a gentleman. I

had a note to the Bishop of the English Church of
Gibraltar. I wonder if it be he. I make bold to
ask. He is happy to say that his church *is* founded
on that rock; but his diocese runs to Constantinople.
He is about to go to Corsica, where a church of his
persuasion is about to be inaugurated. He proves
an agreeable companion. We need some cheer. The
rain begins to dampen the deck before we sail; but
soon the stars appear, the moon struggles through
the clouds; the dark mountains above us on the shore
lose some of their gloom. The clear obscure of the
night is gilded by the glancing lights from the long
line of gas jets along the Promenade des Anglais at
Nice. Further out, as we enter upon the open sea,
the glaring and meretricious lights of that bold, but
not big nation of Monaco, are reflected upon the
waves and broken into myriads of sparkles. Soon the
terrestrial lights give way to the celestial luminaries,
and we give way to sleep.

We awake in the morning at seven, after a nap
of twelve hours, within an hour of the north-west
coast of Corsica. We are opposite to Calvi; but our
vessel, for nautical reasons, to avoid the rough sea,
hugs the coast. It thus gives us a near view of its
crested white mountains, the macchie covered land,
the chestnut, olive, and orange-clad valleys and hills,
and the wonderful fringe of white breakers foaming
among the rocky indentations, whose peaks are decorated with the old Genoese towers. At first the island
reminded me of the Archipelago of Greece. It
looked so nude and rocky—destitute and deserted.
The scenery was thoroughly Alpine, back from the
sea. The interior seemed one magnificent range of
mountains. No plains, no marshes were visible. Corsica has two ranges of mountains. You never lose
sight of its primitive granite. It begins near the

coast at Isola Rossa, and bounds upward 8000 feet, and 'keeps it up,' longitudinally, the length of the island, which is 150 miles. The eastern range is secondary and calcareous, like the mountains along the coast of France and Italy. The granitic range has not been washed away to any perceptible extent. Its grand, grotesque, and imposing forms—twisted and writhed by the fires which heaved them from the ocean—are to-day as they were countless ages ago. The eastern range has been worn. Some alluvial and unhealthy plains have resulted. Brackish ponds and deadly malaria are incidents of the autumn on the eastern side of the island, but not the western. Hence, the western side has been selected by eminent physicians for health stations. The mountains, especially around Ajaccio, shelter that city; and the weak in throat and lung find relief, and often cure.

But I anticipate. We have not yet landed. Nor yet have we reached Ajaccio. First, we must pass within thy gates, Sanguinaires! In which apostrophe I allude to two or three, or more picturesque islands, almost mere rocks, called Sanguinary (wherefore called I cannot guess), which guard the outer harbour. These rocks are neither red nor bloody, but grotesque and brown, and were under a white veil as we approached. The sea covered them with the finest spray from the breakers, which howled among their caves, and hallooed, over their sides. Passing between them the sea is calm at once, as if by magic, and the air is balm. The fine bay of Ajaccio is spread out to the view. It is surrounded on every side by mountain ranges. Many parts are snow-covered, and their sides checkered by shade and sun. There are few houses or villas to be seen. The bay reminds me of the bay of Naples. Others say, it recalls Lugano, in Switzerland; and some, that it reminds them of

Irish scenery. I cling to Naples as the best known portraiture of the features of the bay. If the shores were not so desolate the resemblance would be more striking. The air was balm; not alone, because the exposure was to the south-west, and the bay was so inland as to be protected from the mistral or sea breezes; but because the atmosphere was freighted with the perfume of the mountains. Will my readers in New York, which is on the same parallel of latitude as Corsica, believe that here, at 42° north latitude, the very weeds are the fragrant flowers of the New York conservatory! The very scrubs are the sweet-scented shrubs of the lady of the Fifth avenue! Even the winter air is redolent with a burden of lemon, orange, and myrtle aroma, for which Arabia has no parallel in summer, and the mind of the stranger no conception till he inhales it. Do you remember what Napoleon said, from his rocky island prison of St. Helena? '*A l'odeur seule je devinerais la Corse, les yeux fermés.*' Now, as I write, in the afternoon, in a villa with open window, with a gentle breeze blowing across the bay from the sea, and towards Corte, the realization of Napoleon's words is faithful to satiety. The heraldic bearing on the shield of Corsica once was a Saracen's head, with eyes bandaged. It seems as if there were here such an affluence for one sense, that sight was superfluous. Besides, we have nightingales in plenty, whose song is fragrance to the ear.

As we ploughed up the bay, we asked one of the passengers the question asked by many a stranger before: 'What are those beautiful little houses, so oddly made, lining the shore?' 'These are the villages of the dead.' The dead lived in these towns, as an Irish friend afterwards remarked about these mortuaries. Cemeteries there are none, or but few. These few are not attractive to the living; hence these private burial houses. The funeral custom of the public cemetery here is repulsive. The grounds are poorly kept. The graves are in the form of long trenches, to be filled up from time to time. The coffin is brought into an adjacent church. A crowd attends, with one or two priests; the latter utter a few words, and go off. The coffin is then hurried, on a trot, Mahomedan fashion, to the grave—the crowd following. It is dropped in a hurry. A rush to the town, with great disorder and noise. So concludes the ceremony. No wonder that private burial is preferred. While among the mountains we found many of these odd places of domestic sepulture. Each family which can afford it, has a private mausoleum. It is nothing here to decorate the little garden around the tombs. The cypress, evergreen flowers, and shrubs, which no winter strips of foliage or fragrance, by their own energy of vegetation, and without even the care of the bereaved, preserve in picturesque and perpetual sweetness, mournful memories of the loved and lost.

But our vessel moves on. Soon we are at the place of landing. There are no docks. The anchor is dropped. A throng of boats surrounds the vessel. It seems as if there were 500 people on the shore—an eager group. We are the objects of scrutiny and sympathy; the sympathy has some reference to baggage; for while no carriage, dray, or wheeled vehicle, not even a wheelbarrow is visible, there are plenty of lazy-looking people ready to carry, for a sou or so, our 'traps' to the hotel. We walk up the leading street, past the allegorical monument upon which Napoleon I. stands in marble, draped in a toga. Around it are some score of women. From the spouting mouths of the four lions at its base they are filling their earthen jars with water. They form quite a classic group of antiques, but would look better if they were not so dirty. We have but little leisure to admire them in their mandiles and fandettas, as the immediate question is—who shall get to the best hotel first. Accommodations are limited, and on steamer days especially so.

One thing the stranger, even on the most superficial glance, will notice—not the street, with double rows of orange-trees for shade; not the little railroad, carrying stone to the shore for the jetty in process of construction, followed by a small boy to blow a horn — but the signs, emblems, and influences pervading the town. Bonapartism is the presiding genius of the place. Ajaccio would not be accounted much, except for its bay and healthy situation, were it not the birthplace of the most remarkable man of the past thousand years. It has a fine gallery of paintings; but that is a gift of one of the Napoleonic connections, Cardinal Fesch. It has a beautiful marble chapel; but that is a mausoleum of the Napoleon family. It has an old cathedral, where Madame Letitia was accustomed to take the young Bonaparte to mass. It has some fine villas; but chief among them, which we visited, was that of the Princess Bacciochi (still Bonaparte), which that lady, just dead, has bequeathed to the young Prince Imperial. All the streets bear Napoleonic names. The cafés, also, are named after Jerome, or some others of the family. There is a street called after the King of Rome. But chief among the souvenirs of the great family is the house where Napoleon was born. It is situated next door to a hatter's. I looked in upon the hatter at his work; looked through the grated windows of his workshop, and there, unconscious of his ennobling proximity, he was beating away at his trade.

The house where so many princes of the earth first drew breath is next door to a hatter's! 'We shall all meet at the hatter's,' saith the homely proverb. But let me be more precise.

Passing up a narrow street in the heart of the town, you come upon the Place Letitia; then upon a little open spot, ornamented with tropical palms, shrubs, and flowers. This plot was opened and enlarged by the mother of Napoleon, by tearing down a house in front of her own. She used the house on the left side of the plot as a stable and coach-house. The woman who showed us the premises was formerly a waiting-woman of Caroline Murat. She lives in the old stable opposite the hatter, which is neatly fitted up. Upon the front of the Bonaparte mansion there is an inscription, that within this house Napoleon was born. It gives the date, the 15th of August; so that if tablets of stone are evidence, as they often are, according to Greenleaf, of old dates and events, here is a record to satisfy enquiry, or rather to provoke more discussion as to the great day, so soon to be celebrated on its hundredth round. The house is large—four stories—and of stone, evidently one of the best of its day. It is hardly excelled, even yet, in Ajaccio. It shows that the family were 'well-to-do' in the world. In fact, Madame Letitia's family (the Ramolinis) were rich, and had many estates upon the island. When she died, she bequeathed her furniture to members of her family. The present Emperor has collected most of the articles, and replaced them in the house which he now owns. Our conductress told us what each object meant, and whose room this, that, and the other was. The second floor was Madame's bed-chamber, dining-room, and salon,— quite commodious. The floor above it was devoted to the sons. Here was Napoleon's room and his

bureau! Yonder, Joseph's! There Jerome's! The
floor above that for the boys was occupied by the
daughters. There was a very small terrace looking out
from the reception-room, with flowers in pots—all in
bloom. This reception-room contained five little
square mirrors, as high as your head, running down
each side; three other mirrors near them, and one
grand mirror at either end of the room. In fact, we
counted some fifteen mirrors in this room. The Napoleons were well provided with looking-glasses. We
cannot say as much for any of the other houses we
have seen. What effect these mirrors had upon the
young Napoleon,, I will leave, with other *reflections*
to the reader. Two brackets for wax tapers hung
below each little mirror. The ceiling was ribbed.
It had been newly plastered, with some little effort at
fresco. The mother's portrait, very finely executed,
hung over her dressing-table. It presents her as a
splendid woman. The cabinets were inlaid with
every-coloured marble, and very antique. Some rare
objects sent by Napoleon from Egypt were displayed.

But the chief attraction was the bed on which Napoleon was born! It is a wooden, rickety affair; near
it is the sedan-chair of the Madame. It is well-
known;—well I will put it—politely—in French, that
—Madame Letitia *surprise à l'église par les douleurs
de l'enfantement le* 15 *Août,* 1769, *fut rapportée a
son domicile!* In this very sedan the mother of the
great Emperor was borne from the church, and on
this very couch, the 'little corporal' first drew his
breath. From this bed to St. Helena; from the
capitol to the Tarpeian Rock! Fill up the gap—
Arcoli, Austerlitz, Waterloo! Here, in this house, he
passed his time, playing at soldiers with his fellows,
and mourning because he had no moustache—till, at
the age of fifteen, he entered the Military School

at Brienne. He returned home for vacations, and mixing his young ambitions with the daily round of boyhood pleasures and walks, he, at the same time, imbibed from the peculiar characteristics of the Corsicans, those feuds and feelings, which his grand, gloomy, and marvellous after-life illustrated.

Just above where we lodge, at the end of the avenue—where, if you go, you may probably see a company of lively ladies and gallant officers playing croquet—is a grotto (of which I present a sketch) formed by boulders, and surrounded with foliage, and musical with birds, celebrated as the favourite study-spot of the young Napoleon. The grotto commands a beautiful view of the bay and the snow-clad mountains around. From it can be seen on a clear day, the island of Sardinia. Gardens of oranges, from which we are permitted to pluck at pleasure, cover the slopes near. Hedges of cactus (cactus opuntia) —cactus piled on cactus; the famous macchie, so sweetly scented; the arbutus, the myrtle, the olive, and every kind of tree and colour of flower, grow in the air of winter. It was within the circle of such influences of sun, sky, land, and water, that the young Napoleon formed his plans of life. How many of them failed, or how many of them were realized, we can only guess. One thing remains to be said : that in the hurly-burly of his active career he never forgot Corsica. He always intended to do more for her than he did. His last thoughts were about his native isle.

In Ajaccio, you cannot escape the Napoleonic impressions. They follow you whenever you go out; and they enter with you into every public place. If you go to the Hôtel de Ville, which contains a library of historical interest—at least to the Bonapartes— you will see the picture of a lawyer—Carlo Maria

Bonaparte. He is a handsome man, of elegant appearance, fit to be the bridegroom of the belle of Ajaccio —Letitia Ramolino; and fit—if any body is fit—to be the father of Kings and Queens. Go into the Collège Fesch—named in honour of Letitia's half-brother—and you will find the pictures once owned by the Cardinal, and presented by Joseph Bonaparte as lately as 1842; and in the library, a bronze of the Cardinal. If you would see the tombs of the Cardinal and his half-sister, you will go to Rome for them in vain. They were removed to Ajaccio. The remains of Madame Letitia repose, beside her brother, in the vault of the chapel in the Rue Fesch, near the College. The Cardinal began the chapel, and Napoleon III. finished it in 1859. It is a stone building in the form of a cross, with a dome over it. The inscription in Latin over the tomb of Letitia is 'Mother of Kings!' Everywhere in Ajaccio, among living and dead, Napoleon the Great appears. He may have been the Prometheus of Byron's ode, chained to the rock, or like that Tartar prince, caged; or so much worse than Charles V., that he took a throne, instead of abdicating one; but Byron's muse failed in her vaticination. Whatever the premises may be, the conclusion is not verified :—

> ' If thou had'st died as honour dies,
> Some new Napoleon *might* arise.'

A group of kings may be seen upon the fine, roomy, open plaza, called the Place Bonaparte. Here the first object that arrested my attention, after I emerged from the hotel, on our arrival, was an equestrian bronze statue of Napoleon—colossal and grand. At the four corners of the monument—*on foot and below*, and draped in Roman vestments, are his four brothers. They all look off to the sea, as if they

were in, but not of, Corsica; as if their musings
were pre-occupied with other lands. This fine me-
morial was dedicated by Prince Napoleon in 1865.
He made, on its inauguration, a fierce democratic
oration, which gave offence, as it was alleged, to the
Emperor.

In glancing at Ajaccio, I have thus far written the
impressions of one or two days' sojourn only. I shall
return to it again after our visit to the interior, whither
I am about to take the reader in the next chapter.

CHAPTER V.

AMID THE MOUNTAINS OF CORSICA—A STRANGE, WONDERFUL LAND.

> 'By anemone and violet
>   Like mosaic, paven:
> And its roof was flowers and leaves
> Which the summer's breath enweaves,
> Where nor sun, nor showers, nor breeze,
> Pierce the pines and tallest trees.
>   Each a gem engraven,
> Girt by many an azure wave.'—SHELLEY's 'Isle.'

IT will require a long chapter to describe, even compactly, the views, experiences, and impressions of our four days' visit amidst the mountains around Vico, Corghese, and Evisa—amidst rain and snow, sunshine and cloud; among forests of pine and ilex (live oak); everywhere the everlasting evergreens and efflorescence of Corsica. A strange, wonderful land—half Oriental; tropical, yet with snow-clad mountains; so wild in parts that the wild boars and wild sheep are hunted in forests where the pines grow 30 feet in circumference; so tame and sweet in parts that even the macadamized roads are covered with the sweetest and smallest clover; so sea-surrounded that at no point do you escape the impression of the mobile element; so grand in its mountains that the Alps scarcely surpass it in magnificence.

I am free to confess, that during this fascinating journey I forgot that my search was for Sunbeams. Reaching a wintry altitude, it seemed that the nearer

I approached to the Sun, the less I saw of His Majesty. The truth is, that the month of April or May is the best month for the mountains or interior of the island. No invalid should travel here in the month of March. As our driver said, March is the month for caps and bonnets, and she changes them daily with the fickleness of female caprice. It is the only month, however, which tries the perennial geniality of this climate. The mountains now seem full of clouds, which are a reservoir of rain and snow. The clouds above them may hang heavily; when all at once the sky and sea are changed, and we have the air and shine of an English June. One thing, however, never occurs to mar our enjoyment: it is not cold; never, unless you ascend very high into the snow-mountains. That you can do; for the roads are excellent to every part of the island.

Before we start, it is best to procure one of the departmental maps, issued by authority of the French Government. The French engineering is nearly perfection. Not only so because of the roads which the engineers construct, but because they publish their labours in such perfect style and with such abundant detail. Arrondissements, communes, post routes, bureaux of letters, big towns and little towns, Roman and Moorish ruins, châteaux, the canals of irrigation, and even the population; but, most useful of all, the routes are marked with precision. The Corsican roads are of three kinds: the imperial, the departmental, and forest roads. There is not much difference in the quality of these roads, strange as you may think it. The English roads are not better. Except here and there where the big timber carts have rutted them, they are superlatively excellent. The Central Park Commissioners of New York, with all their police, taste, vigilance, and skill, have not made any better roads for equestrians or carriages than we found in the very midst of mountains here, four or five thousand feet above the sea. This is owing mostly to good engineering. It is due somewhat to cheapness of labour and proximity of good stones for macadamizing. The water is drained skilfully and runs harmlessly from or under the road.

Yet, with all this expense and care, comparatively few, as yet, use these roads. On a ride of four days we met not one carriage. The diligence goes; donkeys and women—(excuse me for the connection)—bearing burdens, and the famous tough, ill-kept, but spirited Corsican pony—these are to be met with; but it is such a rarity to meet a carriage that when we pass through the mountain towns all the population turn out to see!

This may account for the trouble we had in obtaining a conveyance. At Ajaccio, but one man, Jean, had carriages, and he had but few. These were already let. We found what is called a *breck* at another unpretentious stable—owned by one Rivies. It is a cross between a small uncovered omnibus and a jaunting car. We paid for it twenty francs a day, and had to 'eat the driver.' When I proposed, in a horrible way, *manger le cocher*, that humble person standing near—Caggini, by name—was not merely pale, he was paralysed! He thought it was an American custom. To give drink—*pour boire*—is a common and rather expensive custom all through France; but we had improved on it. I do not complain, however, of our vehicle, and I found Caggini very brisk and serviceable. We had a very pleasant day to start, all Sunbeams. Our first route lay along the sea from Ajaccio to the north. We took the Imperial route, which has nearly been completed round the island. It has only one gap, soon to be filled. We started on the rise from Ajaccio. As we gained the heights, we perceived a good deal of cultivation. Besides the sheep-pastures, there were patches of bright green, which we find to be flax, and vineyards in plenty along the terraced mountain sides.

The sheep are nearly all black; one white for twenty of that hue. The white are small, and have coarse wool; the black have a sort of hair—much used here and elsewhere for mattresses. Their hair is long and nearly straight. I might call them a little "conservative." The mountains and valleys were dotted with them as thickly as Carlyle's page with capital letters. A shepherd, here and there, was on the watch, with his dog. Occasionally he would hurl a stone at the flock to remind them that they were under supervision. Goats were nearly as common as sheep; but they looked and acted more like vagrants. Indeed, the goats have much to answer for in Corsica. They have spoiled the land. They live around among the macchie or brushwood, and are voracious of every green thing. They furnish nearly all the milk. A cow here is a curiosity. We like the goat's milk, especially the broccio, or curd boiled in sweet milk. Among the picturesque objects to be met with upon the roads in the mountains, are the processions of country women, bearing upon their heads large boards covered with little dainty baskets, in which the broccio is moulded and borne to market. It was not uncommon to see the milk dripping, in unctuous rivulets, from the baskets about the handkerchief-clad heads of the bearers.

Have I mentioned the moufflins, or wild sheep? They are very common, and have been much hunted. They are larger than the tame sheep. They are more like deer than sheep. Their horns are thick and bend forward. You can easily separate them in

your judgment from the goats. They are thoroughly
conservative—as to hair; but their colour is mixed—
brown, black, and white. They have a frizzled mane,
and an eye like that of Genius, quick and fine. They
are hunted by dogs and men. When caught they
are easily tamed, and become pets with the children.
They are as timid as hares in the woods, and become
as affectionate as dogs, when domesticated. They are,
when caught young, turned over to the goat to feed,
and they are not tender of their foster mother. They
often suck with a dash of ferocity that draws blood
instead of milk. Indeed, everything in Corsica has a
wild and fierce style. The Vendetta has inoculated
the animal life.

On our upward way we perceived villages upon the
mountains, perched high, like all the little towns on
the Italian and French coast; so high, that during the
invasions of the Saracens, they were comparatively
protected in their rock-bound strongholds. We meet
on the road, occasionally, women carrying on their
heads their burdens of broccio, wine, fruit, or faggots,
and invariably knitting as they walk. Now and then
they ride upon the donkey, and ride—will you believe
it?—*à califourchon*, astraddle. Indeed, I met, upon
a donkey, a man riding upon a side-saddle and a
woman behind him astride on the meek little bearer
of burdens. Our upward route was splendid, having
now and then entrancing glimpses of the sea, which
was lashing itself against the rocks, and blanched with
beauty. Alata on the left,—Appietto, farther on, and
the almost unpronounceable Calcatoggia village, show
that somebody does live hereabouts, though cottages
and huts are scarce outside of the hamlets and in the
country. At last, after some dozen miles of ascent,
we reach the chapel of St. Sebastian. Here we have
a famous spot for sea views. We obtain a glimpse of

the white-clad mountains on our right, into whose heart we are to penetrate, and whose glistening scenery beckoned us thither. There is one peasant living near the chapel. His house is a jail-like, stony edifice. The smoke comes out of the windows and doors. The cactus is abundant. The 'proprietor' is, as all proprietors are here, clad in cheap brown velvet cloth. He seemed utterly at leisure. The land is cut up into little patches, sometimes separated by stone walls or hedges of the macchie, but, on the whole, it looks desolate. The little dashes of green which smile as the sun glances upon them through the changing clouds, give some relief to the general desolation.

As we wind our way round mountain sides and over the valleys and under the shadow of overhanging rocks, we discern, in cultivated places, workmen in gangs, either digging round the olive, making terraces, or working the soil for the vine; or, what is very common, making holes in the ground to hold the rain, to assist the work of irrigation. Indeed, the vineyards, which furnish the full-bodied wines of Corsica, are everywhere perforated with holes. Every other hole has in it a labourer. As we go by, he rests upon his spade, and wiping the sweat from his brow, respectfully lifts his Phrygian cap, with a '*buon giorno.*' If he were before the footlights he would make a capital personation, in attitude and costume, of the grave-digger in Hamlet. But who are these workmen? Not Corsicans. The small 'proprietor,' who wears his velvet clothes, and owns a few chestnut or olive trees, or his flock of goats and sheep; and his more wealthy neighbour, who has his vineyard or lemon grove in some sequestered valley, are not the workmen of this isle. These gangs of labourers are Lucchesi. They live in Italy. They come over to Corsica for a few months, make and save a few

hundred francs, and return to their homes. They
work for two francs a day. In their industrial and
economic migrations, they are like our Chinese ;
and are equally despised by their superiors. In fact,
they are the only labourers of this land, so favoured
by sky and sun, and so neglected by the ' proprietors.' What work the Lucchesi do *not* do here,
the women do.

Before leaving St. Sebastian, the ladies of our party,
not content with glances at the sea through the magic
aerial mountain prospective so eulogized by Ruskin
and so illustrated by Bierstadt, led me, not unwilling,
along a steep track for a mile beyond the chapel on
the road, to a point overlooking the coast and almost
overhanging the wild foam beneath. There, amidst
jagged and ragged rocks, ripped and torn by the mad
elements from a dateless epoch, we enjoyed view upon
view, 'splendour far sinking into splendour without
end.' Far along the shore to the south, and reaching
along the Gulf of Sagone to the north, even as far off
as the Greek colony of Carghese (which we afterwards
found lost no beauty by being near to it), the eye
could see, in the clear air, made more clear by the
bright cerulean of the water, all the indentations of
this wonderful coast. The sea was not calm. It made
me think of Tennyson's line :—

' League-long rollers thundering on the shore.'

The immediate surroundings of this romantic spot
were beyond all expression wild and grand. At the
extreme point of the promontory there was an amphitheatre, resembling in form the Cirque or Oule,
peculiar to the Pyrenees, except that the latter is a
basin for water, while the former is as dry and warm as
the rocks can be under winter sunbeams. The walls,
and especially the varied shapes of loose and scattered

rocks, give the impression at some distance of human handicraft, and even human beings. The semi-circle might, with no extravagance of fancy, be likened to a grand granite Congress. True, these petrified Congressmen were as silent as Congress is not. Their postures were rather stiff for eloquent gentlemen. Their heads—I speak with parliamentary respect, under the rules—were hard. We picked out of the group the notabilities. There was one half-boulder, half-granite sort of member, whom we all agreed was a general as much renowned in debate as in war—in discussion as in concussion. With all regard to the proprieties, I clambered on the top of this honourable member's bald head, and undertook, under the five-minute rule, to give him voice. The surf below bellowed its applause. The lizards, which were lobbying around, came out to see. An occasional goat looked down from the galleries of rock above and seemed to shake his head and beard in grave acquiescence. My time expired. Down came the gravel and down came I. The ladies called the previous question, which was 'lunch,' and we were soon off to the chapel again. Thence, for the mountains, striking eastward for St. Andrea upon a forest or bye road.

Even this road gave our Corsican ponies but little trouble. We reached the small village of St. Andrea; and having bowed to the old curate standing before the church, we were greeted heartily by him in return. Learning that we were American, he tendered what we have invariably received in this island from the Catholic priests, the utmost of courtesy. They are as scholarly as they are urbane. Indeed it astonished us not a little, being such strangers, to receive such unbounded attention. 'Would we alight? Would we enter his church? What do we think of Corse? He eulogizes the wine of St. Andrea. 'Would we do

him the honour to try it?' We did. Under the
convoy of some twenty children who had gathered
about us with inquisitive eyes, we drive through the
little village, down a narrow street, so narrow that our
hubs grazed the houses on either side, and are met
by the curate and a delegation of his flock. One of
the latter bears a platter with two bottles of wine and
three glasses. The white wine is for the ladies; the
red—a full-blooded, spirited wine, emblematic of the
fiery qualities of the Corsicans—is for the gentleman.
Healths to the priest; to America; to Corsica; and
blessings from the heaven above, so bright and so
near! May it ever be near to the venerable father!
Before we go, we desire to present something substantial for the wine and kindness. Our douceur was
offered to the curate: 'No, I offered the wine with
my heart.' Then to the cup-bearer, 'Oh no, monsieur! we are not poor.' Finally to the church we
offered it, and it was accepted. With many kind
farewells, we were again on our upward way. We
meet one of the guards of the royal forests, gun on
shoulder, and dressed in green. We talk with him.
He will not drink. He is on duty. He looks quite
like a picturesque forester.

All along this route the sides of the road are white
with a fringe of marguerites. The shrubs furnish a
surfeit of perfume as evening comes. The bells tinkle
upon sheep and goat in the valleys far below. Our
afternoon is one oft-repeated exclamation! Magnificent! glorious! We reach the chestnut groves;
but they alone are leafless. No winter sunbeams
clothe their branches. Then the forests of ilex or
evergreen oaks appear. These, like the rocks, are
old, storm-worn, and twisted. Each one is a picture;
almost a grove of itself.

It may not be amiss here to say a word about the

chestnut. It is the bane and the blessing of Corsica. It grows in the higher latitudes. It requires no care, as the olive does. It never fails to yield a crop. It is more plentiful on the eastern side of the island. In fact, there is on that side a circle of land, called Castagniccia, or chestnut country. There the people live almost entirely on chestnuts. Where the chestnut forests abound, a little olive oil, a little wine, a few figs, and sometimes a kid or a wild sheep complete the diet. To this chance food, the chestnut, picked without labour, may be attributed much of the improvident, lazy, and independent habits of the Corsicans. The dry chestnut is given as food to the horses. They like it. It is hard, but they grind it in their mouths as a sweet morsel. The people use it roasted sometimes, but generally made into flour. It is said that there are twenty-two dishes made out of the chestnut in Corsica! The chestnut cakes we had at Cauro, from an old brigand hunter, were delicious and nutritious. They are baked in square, flat pans, and have the colour of our buck-wheat. When the Pisans, Genoese, and French undertook to subjugate Corsica, the 'Chestnut Boys' beat them. Their rations cost little. Hemmed in by mountains and rocks, or hiding in caves, while other parts of the island succumbed to the yoke of the foreigner, the independent chestnut-eater was un-subdued. It is not a new kind of food here. It is as old as the earliest traditions of the island. When Æneas deserted Dido, and when his companion, Corso, a bad young man, from whom the island is named, eloped with Dido's niece, Miss Sico (hence Corse-Sico), and when the young people were followed hither by the irate brother of the bride, they celebrated their honeymoon in a chestnut grove, had baked chestnuts for breakfast, chestnuts *à la poulet* for dinner, and chestnut cakes for supper. The

mother of Napoleon, it is known, had a dairy, when at
Paris, for goats' milk and broccio.   Doubtless the
young and old Napoleons preserved this family affection for the food of their native land.   There is a
good deal of fight in chestnuts.   The brigands, who
have hardly yet been exterminated, had a constant and
ready supply of this home-made, belligerent aliment.
They retired on it to their fastnesses and defied starvation and surrender.   Perhaps the world is indebted for
its Bonapartes—to chestnuts !

Above the region of chestnuts, is that of the live
oak ; then the big larches, 170 feet high ; then the
beech and birch ; and then the ever-during snows.
In saying so much about the Corsican trees, it would
be unfair to omit the olive.   The tree does not grow
here so large as on the Riviera coast, but it is very
productive.   If one tithe of the expense and time put
upon the olive on the coast were bestowed on it here,
the yield would be immense.   The olive requires care
and skill.   These the Luchesi, who are mere labourers,
do not give.   All the care given to the olive is by the
women.   The olive requires water, and water here can
be had only by directing it into reservoirs and thence
to the roots of the trees.   The olive is indigenous to
Corsica.   Birds bear the stones about the isle.   These
little olive-planters are almost as thick as the leaves
amidst which they hide and carol.   When these bird-
borne seeds grow up they are called 'sauvage olives.'
Not many years ago, the authorities took a census
of the savages.   They then numbered twelve millions.
Two years ago Corsica showed at a fair 250 different
specimens of olive oil.   Were this land in America,
with its sweet sun and energy of soil, as yet unworn
by tilling, what a revenue it would bring !   How
many of these young 'savages' would be civilized !
How they would set off—æsthetically speaking—with

their sombre green, the pale pink blossoms of the almond and the gorgeous pink of the peach! How the rich verdure and golden fruit of orange and lemon would contrast with the foliage of that tree which has been made sacred by its memories of Olivet and the Holy Land! How the endless macchie, the myrtle tree, the arbutus, with its red, ripe berry, and the laurustinus, would make this island a magnificent bouquet worthy to be laid before the Olympians! Yet many of these fragrant and beautiful plants are being made into—charcoal. The bright, blue flags of smoke which float here and there upon the mountain sides, proceed from the carbon-makers. The heavy-laden carts we meet have their sacks full of black diamonds—for is not the diamond of carbon? The cuisine of France, so celebrated, is lighted by the coals which are made from the most aromatic plants known to the sun and earth! In fact, the sunbeams I have searched for and found, afford, in one analysis, the fuel for cooking the elegant viands which furnish the entertainments of millionaires and kings!

There is little work in the Corsican; but how much he might produce from his isle! Out of the two millions and a quarter of acres of which Corsica is composed, it is estimated that only *one six-hundredth* part is under cultivation! In a land where the seasons are one—even as the sun is one—and where perpetual vegetable life is the law, the great body of the 250,000 people of the island are content with chestnuts and the memory of Paoli, goat's curd and the glory of Napoleon! Instead of soldiers everywhere seen, we might see fewer goatherds, more shepherds, and still more labourers in vineyards and around olives.

In the elder day, Rome was mistress here. There are innumerable traces of her power and greatness. Witness the ruins, coins, aqueducts, and theatres!

This isle, then, had 2,000,000 of people. Now the
labourers are imported, and the French government
pays ten dollars to help Corsica, while she gets one in
revenue! Yet the isle is pre-eminently rich in unde-
veloped wealth. Not to speak of wines and woods,
olives and oranges; marbles and porphyry—precious
stones, incident to its mountains—are here found in
abundance. Lead is found at Barbaggio. But there
are two things lacking—capital and enterprise. Now
that Corsica is recommended by eminent physicians
as a sanitary resort—not as the rival, but as an adjunct
of the Riviera of Italy and France—travellers will
come. They will observe. They will print their
observations. Already my eminent friend, Dr. Henry
Bennet, in his volume of ‘Winter in the South of
Europe,’ has called attention to facts and deductions,
showing the advantages of Corsica for its climatic and
remedial influences in some pulmonary disorders.
Corsica owes him a lasting debt of gratitude.

Hitherto the island has been so far apart from the
ordinary routes of travel that few have observed, even
what has fallen under my eye. I think the very
wildness, not to say, reputed inaccessibility, of these
mountains, with their great forests and ready sub-
sistence, have had much to do in forming the Corsican
character. They have left Corsica fifty years behind
its neighbours. The people are made hardy, independent, and defiant, but vengeful and lazy. But
they were full of heroes. Della Rocca, Sampriero,
Paoli, Napoleon—a constellation—and, I may add
(for I have seen him, yet alive, in his own garden,
where his great General Napoleon planted lemons and
wandered, when a boy), General Sebastiani,—these
illustrate the martial daring and obstinate character of
the Corsican. The people, naturally, are very proud
of these names; and but for *Napoleon,* they say, their

soil would not now be a part of France. Indeed, it is common to hear them say: France did not annex Corse—Corse annexed France!

As I think over the salient points of the Corsican character, I forget that we are on the road. Night is coming on, and clouds and storms are gathering over the mountains above us. We are several hours from our destination. We are nearly 3000 feet above the sea, and have just in view the town of Sari. We meet a "solitary horseman." He is coming down the mountain by a by-path. We inquire of him our distance. "Three hours yet to Vico." He dismounts from his pony; asks me to ride it. Without any bridle, only a rope halter; with a saddle almost as big as the pony, and stirrups as big as the saddle,—I mounted. My companion is a "proprietor"—has a vineyard of two acres—is curious about America. I explain to him the homestead law. He looks incredulous. One hundred and sixty acres! and all that, for next to nothing! He shrugs his shoulders. He points out to me the mountains, by name. To the east is the range of Monte Rotondo. It is the highest in the island, 9000 feet! It is surrounded by other mountains of less altitude. The mountains we are now ascending are covered with vegetation and forests. The range of Rotondo is made up of pinnacles, towers, castles,—as grand as anything I have seen in the Alps. If not, they seem so; from our having such grounds of vantage for observing them. I ask my conductor of the pony, half playfully: "Why is it, that here on this mountain, we have oaks and chestnuts; the laurustinus and purple cyclamen, ferns, and violets—foliage, flower, shrub, tree—all in such tropical profusion; while right yonder, as if in reach of our voice, is rock and snow—desolation, sublime desolation?" He gives me for answer, with much seriousness: "It is the caprice of the Eternal Father." An answer worthy of the scene.

From the spot where I left my pony and his " proprietor," I counted fourteen conspicuous peaks, each white with a "diadem of snow." Above them hangs the threatening storm, in rolling masses of clouds. The air grows cool, then cold. The wind blows wildly. We are summoned to walk ; for our ponies have come to a halt. We can barely see the thread of our road, winding around and above and before us. Soon the lights of Vico shine afar like stars. The convent is reached and regretfully passed, for we had letters to those in charge of the institution. Soon we are at the hotel of Vico, bearing the imposing title of Hotel Pozzo di Borgo. It is named after one of the illustrious families of Corsica, represented at Ajaccio, and very wealthy. The hotel itself is kept by one of the family. It is the same family whose escutcheon may be seen on a Genoese palace in Ajaccio. That palace was built by the famous diplomatist Carlo Andrea di Borgo. One of the Borgos, Carlo Maria, was, in the earlier days of the French Revolution, a rival of Joseph Bonaparte for popularity in Corsica. He was a lawyer ; quarrelled with the Bonapartes ; took the side of Paoli and the English against the French, and was made Procureur Général under the brief English rule. As the Bonapartes rose he followed their star with true Vendetta hate. He entered the Russian service ; and, on the field of Waterloo, saw the fall of the rival family. He died, in 1842, very rich and distinguished. His relatives keep the hotel at Vico. It is curious that here there are the strangest contrasts of fortune and position in members of the same family. Some are very wealthy and eminent, some are hirelings and servants. The relationship may be recognised by both ; but the pride of the poor relations is indomitable, and there is no intercourse. The mother of this Vico-Borgo family did the cooking and the daughters the work for the inn. The latter received us at the door of the bleak, three-story, stony hotel. The invariable handkerchief was upon their heads. We enter and find traces of refinement. A guitar is peeping from a closet and an embroidery frame is near. We are met, on entrance, not alone by the daughters, but by a bouquet of unpleasant odours. I cannot commend the fragrance of the Corsican hostelry as I do that of the macchie. Soon a fire greets us. These are not the radiant beams we are in search of; but they are comforting. We change our wet garments and prepare for our supper. The rooms are not plastered; only flooring for ceiling. The windows have a touch here and there of fractured glass. The ventilation, for an invalid, is rather too free. The dinner comes : chicken, kid, beefsteak, no butter ; but the broccio does its office. We slept soundly, and were awakened in the morning by the crowing of chickens in the—hotel, and by the sound of the Sabbath bells reverberating through the valley from convent and church. The storm has gone by ; but the weather is only changing its cap. I must seek advice before venturing further into the mountains.

We had started from Ajaccio for the great forest above Evisa ; the forest where the wild boar and the bandit, the great larches and the snows, hold their sway. We desired to see the celebrated trees, known as the King and Queen. This forest is some 4700 feet above the sea. It is known as Valdoniello, or the Black Forest of Corsica. Corsica is written down in ancient lore as shaggy and savage. Theophrastus, Strabo, and others, mention its rugged inaccessible nature, covered with dense forests, whose excellent timber has been celebrated in all ages. These classic

writers illustrate the Corsican character by the ruggedness of these wildernesses. Although they attribute to
the people a rude sense of justice (La Vendetta ?) they
regard them as wilder than the beasts and more un-
tameable than their lands. The truth is the Corsicans
did not make good slaves. It was impossible to
accustom them to domestic habits. Their mountains
and forests, which have survived all the changes of
man, still remain to attest these ancient accounts.
Immense pieces of timber, drawn by some half-dozen
or more mules on some sort of "timber wheels," the
timber dragging its long length upon the ground,
were frequently met by us on our journey thither.
They furnished impulses for us to go up still higher.
We were told of one tree, already felled and hewn, so
large that there was no team big enough or lumbermen
enterprising enough to haul it out. It would bring
2000 francs in Genoa. If only it could be taken to
the sea, it would almost supply timber for a ship. A
Maine or Michigan lumberman would not long hesitate to undertake the job.

I had a note of introduction to the doctor of the
village—Dr. Multido. I called on him. He was a
young man of fine presence. He had been educated
at Marseilles, was a native of Vico, and had just inherited a fortune of 200,000 francs from his uncle, the
curate. He gave us all the information we desired,
and advised us to go as far as Evisa and to watch for
the storm. After a hearty breakfast, during which we
were waited upon by a fat girl who brought the "things
in" on her head and who did not look as cleanly as
her glasses and spoons,—after many good wishes from
our hostesses and a purchase of their embroidery, and
with many misgivings and amidst a crowd which must
have comprehended half the town,—our driver snapped
his whip at the ponies, and we were off. So was an

old man off—his donkey. The driver,—as all Corsican
and French drivers invariably do,—had cracked his
silk into a series of snaps. These extravagant noises
had frightened the old man's donkey, which had shied,
and the old man had tumbled. The discourteous
crowd roared at the misadventure. We stopped our
" breck," helped the old man on, and, amid the renewed adieux of the crowd, we dashed up the narrow
streets and soon were again upon the mountains.
March had indeed changed her cap. Our view was
clouded and a drizzling rain began. While in this
predicament, and still undaunted, a singular caravan
approaches. Down the mountain, upon Sunday too,
rush an excited throng! Forty men are drawing a
cart, by ropes, on which is a splendid millstone! A
soldier is captain of the gang. A boy on a pony,
decorated with ribbons, is dashing with them. Pell-
mell they go, flushed either with wine or with the
occasion. They are all in high glee, singing a rollicking song. Is this a merry-making of Ceres? Perhaps
the stone was cut in some place inaccessible to donkey
or horse? Perhaps it is a fête of the village handed
down from the time of Dido's niece? Far down the
mountain, till the throng passes into the village, we
hear the shouts come up ; and they are in strange
dissonance with the Sabbath bells which peal below
with sweet and sad vibrations.

Our ascent was not easy. Our driver said that the
horses were fatigued. Even chestnuts had not tempted
them to breakfast. We descend from the breck,—to
walk. We find, even here, violets and a flower—a
delicate, purple, bell-shaped flower—nameless to us,
but very sweet, growing in the very road. The wind
blows coldly. The tall ilex trees wave like feathers in
the blast. Magnificent groves—the vestibule to the
great forest—now appear. We reach the top of the

mountain in mid-afternoon; and, *facilis descensus*, we galloped like mad down into a valley; for though Evisa has to be reached by an ascent, still the last stage of the journey is down hill. We had a sight of a striped wild boar on the side of the hill, among the brush. So had the ponies. This sight helped the ponies more than the whip. The wind roared, and the ponies ran. The writer, being an invalid, in search of—sunbeams,—was huddled and hid under manifold wrappages by gentle hands. Our indefatigable driver braved the storm without overcoat and brought us safely through. We met but one human being on this route—a rough-visaged, brigandish-looking person in a coarse blanket, made dexterously into a capuchon, under which his diabolic face was hidden. He, too, was courteous, and showed us a short cut into Evisa. The town seemed deserted. The storm had driven even the chickens under the rocks and into the houses. Here we found the heartiest welcome from M. Carrara, who gave us his two best rooms, and his blushing wine, produced by himself in his vineyard at Porta by the sea. Although we had letters to M. Carrara, he had no advice that we were coming. When we asked what he could give for dinner, he replied, "Un poulet, c'est tout." Well, chicken is not bad. We suggested a variety in the shape of an egg, or omelet. "Oh, yes." We further suggested: "What will you do for us with only one chicken. It is nothing." This urged him to an adventure on roasted chestnuts. A lunch on Porta wine, pure as the blood of a goddess, and roast chestnuts, bursting with white seams and hot, to a hungry traveller! Can there be more cheer for hungry and storm-tossed travellers than we had of Carrara's poulet? We shall see. The storm comes on, the hail beats, the snow fills this upper air. The sunbeams do not appear; but what is more grateful, the dinner does!   Here is our menu in the wilderness of Corsica :

 1st Course—Chicken soup.

 2nd Course—Poulet bouilli.

 3rd Course—Poulet au riz.

 4th Course—Poulet au champignons.

 5th Course—Pommes de terre, fried in pig's fat.

 6th Course—Une crême des œufs, (embryo chickens.)

 7th Course—Broccio.

 8th Course—Dessert, raisins, grapes, almonds, chest-nuts, all fresh and racy of the soil.

 9th Course—Une tasse de thé.

We crowed over that dinner—at least I did—and the ladies hid it away as a hen does her chicks under the wing.   The females of our party discovered the china closet open and the neatest of china tea sets. We called for our tea in these delicate cups.   But in Corsica they excel in chickens, but not in tea.   They use cold water and only half draw the tea.   We had been served with silver, too!   The curious female asked how many families could boast of china and silver!   " Only one or two in Evisa."   Remember our host is of the great Borgo family ; and Evisa is a village of 1500 inhabitants.   We discussed our forest trip ; but M. Carrara said it was impossible.   We reluctantly gave it up.   We were wise, for we learned afterwards that the forest roads were blocked with snow.

When we rose in the morning the ground was covered and the mountains had on white caps.   We waited till afternoon before we ventured further on our travel.   It lay this time down toward the sea, to Porta and to sunbeams.   This route was a continuous descent.   The Corsican driver fairly rattles you down the mountain.   His sure-footed ponies never leave the path.   There seems to be no end to the tourniquets on this zigzagging road.   Past beautiful cascades ;

and pictures of green and brown—foliage and rock;
past the most tempting flowers—all alone, upon this
road to-day, we push down, down, toward the coast.
At last we reach Porta, and are welcomed by M.
Ruelle. Here are to be found iron-works and saw-
mills; and here we saw droves of Lucchesi working
immense gardens for the cedrat, a kind of large lemon
used for confectionery in France. There was no other
house in Porta but Ruelle's. He invited us to stay
all night. Declining that, we, however, accepted an
offer of his rare wines. We walked from his house to
the coast. The sea was tempestuous. It was driving
in foam over the Moorish tower upon the rocky
eminence at the mouth of the River. We cast our
horoscope for the day, and concluded to move on to
the Greek colony of Carghese.

I would fain have rested here at Porta, if only to
enjoy the luxury of the sky and air after our stormful
experiences in the mountains. From the region of
winter, where the hail dashed into our abode and
startled chickens, cats, and human beings by its wild
saturnalia, down, in a few hours, to the region of sun
and olives, oranges, lemons, and all beauties of vegetable
life,—this was worth a respite for a night, if only to
indulge in the reflections incident to the vicissitudes
of our one day. Think of it! We had been within
sight and almost within a few hours' reach of Monte
Rotondo. From that point, as I am 'informed and
believe,' on a good, clear day, the whole coast from
Marseilles to Naples is visible! To say nothing of a
coup d'œil of Corsica, Sardinia, Elba, Monte Cristo,
and Caprera, the human eye can play from the Valley
of the Rhone, with its grand and castellated mountains,
over the tall white peaks of Provence and Savoy,
capturing the strongholds of Toulon and Ventimiglia;
saunter amidst the orange gardens of the Duke of

Vallombrosa, at Cannes; surmount the Turbia Mountain behind Nice; clamber, without aid of donkey or guide, through the defiles above Monaco and Mentone; linger with Doctor Antonio and his love around the palm trees of Bordighèra; get a straight view of the leaning tower at Pisa; catch a glimpse of the palaces and churches of superb Genoa; and follow the Maritime Alps, which hide the mulberries and vines of Lombardy, until the Valley of the Arno leads it by a silver thread through labyrinths of beauty to gorgeous Florence; thence the eye may roam to the Tiber, with a glimpse of the Pantheon which ' Angelo hung in the air of St. Peter's;' and with the aid of a glass (I do not speak of the fiery Corsican vintage) rest upon the cones of Mount Vesuvius or float upon the richly-tinted and ever-sparkling waters of the Bay of Naples!

I do not know that this is an overdrawn picture of the fancy. Monte Rotondo has been ascended. Judging from the account of that ascent, it is practicable and will repay the exertion. The ascent is from Corte; by a bridle-path up the roaring gorge of the Restonica, through quarries of black limestone and marble, and into the granite region. Passing a few chestnut trees, the utter desolation which the peasant described to me as the caprice of the Eternal Father, begins. A few pines are to be found amidst the precipices and turrets, and other rocky phantasies of the old fires. The clear waters of the Restonica, which washes the boulders almost as white as the snows, still guide you until a pine forest is found nearly at the source of the torrent. You cross the stream, ascend a gorge to the cabins of the goat-herds, where you find milk, hard bread, and kindness. Resting all night, you must be up and *a-foot* to clamber still higher over the loose stone; thence by the Lake Rotondo and several other smaller lakes, into the region of snow.

The traveller is recommended to be at the summit by sunrise, as the whole island, 114 miles long—with all its sister isles and the coast from Civita Vecchia to Toulon—is to be seen. So says the account. This was not tested, however, by my actual observation. *A priori*, it would seem reasonable. I ought not to write of what I did not see. I reserve the rest of my *real* mountain experience for the next chapter.

CHAPTER VI.

BEFORE starting again, on our mountain journey, it would be well to understand better the historic associations of the isle. The origin of the name, as I have given it, is rather apocryphal. But we are not to be too particular where all is traditionary twilight. Another derivation is from Corsus, a son of Hercules. This is supposed, by some, to be a more ancient name than Cyrnos, the ordinary Greek appellation for Corsica. According to Fabius Pictor, Therapne was also a name given of old to the island; indeed it was the oldest name. A traveller along the shores of the Western Mediterranean, as far even as the pillars which bear the name of the demi-god, must be struck with the incredible 'labours' which Hercules is reputed to have performed. Not alone Africa and Spain, Gaul and Italy, but all the islands of the blue sea, have earliest associations with the divinity of Might and Muscle. Are those exploits an allegory? Does it not require enormous physical strength to overcome the obstacles to the occupation of a new country? What monsters were there not to exterminate in the pre-historic time? To Hercules, all the force and valour necessary to make the land habitable by man are attributed. For an unlettered people, in its nonage, this allegorical arrangement is convenient: superstitiously transferring to a single divinity the collective efforts of the founders of their race. It saves the making of books—which is 'a weariness of the flesh'—and, at the same time, honours the origin of the people with the divine ichor. Sallust politely derives Corsica from a Ligurian lady—Corsa. But I prefer the romantic derivation already given from Dido's darling niece. Africa, and its Phœnician colonists, had something to do with the earliest history of Corsica. The Phœnicians colonized the island. It took the Romans a century to eradicate them, and to capture it. They began their conquest 260 B. C. Marius founded Mariana and Aleria.

Since my visit to the island, I have seen Mr. Murray's little Guide to Corsica. In it he devotes scarcely thirty pages to the island. In conversing about the matter, he regretted the meagreness of the volume. I would hardly presume to add to or subtract from anything in his valuable compilations; but from one remark I dissent—that there are no other classic associations with the island, except those of the exiled grumbling philosopher, Seneca. I refer to the Dictionary of Greek and Roman Geography, edited by Dr. William Smith, Part VI., under the title 'Corsica.' From the score of authors quoted, we learn all about its origin, name, products, forests, honey, wax, fruits, and history. We learn that it was never really subjected to Rome, but was in chronic revolt. The names of Marius, Scipio, Otho, Belisarius, and Totila, are associated with attempts to conquer it, down to the advent of the Saracens; but the most that was made out of the stubborn island was a little tribute in wax and honey. Thus we infer that the isle was ever fragrant. Again, we read that the Corsicans were longlived; and this is attributed to their use of so much honey. Again, this explains what we have seen here every step, the floral opulence of the island.

After Rome tried to hold the island, came Greek, Moor, and Goth. It is Africa and Spain over again, with their vicissitudes. A feudal aristocracy—called *signori*—sprang up in the middle ages. In the eleventh century Pisa obtained a footing ; then Genoa, jealous and belligerent, in the fourteenth century. The war between Pisa and Genoa for Corsica is illustrated by the biography of Della Rocca. Until 1768 — the year before Napoleon was born—the Genoese held the island, but always under a protest, with a fight on hand. Out of these fights came the heroes of the island. Their line ended with the splendid name of Paoli. In 1768, Genoa sold Corsica to France. In 1769, France defeated the Corsicans at Pontenuova. France held it, unsubdued in spirit, until 1786. The French Revolution came, and fused all elements, and to such an extent, that Corsica gave to France an Emperor—Napoleon ; and that too, from the family of a patriot, who was the Secretary of Paoli! But why rehearse these events ? My duty is in the sunlight, and on the road.

In closing the last chapter, we were on our way down the west coast. We are making for La Piana. The road soon leaves the sea and ascends again. On this route we had splendid views of mountains and sea-coast. All the fauna and flora were to be found, the former petrified by fire or worn by water in the many-shaped rocks, and the latter dropped deftly into every nook and crevice where water ran or birds flew. One of the attractions of Corsica is its endless variety of scene. We have bird, beast, and creeping thing fashioned by the accidents of the volcano, which heaved this isle out of the blue sea into the blue sky. And after the fires had done their work, the waters gave forth genii to shape the rocks and mountains into statuesque and grotesque forms. Certainly Seneca

either did not leave his tower at Calvi to go into the
mountains, or else he was cross and unphilosophic,
because of his exile : or else he was without refinement or taste for natural scenery.   If not one or all
of these, he would never have traduced Corsica so
outrageously as he did in his *de Consolatione.*   He
was exiled under the reign of Claudius, and remained
in Corsica six years.   His tower is one of the monuments of the island. He even abused the water of
Corsica.   We know that in this he slandered the
island.   He described the island as utterly fruitless.
That is hardly true now, and then there was more
cultivation than at present.   If there is one thing
more than another which Corsica produces, and
readily, it is fruit of every kind.   Indeed, the orange
and lemon are produced the next year after planting.
Seneca says the island had no charming trees.   He
had never seen the evergreen oaks near Evisa, or the
splendid larches of Aitone, or the chesnut groves of
Castagniccia.   They are indigenous, and must have
been on the island eighteen hundred years ago.

Again, this philosopher says : ' She produces nothing which other people seek, and is not able to
nourish those who cultivate her.   What more has she
than rocks?   Where are to be found more priva-
tions?   Where does man suffer more?   What is more
horrible than the aspect of the country?'   I have no
patience with this old slanderer.   At the very time
he was writing this libel, Rome was drawing on Corsica for many varieties of marble—the white equal
to that of Carrara, the verde antique, whose beauty
now decorates the Sistine Chapel — for grain and
fruits and fish, and many other supplies.   The fish
in and around the island are those which Horace
celebrated in song, and Lucullus served upon his table.
I have seen the nets drawn upon these shores with a

miraculous draught at times, and seldom with a 'water haul.' The fish are gorgeously coloured—having all that the prism can give of hue and all the delicacy the epicure could desire. Green and gold fish, fish all a-flame, fish grey and silver, fish striped with bands of beauty, fish with fins and horns, and fish with black and white spots, and in fine, the devil-fish itself—are here used for food. Here is to be found the chanticleer of fish, *coq de mer*, with brilliant and blue wings; hard to catch, for he can fly out of the net. This chanticleer of fishes is a searcher after sunbeams; and that he finds them in the denser element, his vivid hues attest. Crabs are found here; oysters are plenty at Bonifacio; the languste, or lobster, is common; sardines, mackerel, soles, and whatsoever else is delicate for the taste and suitable for food. Seneca had none of the Apostolic—or Waltonian—predilections, or he would not have abused Corsica.

The people may have been then what they are now, rather independent and indolent; but why should Seneca depreciate what is the glory of Corsica—her rocks and mountains, which he calls horrible spectres—unless the cunning old courtier was home-sick and wanted to be about the palace. Doubtless he deliberately exaggerated the inconsolable state of his exile, in order to interest the Romans in his favour by drawing on their sympathies. He accused the Corsicans of being revengeful. He wreaked his spite for his exile on the people and the land; and even on the rocks. To him the rocks were frightful. Perhaps he had seen some of the grotesque and strange forms cut in the rocks to which I have just referred. May be he had visited the region of Monte Libbio, whose range seems like a wall of ruined towers, guarding one which looks like the hooded head and form of an old woman, and known as 'La Sposata.' Perhaps

he had been to Rocapina, upon whose headland the
figure of a lion reposes, head up, one leg down, looking off to Sardinia—a wonderful piece of natural
statuary! Perhaps he had seen the rock at the gate
of Ajaccio, known as the boot of St. Peter. Perhaps
he had seen the anvil-shaped Col dell' Incudine, or
the finger-like promontory of the northern end of the
island ; or, perhaps, he worked up his frightful picture
of this charming isle as the German philosopher did ·
his camel, out of his own consciousness. One thing
is sure, he could not have visited this part of the
island where we now travel ; for it would have extorted
even from exiled cynicism a tribute to its cultivation,
charm, and resources.

Not alone are these rocky shapes or mis-shapes so
common as to attract the wayfarer; but upon them
are hung, like garlands, many a beautiful tradition or
weird story. There is one mountain, Talafonata, which
has a hole in it, into which the sun when he gets up,
tired with labouring over the Corsican alps, creeps in
about noon for a siesta. The people of Niolo say,
that the devil—who is represented as an agriculturist,
—perhaps because agriculture is held in such low
esteem in Corsica,—was one day ploughing with his
team of oxen on Campo Tile. St. Martin was coming
over the plain ; and as he had prejudices against diabolism, he tried to exorcise the devil. A quarrel
ensued. While Diabolus was engaged in using harsh
language to the Saint, his plough struck a rock.
I should not wonder at that at all, nor that the
plough broke. The devil attempted to mend it, but
could not. Getting as mad as the devil—does at
times, he hurled his hammer high in air; it knocked
a hole in Mont Talafonata. When he turned to
look after his 'cattle,' he found them, by some
Medusa-like witchery, turned into stone ; and we saw,

or fancied we saw, several yoke of petrified oxen on
our way to La Piana.

When we reached La Piana, in the afternoon,
the whole population, including some hundreds of
children, came out to look. As we walked out to
look at the sea, under the guidance of the mayor,
they followed. He used his bâton to drive them
back, in vain. After our view, we were invited into
his house, and were there hospitably entertained. It
was cool, stony, and clean. It had no cushioned chairs
or sofas, and no glass in the windows; but it was
hospitable and open.

We had no sooner entered this town than an apprehension seized us that we should have to roost out
among the chickens. On looking at our memoranda
—for no guide-books give us information—the only
inn mentioned was put down as 'bad.' We had, on
our entrance, luckily passed by it. One house in the
main street looked comfortable. We ask whose it is
'It is the curé's.' We are invited to occupy it, all
thanks to the curé. We mount to the second étage
by a steep wooden stairway, half ladder and half steps,
and through the narrow and dark stony passages we
reach the third floor. Our horses are stabled on the
ground floor. The street is, happily, wide enough for
our carriage to repose in over night. We have one
chamber, and an improvised couch among the fleas in
the 'Salon de Réception.' I must say that while
Corsica has many virtues, and good food is one of
them, the abodes are comfortless and dirty. A bright
young boy, belonging to the hotel, and who would
take no money for his courtesies, escorted us about
the town after our meal. The curiosity to see us
was still intense; throngs followed us in the streets,
and every window was adorned with female heads. It
had been bruited about that we were Americans, and

one citizen asserted that he had had the happiness to see the writer in Washington during the war. As that was not unlikely, we had an interview with this happy citizen. I wondered what could have taken him to America. He told us that he had been to New Orleans on a venture, and was caught there by the war. On trying to get North, he had been captured. As he could not speak English, he was taken before a general, whose name he could not recall, but he was a great astronomer! 'General Mitchell?' I asked. 'Si, signor.' General M. had questioned him, and finding no evil in him, let him pass through his lines to the North. This incident made him quite a hero. Under his conduct we visited the little church, of which the population were proud, because the altar had been given by the Empress Eugénie. The curiosity of these kind people may be pardoned for their hospitality. At dinner, the curé, whose chamber was opposite ours, sent his compliments, and a present of fresh and toothsome prunes, and then some wine. He spent the evening with us, smoking and talking. 'His people,' he said, 'were poor, but *rich*, as they did not require much.' Our beds were hard, though clean. Mine was designated as 'un lit de mariage.' Wherefore? Because the pillows were trimmed with embroidered ruffling. The sheets were very coarse, so as to feel cold.

While enjoying our after-dinner smoke and gossip, quite a scene occurred. In rushed upon us an insane woman, little, thin, and weird. The Madame of our party seemed to be the object of her attention. Before we knew it, the witch was down on her knees before Madame; and before we could understand what she was doing; she crossed herself, then made the sign of the cross on the forehead of Madame. Her incantations were in patois, and not to be understood.

We said to her, in Italian, that her language was unintelligible—that we could not understand her: 'No matter,' she replied, 'the good God does!' Finally, we were given to understand that she was divinely directed to have the benediction from the American Madame! Equal to the occasion, the Madame gave it with a touching pathos and out-stretched palms, that seemed to 'satisfy the sentiment,' and which, as an artistic touch, any histrionic Heavy Father would approve!

We heard no more of her, unless it was her voice, whose wild crooning awakened me at day-break, with a song so unearthly in its intonations and so sad in its moaning, that it made the heart ache. The tears seemed to be in her voice as she sang. This was a lyric sample of that national music of which we had heard—a lament much in vogue in the days of the Vendetta, and not yet obsolete. Sometimes, in times of bereavement, these elegies are yet improvised. Their simplicity is very touching. I have read one song, sung by a young girl of Ota, who rejoiced in the name of Fior di Spina, or hawthorn flower. She had killed a lover who had refused her marriage. The burden of her song was that she was very unhappy, though courageous, and had to rove the mountains and wild paths among the bandits, whose brave hearts would protect her. The recitation of our songstress, as we learned, had reference to a brigand and a peasant girl. The girl had been out on the mountains gathering faggots, and had loaded her donkey. She was hastening home as evening came on. It was in the time and region when and where the famous brigand Serafino—(I think I have seen him on the theatrical boards)—lived and preyed. Apprehensive of Serafino, her fears were soon allayed by the appearance of a gentlemanly gendarme. They walked, and

talked, enamoured and enamouring. They chatted of
the terrible Serafino. I need not say that the soldier
and Serafino were one. He saw her home—her and
her donkey—but the maiden 'ne'er' saw him more.
She pined in monotones of lamentation for the love
of her brigand! There! is not that a pretty story
by whose music to awake in a region so renowned
in brigandage? By the way, the last story told of
Serafino has its scene near Ajaccio, on the road to
Sartene. A captain of gendarmes was traversing
Campo di Loro on a pony. He was overtaken by a
peasant on foot. The officer dismounted to fix the
girth of his saddle. The peasant, allowing the sun
to glance upon his dagger, civilly remarked, pointing
to his own boots, 'Are these fit for Serafino?' Exit
Serafino on a pony. Exit officer, bootless!

Since penning the above, I have heard the songs
of the Moors and the Spaniards. I declare that the
similarity of tone, the drawl of lamentation, the soul
of sorrow, which follows my memory of Algiers
and Andalusia, seem but the echoes of the lyrics
of Corsica? Have the Saracens set the Mediterranean to a music of their own?—These voceros,
are they the language of the dead nations, whose
ghosts haunt the shores of this historic and poetical
sea?

Most of these *voceros* are laments over those slain
in family feuds. They are the poetry of La Vendetta. Happily for Corsica, this barbarous system of
rude justice is nearly extinct. Its principal seat was
in the centre of the island, around Sartene, Vico,
and Corte. Persons are known to have been shut
up for years, close prisoners in their own stone
houses, with only shot-holes open; the windows being
stuffed with mattresses. The prisoner was always
on guard for his life, from the bullet. His relatives

tilled his land, or cared for his flocks, but did it under sentinel. There are instances where men have lived in this self-immured jail for a score of years. In 1853 the Government forbade the carrying of arms, in order to suppress this system, but the prohibition failed. Then the system of mounted mountain-police was established. It has done much to eradicate the evil. The Catholic priesthood and religion have been more efficient agents of reconciliation. I heard of many instances where these agents were of inestimable utility. The law of revenge is so alien to the spirit of Christian kindness, that to establish the latter is to extirpate the former. At Sartene there had been a feud between the Rocca Serra and the Ortoli families since 1815. This feud grew out of political differences. One clan called themselves White Bourbonist, and the other Red Republican. When Napoleon became Prince President, a public reconciliation took place, and the children of the two houses, in happy innocence, were allowed to dance together! With the extirpation of La Vendetta brigandage dies, as the brigand is not so much a robber as an outlaw. He is driven to the rocks and mountains because the avenger of blood or the Government has been on his track. Strangers are not in any danger from the brigands of the isle, if any remain. The principal cause of these vengeful raids and murders is the old passion—love. An insult or a wrong to a female relative, and La Vendetta takes it up. Often the woman herself, with her bright, little, pointed dirk, with *Vendetta* burned on one side, and *Mort* on the other, pursues her vengeance, like a fury, to the 'sticking point.'

Sometimes there is a comic side to these feuds. On the eastern side of the island, the inhabitants of two villages were celebrating a religious *fête*. Their pro-

cessions became confused by the carcass of a dead
donkey in the way. The inhabitants of each village
accused those of the other of placing it there as an
insult. A mortal enmity and stubborn conflict ensued.
The two towns, Borgo and Lucciana, were held in
blockade by each other. The contest was, who should
keep the carcass. The donkey did more travelling
in that asinine war than when in full sonorous life.
It was carried from village to village with many sanguinary conflicts. Once Borgo placed it in the church
door of Lucciana, and once Lucciana hung it as a
trophy on the steeple of Borgo! Finally, to prevent
further bloodshed, the Mayor of Lucciana 'digged
a pit,' and hid therein the *teterrima causa belli*. A
burlesque poem has been written to celebrate this war
of the donkey. How much moral there is in the
lesson, for nations outside this fragrant island!

The lament of the crone did not awake me at the
hotel. I was already awake. It saluted my aroused
sense. Other lively matters surround the traveller in
the Corsican mountains besides vociferating females
and supposititious bandits. If the fleas were only as
harmless as the brigands are now-a-days in Corsica
travelling here would involve less trouble. Our Italian
companion was frequently heard to ejaculate, with
polyglot volubility, during the night, from her
populous couch, '*Mon Dieu!*' '*Sapristi!*' '*Ach!
Der Deufel!*' '*Corpo di Bacco!*' '*Zounds!*'

Morning comes. The lament had hardly ceased,
before it seemed as if all the population were out and
about. The bells begin to ring. The chickens in
the house begin to crow and cluck. The donkeys
begin their dissonant noises. The interminable chatter
of voices is heard in the street below. We could
not sleep, even if fleas were not; so we arise and
walk out to find some cigars. We find them in a

shoemaker's shop; his wife makes them herself. We perceive upon the piazza—40 by 60 feet—about two dozen of the principal citizens—including the Mayor and the travelled man. They are walking by sixes backwards and forwards, and talking very loud, as if they were all mad. They lift their caps altogether, and the travelled man astonishes his companions and myself by saying 'Good-bye' for 'Good morning!' I passed round the cigars. They were received with exemplary politeness. After breakfast they came to see us off, the kind priest among them, and we were soon upon the high-road, flying down and toiling up the mountains which skirt the coast. Relics of the Moorish invasion and of Genoese rule—old castles and towers of look-out—cap the pinnacles along the sea. Soon we reach the 'Plain of Silver.' It was so named because richly cultivated with flax and barley, olive and vine, sheep and goats.

We met here some peasants with guns. They had licenses, I suppose, as that is the law. The Vendetta must be suppressed, and guns are prohibited, except under special license. The game is plentiful. We saw the red-legged partridge and the hare. Blackbirds, called merle (not our blackbird of America, but a bird quite in vogue here for the table) are sought, not only for sport but as a business, to be potted for commerce. The crow here is partly white, and we saw plenty of them. Hundreds of peasants, women and men, were scattered over the plain, picking out the weeds and bad barley from the young grain.

This plain is the precursor of the Greek colony of Carghese, which we have resolved to visit. Certainly the environs of Carghese bespeak more of industry than the Corsicans bestow even on their fat plains. The Greek colony has been here for over four hundred years. I have looked in vain through the meagre little pamphlets on Corsica, into the French geographies, and guide-books, intended to instruct the stranger about this island ; I find no allusion to this colony or its history.   But from personal observation of the people and their habits and features, and from a comparison of my own observations when in Greece, I have no doubt that Carghese is, as represented, a Greek colony.   These Greeks, flying from the tyranny of the Ottoman, landed on this promontory, which encloses the Gulf of Sagone.   They have made it rich in production, if not classic in cultivation.   For many years their descendants refused to intermarry with the islanders.   They preserved their religion, their dress, their language and customs ; but their seclusion is giving way, as I should judge, by what I learned during our stay.

We were hardly ready for this Attic colony before we were in its village.   It lies under the brow of the mountain.   Our eyes and minds were employed as we rode along the coast, in viewing the ever-splendid sea, all white with the recent tempest, and lashing itself into mist against the rocky shores.   The cultivation of the lands is superior.   The prickly pear is common, but most the vine, for which these Greeks are celebrated, abounds on the mountain sides.   Our premonitions that we were approaching the Greeks were, the *Greek* cross on the road side ; and here and there fountains, where Grecian maidens and old women, with their classic water-jars, made quite a tableau. These are the children of the great old Greeks of whom we have read.   The nose, the eye, the mien —though they are of peasants—indicate the physical elegance of the Grecian.   We drive into town.   There, as if on parade in the street, and marching with a French officer, is the venerable Greek priest.   He fills your idea of a Druid, under the most ancient of oaks. His beard was half-way down his body, and milk-white : his hair is long and white ; he wears a long, black robe, and black sombrero of immense periphery of rim. We drove to the inn. There we find a Greek landlady, very proud, indeed, of her Grecian blood, and of two Greek daughters, very beautiful. Their faces might have been chiselled by Phidias. The son of the landlady, however, has married a black-eyed Corsican girl, who can talk French. She invites us to the hospitalities ; but, unlike the Corsican and Catholic people, she charges for them with an exaggerated idea of their value. She is the daughter of our landlady at La Piana. After partaking of the rare wines of Carghese, we visit the Greek church, and luckily, while prayers are said by another Greek priest with long black hair and beard. This church is very small, but there is a grand one now building, started by a subscription from the late Emperor of Russia, Nicholas. The process of erection does not go on with much alacrity. The church we visited was slightly decorated. It rejoiced in some pictorial daubs of saints. The neighbouring Catholic church is much more prosperous ; and I should say that the Latin race and its religion was fast absorbing the remnant of the children of Epaminondas and Pericles, left so forlorn on these distant shores.

I cannot say much for the intelligence of these descendants of Socrates and Plato. Endeavouring to explain to some of them that I had sent a despatch under an ocean three thousand miles broad, which was received several hours earlier in America than when it started from France—these people looked amazed. No explanation of the sun's apparent motion, caused by the earth's motion about its axis, or of the telegraphic cable, could make the statement comprehensible. Well, perhaps it would be a poser for Aristotle himself. Wine, figs, raisins, broccio, they understood : and after a good time, long to be remembered, we started again, and arrived at Sagone.

I do not know whether it is from the intensity of the sun's beams, or from want of cleanliness ; but in Corsica there are a good many people with sore eyes. In Sagone the people seemed to be distinguished by being one-eyed. The landlady, and the girl who helped her—almost everybody but curate and custom-house officers—were one-eyed. The blacksmith was a genuine Polyphemus. But all—officers included —had an eye-single to our comfort. No pay would they accept. They pressed us to remain ; forced on us wine, cheese, and bread ; and after many regrets, we left them to dash through four hours of descent into Ajaccio.

From this round trip, which I have described, made at an unpropitious season, but made very pleasant by unceasing hospitality, my readers may glean something about this island not to be found in any other way, or in any book. We propose to make visits to other towns over the island, over the mountains again to Bastia ; through Corte, the ancient capital ; thence to Calvi ; thence again, after returning to Bastia, to the South, through the chestnut land, to Bonifacia. But this is in reserve for a better month ; and, perhaps, from the other side of the island. Bastia can be reached from Genoa, or Leghorn, in a few hours of boating and in a smooth sea ; for Corsica is a break-water against the storms from the South and West.

CHAPTER VII.

THE CLAIMS OF CORSICA AS A HEALTH RESORT, ETC.

" Almost all patients lie with their faces to the light, exactly as plants make their way towards the light."—*Florence Nightingale.*

IN the preliminary chapter, I have en-deavoured to explain the title of the volume—'Search after Sunbeams.' I have pursued this search, with some days of failure, in the mountains. But on returning to Ajaccio, whose situation and surroundings are at once beautiful and salubrious, I might be pardoned for trying in a few pages to attain a more valuable object than the mere description of scenery or people.

It has been a great desideratum with physicians to find a climate which is dry, warm, and stimulating in winter for patients suffering from pulmonary troubles. Any doctor of eminence in Europe or America, who has kept pace with the advance of his pro-fession, will advise, not that his patient should go to Nassau—that is too hot; nor to Cuba, for the same reason; nor the Island of Madeira—that is too moist, and, while it mitigates suffering, does not cure; but he would advise a climate which is at once stimu-lating and mild.

Ajaccio, like Nice, Cannes, Mentone, and other towns of the coast, is screened by the great mountains of the island. The town is especially protected from the north-west by a spur descending to the sea from the great range into whose bosom we have travelled. It is warmer than Nice. No harsh winds blow unless you go, imprudently, as I did, into the very mountains in February. The atmosphere is still; the weather is constantly fine. Although we are here in the worst month and have ventured into the high latitude of the mountains, and roughed it generally, yet we have found but little difference between this climate in March and that of New York in May.

The number of aged people we have met with in Corsica is incredible. Joyce Heyth would have lived here to the full measure of Mr. Barnum's apocryphal statement. Indeed, there does live, up in the beautiful country which we have visited, near Vico, in the valley of Liamone, in the hamlet of Murzo, the veritable old woman, Angela Pietro, aged 110 years, who was the servant of the mother of the first Napoleon. She was with Madame Letitia at the time of the Anti-French émeute at Ajaccio, when the Bonaparte family had to 'run for it.' She is nearly blind, but has some memory, and talks of the Great Napoleon and the events of those memorable days with the volubility of Joyce, if not with her imagination.

But, with all these healthy elements, one thing Corsica has not, as yet, but soon will have—good and comfortable accommodations for the invalid and tourist. One thing, however, it has, which makes up for the lack of these—a fresh, glorious scenery of mountain and sea, whose ever-varying charms I have tried, I hope not quite in vain, to picture. But these attractions, however charming and however painted, are but whip-syllabub compared to the substantial benefits which, like good nutriment, help to build up the debilitated frames which have inherited or inhaled the poison of consumption.

There will always be one impediment to Corsica

as a winter health resort. The invalid must cross in winter a stormy sea. He may become port-bound, before he can either reach his haven or leave it. An American friend—a lady—was kept three weeks at Ajaccio before she could get away. This was in consequence of the storms, and the uncertainty of the arrival and departure of the Nice and Marseilles vessels. Besides, the steamers propose to run only once a fortnight. The accommodations in Ajaccio are poor; the apartments are not 'glorified' by the sunbeams. Most of my time was spent in lying around, over the sun-warmed rocks of the bay, reading and writing, or gathering the tiny exquisite shells which fill the nooks of the rocks and the sand of the beach. We found hospitable and comfortable quarters with Dr. Ribton, in cottage No. 1, along the promenade or Cours Napoleon. Our windows looked out upon, and across the bay—a perpetual sparkling joy! The view reached to the mountains of snow beyond its shores—and the breath of all the perfumes of Arabia seemed to come from these mountains on the approach of evening. We were happy in having such apartments with such views. The hotels of the town—some two or three—have sunless and dirty rooms; the servants are ignorant and not by any means neat in apparel or habits, and the prices charged for such poor accommodation are exorbitant. These are the offsets to the attraction of Corsica. One thing is to be said, and said in a hundred ways, that the main thing, for which Corsica is to be visited, is its mountain scenery; and this as well by invalids as healthy tourists. The proper time for this is in May, when the snow is all off the interior, when the forests and peaks may be reached and the vines and chesnuts are coming out in green.

The climate of Ajaccio may, for some invalids—

especially those in the earlier and more curable stages
of phthisis, be too relaxing. One of my American
friends found it so this winter. He left for Capri, to
test that place; he had spent most of the winter in
Mentone, with the best results. He needed the dry,
invigorating climate of the Riviera. A winter climate,
like that of Ajaccio—temperate, sunny, and rather
moist, with the uniformity of Madeira, and without
the heat of Egypt, or sirocco of Sicily, without the
bad drainage of Naples, or the unpleasant river-bed
of Malaga—would be more suitable for the advanced
and almost remediless stage of the disease. Life might
be happily prolonged under such a sky; and under
proper hygienic conditions, it might be saved. But
as the conclusion of my judgment, and as the result
of a long search after the 'beams' which have the
most balm, I yield my preference to the Riviera and
Mentone. Yet as I leave Ajaccio, where I have
received so much happiness and strength, I cast a
lingering look behind. If I visit it again, I shall
regret that social disquietudes have driven my friend
Dr. Ribton even to Capri.

I close this chapter on the steamer from Ajaccio.
It is a bright sunny morning. Three or four hours
and we are at Nice; we shall then be on solid continental ground again, and nearer home. We have
had a rough night on the sea; but how brightly
breaks the morning! Far off—forty miles or more
—I can discern, with perfect clearness and without
glass, the snow mountains of the Maritime Alps,
making a circle of silvery beauty, almost afloat in the
upper ether! From this distance no shore line is
visible, nor is the lower range of mountains on the
Riviera. The dark range of the Estrelles beyond
Cannes, to the west, is faintly seen. The splendid
vision of these distant cloud-like mountains, is proof

to me of the rather *grandiose* statement in a preceding chapter, viz. :—That from a lofty, or a distant
point, in this pure air, and under favorable conditions,
one may, with a *coup d' œil*, grasp a splendid range
of mountain scenery—almost take in half the shores
of France and Italy !

CHAPTER VIII.

ADVENT INTO AFRICA.

"Gaze where some distant sail a speck supplies,
    With all the thirsting eye of—Enterprize !
    Tell o'er the tales of many a night of toil,
    And marvel where they next shall seize a spoil.—
    No matter where,—their chief's allotment this :
    Theirs, to believe no prey or plan amiss."
                                                    BYRON's ' Corsair.'

THE excitement of leaving a port like Marseilles is not lessened because it is in a foreign land. There is always a sense of the uncertainty of such a voyage over the waste of waters ; but there was a special and emphatic feeling as we passed out of Europe with our prow turned towards Africa. The gateways of Marseilles, constructed out of immense blocks of artificial stone, the rocks and isles about the harbour, and the mountains by the sea, are so bleak and grand, that one cannot leave them behind without admiration, and, if a poor sailor, without emotion. One hundred and sixty-eight years before, Addison sailed out of Marseilles on his visit to Italy. He could not fail to see the classic aspects of the place, and records the olive-trees and gardens, ' the sweet plants—as wild thyme, lavender, rosemary, balme, and mirtle ;' and the deserts, as he terms the rocky places. He found where Ulysses shed the blood of victims, and ' raised a pale assembly of the dead ;' but he gives us no account of the splendid harbour out of which he moved, and which Nature had made for the commercial metropolis of France. Could Addison have seen, what we saw, the hundred ships under sail; and the steamers under way, he might have given more than one quotation to illustrate the initiation of his voyage.

Our vessel was not well ballasted; it had a wabbling motion scarcely endurable. But the beautiful sea nearly made us forget the consequent uneasiness. What a sea it was! The greenish-blue was so bright, the crests so glittering, the offing so ethereal, and the sun sank in bars of red, violet, and gold. The warm air, in spite of all premonitions from within, drew us to the deck, and we began to feel that we were on our way to a sunnier clime. It was like a voyage to the Hesperides, with the anticipation of the sun, bloom, fruit, and joys of the classic garden, and without the Dragon to disturb the promise.

On the next day we passed between Minorca and Majorca. Our glasses failed to reveal much of interest or of beauty. Still the lofty summit of Monte Toro was to us a beacon and a sign.

As we neared Algiers, the sight of that mysterious Afric land was welcome. At first the city was a little triangle of white specks; then it grew into a city whose surroundings, from the Sahel hills to the Atlas snowy mountains on the left, were full of real interest. Then the long low line of coast came in view, and the white square houses loomed up before the vision; then, with the aid of an English resident, we could perceive to the right the dark rocky inlets and forts which the corsairs made their lair; and, finally, we wound into the jetties of the bay and harbour, and were at rest in the tranquil waters of Africa. There is here a harbour or jetty quite artificial, and a hundred vessels, many of them steamers, are within. A green spot within the harbour tells us that the French are here, for we see the soldiers and cannon. The heat is

intense, but it is not that of the sirocco. Nor is the
city wanting in European comfort. The new Hotel
d'Orient received us, and the motley groups at wharf
and dock and in the streets were a continual provocation to our curiosity. The crowds were as many-
coloured as Joseph's coat, and they fought for our
baggage in a very rough way. The pervading impression as I looked at Algiers from the ocean and in
the bay, at its old and new town, its green hills
covered with villas, and its compact town of white
houses, was—how remote from the world is all this!
Who would come so far to see so little? This is all
that remains on this coast to mark the power of the
Barbary corsairs who held Europe in fear so long!

Soon we are assured that we are in the Orient.
Veiled women and bournoused men; flowing robes
and mysterious wrappages; ragged Bedouins driving
donkeys, or sitting on them, their long robes almost
dragging in the white dust of the street; mosques
with minarets and domes, surrounded by slow, lazy,
easy-going people; palm-trees and stately aloes, cactus
and lilac, fuchsias and bignonias; all the sweet flowers,
fragrant shrubs, and stately trees, which are made of
the sunbeams, soon assured us that we were in a milder
clime than we had left. It was the Orient, at the
door of Europe. Forty odd hours of sea-going, and
here we are in the very home of Abd-el-Kader, and
among the children of the desert!

But all is confusion yet. Such a mixed and bizarre
company of beings at every turn must give us pause.
It takes time, inquiry, and study, to classify the confusion incident to a city so Mosaic in its character
and people. The different nationalities and costumes
were like the moving figures at a fair. The thirty-
six years of French occupation has not greatly changed
the shell, much less the kernel, of the Orientalism and

Mahomedanism here prevalent. The more I see of the city, the more I see that French toleration, and its un-Puritanic ideas of social life, have not made much progress in the way of changing the distinctive civilization of North Africa. It is true the French hold this city, and the country as far as the Desert, by the aid of forts and soldiers; it is said, moreover, that there are more stores, cafés, music-places, tobacco shops, and other signs of French life, than in any town of the same size in France. But these are for the sixteen thousand French, who regard Algiers as a place of exile, and hope, when they realise sufficient means, to return to France. Even the peasants have an idea that their home is not permanent. The French adventurers on our boat said: '*Au revoir,*' to their friends. What with danger of outbreaks, and the roughest usage earthquakes can give (they have just experienced one in the province of Constantine), and the parching heat of summer, the prosperity of Algiers is likely to run hand in hand with military occupation and expenditure. The old elements— Arabic, Turkish, or Moorish, or, to comprehend all, Oriental—remain intact. True, in Algiers City property has advanced. The city, which used to be visited by pestilence, in the time of the pirates, is comparatively clean and salubrious. Business seems brisk on the quay. A railroad runs as far as Blidah, west, thirty miles. A theatre, quite elegant on the outside, is established. The city seems to be, and is, growing; but it is the hot-house growth of official and military occupation, rather than a healthy, steady, national growth. But is it the less interesting on that account?

I learned, when inquiring about the character and condition of the Hebrew population, that they were the most prosperous. In fact, they have, since the

French occupation, and since they have been permitted to hold property, acquired nearly all the estates
of the city. Going into their Synagogue, we were
received with the greatest courtesy. Dressed in the
black turban, wound around a red fez cap, and in
their dark cloak, gracefully thrown over the shoulder,
and the inevitable loose pantaloons, they seem at once
the best apparelled and most intelligent of the indigenous population. They have been almost as long
here in North Africa as the Berbers. They number,
in the city, over 6000. Their women dress in gaudy
colours, with cinctures of gold, and embroidered
ribbons, and invariably their black hair is hid under
a black satin cover. Their children are beautiful;
though that may be said of the Arab and Moorish
children. I have been in several schools. I never
saw more handsome little children than the Arab and
Moorish. Their eyes are dark and vivaciously ex-
pressive. It is not a gloomy, dead black, but it has
a daring glitter, that spoke of the grand and active
race which brought civilization from the East to the
now dominant Western races. I need only refer to
Buckle's second volume, where justice is done to the
tact, skill, learning, and intelligence of this remarkable
race. They gave us arithmetic, algebra, astronomy,
and chemistry. Their doctors in medicine and scholar-
ship informed the world of mind during the darkest of
the ages. As I looked on these sweet blossoms of the
old magnificence of the Moorish tree, I recalled how
the old stock had weathered the storms of centuries!
How they fought the Spaniards on the soil of Andalusia, to hold their own, amidst the smiling lands,
where every vale was a Tempe, and every Tempe a
poem—where the Alhambra itself arose like a grand
epic, through which resounded the clash of arms and
the songs of maidens! I remembered how more

egotistic nations have themselves pirated on the weak
—Britain in the Caribbean and in India, the Dutch
among the spice lands of the Far East, the French
in Cochin China—not to come nearer—and Spain
herself, with her flag of red and gold—rivers of red
between banks of gold—had pillaged and enslaved a
hemisphere. I remembered all these things and, in
the face of the sweet Moorish children, I forgot the
Barbary Buccaneers, so vilipended by history, who
scoured the seas from the Golden Horn to the Gates
of Hercules ; and for these recollections and in this
spirit of charity, I confess to have heard, with com-
pound interest, these children of the children of the
Orient sing their alphabet from the tablets before
them under the tutelage of an Abyssinian gray-beard,
all black, save his turban of spotless white. I could
see that the schoolmaster was 'abroad' as well as at
home. When we went into Madame Luce's house, in
the crowded part of the city, to see her Oriental
embroidery, what an interesting juvenile group we
found ! Some thirty beautiful Moorish girls, as fair
as any such group in New York (save one of glossiest
ebony), were all at work, sitting on the floor over
their frames, and finishing the inwoven elegance of
those fabrics which so astonish the Occidental lady by
their perfection of needlework. I saw new meaning in
Shakespeare's lines :—

"The hand of *little* employment hath the daintier sense."

These girls—nearly all—even the smallest, of four
years, had the tiny nails of their pliant fingers stained
dark with henna, and their hair coloured into a reddish
wine-colour. This colour of the hair they retain till
they marry. Then it is stained black. When the
hair becomes grey, in their old age, they stain it red
again. One of the children was tattooed over her

beautiful face. So modest, so pale, and so fair, it seemed cruel to pick into her pale cuticle the bluish tints, in shape of star and crescent, and other marks signifying, as our guide Hahmoud told us, the tribe to which she belonged. What a contrast to these little beauties is the shoeblack, whose portrait I give, and without comment !

We visited another group of children, in the crowded part of the city. We trudged through streets five feet wide, up and down, and with barely light from the sky above to find our path, sometimes going under dark archways, to find the home, where children are kept for mothers who go out for the day to their work. At length we found it. The mothers pay only two sous (two cents) a day. The institution is economically and neatly managed, under the direction of the ever-blessed Sisters of Charity. They showed us over the establishment, from their eating saloon and dormitories to the topmost story. The topmost story had in it some twenty girls, of larger growth, making artificial flowers. They all rose and saluted us as we entered. In this company were French and Moor. On our descent, we crossed the court—for all these institutions are located in a Moorish house; it is so convenient. We were ushered into a neat apartment, where were some thirty or more little cradles, with white coverlets, and ready to rock at the first infantile whisper. Over them, upon the wall, were written the names of the kind ladies who ' founded ' the cradle. These are for the abandoned children, born out of wedlock. There is no exclusion on account of race or blood. Behind a sheet-iron door in the wall, with an opening in it for ventilation, is a bed, into which, from the street, the mother or other ' party ' places the child. The moment the child touches the couch, the bell rings. Under the bell

there sleeps a Sister of Charity. She receives the little
one before the tintinnabulation ceases. On the outside,
in the street, locked by day, open by night, is a
window-shutter, and over it a sign, '*Pour les enfants
abandonnés.*' So you see Algiers has one of the
requirements of French civilization which London
and New York have not adopted.

I have said that all the houses of the old town, and
even those occupied by the Government, have the
open court inside and the verandahs around the court,
story above story. The engraving will illustrate this
better than the text. The most superb building of
the old regime—the residence of the Deys, in the
town—is built in this way. It has frescoed walls and
mosaic pavement. The doorways are of marble,
yellow with years. The crescent still gleams here
and there. Upon the third or fourth story is the
seraglio, with barred windows, against which the doves
of the Dey used to flutter and break their tender
wings. This last remark is more serious than was
intended ; for it had for Algiers—where for centuries
the white Christian captive maiden was coerced into
both slave and wife—a sad significance.

No one, from description, can have an idea of the
tortuous, narrow, and dirty streets which one has to
traverse in order to go through the old city. The
engraving presents but a small section, and that
faintly. The lower part of the city is Frenchified.
Arcades are built, like those of Paris. But above the
first few streets, parallel with the Rue de l'Impératrice,
upon the shore, are all the vicissitudes, the ups and
downs, twists and turns, of Oriental thoroughfares.
It would puzzle New York or London street super-
intendents, with their thousand miles of streetage
under their daily eye, to rectify this Algerine system.
But there are two reasons why the streets here are

thus : First, they were made narrow — the houses
opposite actually impinging upon each other at the
top—to obtain shade in the hot summers of the
south ; and, secondly, they were made compact, and
on a hill side, for defensive purposes. All the Oriental
towns, like the towns in Italy and in Corsica which
I have described, have these characteristics, especially
on the seaboard. We ascended through these defiles
to the summit of the town, and came out to the sky
and air, and had at once a sea view. It was a positive
relief to nose, lung, and eye. We came out at the
top of the town. Here was the fort captured in 1830
by the French, after their landing and fighting west of
the city, and which was the most substantial evidence
furnished to mankind, that the Algerine piracy or
polity was deceased, and that France had its grip on
this coast. Here were the old walls—twenty feet thick
—here the old gates of the city and fort, the chains
still hanging as they did when the Deys here held
court. We were shown within the fort. It is an
Oriental establishment, with French improvements—a
large courtyard, and some four or five stories of
porches, arched and pillared after the twisted, spiral,
Byzantine order, surrounding it. The seraglio is at
the top ! Within the court, as we were told, wrestlers
and gladiators displayed their strength and skill, for
the houri who peeped above between the iron bars,
and for the Dey and his eunuchs, who smoked their
chibouques and drank their mocha from the verandahs.
There, too, in a box of a house, about as big as, and
not unlike, a locomotive photographic shop which you
see on wheels, is the room occupied by the last Dey—
Hussein. It was in this he was wont to receive his
visitors and to do business. It is built out, as it were,
from the verandah. It overlooks the court, and is on
the third story. It is now closed. The soldier who

conducted us said, that if all who wanted to see its
inside were permitted, it would not last a month.
We glanced into its window. It is after the Oriental
style. Its ceiling is arabesque. Here the Dey received
the French Consul, who came, in full uniform, to
remonstrate against the non-payment of a debt to
French *protegés*. The old Dey lost his temper, and
slapped the Consul in the face with his fan. It was
'all Dey' with him then. The consul retired without
saying 'good Dey,' and (if I may be again permitted)
the prospects of that Dey were not afterwards so
brilliant. In fact, they were clouded. The French
went after him, and got him. As this is a pivot in
history, as so many terrible battles resulted from it—
I have indulged a little freely in some artistic touches
to represent the scene to my reader. If I have not
done justice to the Dey in the sketch, history has, and
the French have also. I have seen the splendid
pictures of Horace Vernet representing the wars of
Algiers, especially that grand tableau at Versailles,
where Abd-el-Kader is represented as taken; but I
confess that they were not the originals of my idea and
picture.

The grand house where the Dey lived with his
hundred wives, and where he is supposed to have
revelled, and which he ordered to be blown up in
1830—to save it, is held by the French artillery.
The very mosque in the fort is used as a barrack.
Around its porphyry columns and under its ample
dome, the French soldier sleeps in his iron bedstead,
and may be seen—as we saw him—sitting about the
holy places cleaning his uniform, without seeming
to care whether he looked towards Mecca or Paris.
The crescent pales before the cannon of modern
civilization.

As I have taken the liberty to illustrate how Algiers

lost the Dey by a blow from that personage, with his
chasse mouche, upon the infidel head of the French
Consul, it would be more complete to append a brief
history of that event. The citadel, or Casbah, is
already before the reader's eye in another engraving.
It is some 464 feet above the sea, and overlooks the
splendid mosque of Djama, and the port and city.
Here the Deys lived in perfect security from popular
violence, under the guard of Janissaries; here were
prisons and beheadings. In the interest of civilization,
and for some good purpose, Hassan was moved to cuff
the Consul. Before this scene transpired, France had
the Algerine coral franchise and paid a fixed sum for
it to the Dey. 'Without consideration,'—he enhanced
the sum. When the French paid the extra 3000
francs, he perfidiously allowed others to poach on the
coral manor. When the Consul protested, that officer
was fined 100,000 francs. An 'unpleasantness' arose.
In 1827, the Consul, M. Duval, still feeling unpleasantly, assumed to protect two Algerine Jews from the
rapacity of the Dey. Seven million of francs was the
sum which the Dey desired to confiscate, and which
was due to the Jews. An interview, a quarrel, hot
blood, and a blow! When France desired reparation,
and sent a minister with the demand, it was denied.
The ship of the minister was fired on as it left the
harbour; hence a hostile French fleet from Toulon,
and hence battles innumerable and sanguinary, from
Constantine to Oran, and from the white marble gates
of the Casbah to the mountainous portals of the
Desert.

We dread going through the old city again on our
downward tramp to our hotel. Our guide, Hahmoud,
takes us to the top of the hill overlooking the west
side of the city. Here is to be seen, across the valley,
the cathedral, called Notre Dame d'Afrique. It looks

like a mosque, and is exquisitely proportioned and elegant in its airy architecture. It has been built since the occupation. Indeed, all the churches and the synagogue, even, have the Byzantine style. They look like mosques outside and inside, except this: that, whereas the mosques have no paintings—nothing but the carpets and matting on the floor, and a few mottoes from the Koran on their walls—the French churches are richly decorated. One of the latter in the city has an attractive picture of the transfiguration.

We visited three mosques. One of them was most interesting. It was built by the Turks 500 years ago, and has in it a splendid copy of the Koran, the gift of the Sultan. It is called Pêcherie. It is in form like the Greek cross. We take off our shoes and boots, and slide around over the matting in a comical way. I have on some red slippers, which Hahmoud provided; but one of them is so irreverently large and clumsily Christian as to lose itself. Its brother slipper is proceeding regularly and reverently when arrested by Hahmoud, for Hahmoud is particular here. He is half Turk and half Arab. Fountains are in the mosque, where the bare-legged faithful lave their legs before they cross them in prayer, or lie down to sleep. We see ungainly human bundles lying in corners of the mosque, looking like sacks of grain, so motionless is their slumber under their rude burnous. I said to Hahmoud, 'Is there no rule to prohibit these lazzaroni from sleeping in your sacred temples?' I said this with something between the sacerdotal and police tone. I think Hahmoud conceals much waggery under his turban and beneath his long red silk sash and broidered jacket, for he said demurely: 'My honourable friend' —he calls me that, inasmuch as I told him that we were in America, all sovereigns, wearing crowns and bearing sceptres—(this metaphor is one of my weak-

7

nesses)—"my honourable friend asks me why these
faithful unfortunates sleep in the mosque? I ask him,
"do the faithful never sleep in the churches of the
Christian?"' Finding his honourable friend mute, he
added, 'Our mosques are ever open for prayer. We
know not whether our serene brethren (meaning the
snoring Zanies) may not be overcome with prayer,
nay, actually in prayer. Allah is great, and Mahomet
is his prophet!'

The doors of this mosque open into court gardens,
where are tropical trees. Starlings fly in and out of
them, and sing as they fly, and fly even into and
through the mosque, and among the Byzantine
columns and arches. Their cheerful treble brings
sunshine to the gloomy dome, and accords somehow
with the hoarse undertone of the Huck Hassab, or
teacher of the Koran, who sits cross-legged in one
angle of the mosque teaching in low monotones a
company of youths preparing for the gospel, according
to Mahomet. One of the youths is his own son.
Hahmoud says that he knows the Koran already by
heart. As the questions are asked, now and then the
youths bow and handle their beads. They are receiving
instruction in the ritual. Occasionally the teacher
smiles, and once he laughed—gravely. He stops his
laugh suddenly, as well as seriously. The Arabic
humour has been accounted rather stern and moral.
I seldom see much hilarity among the Moors or
Arabs. The Kabyles are rather jocund. This Moorish
priest and teacher shut down the breaks on his
jocoseness, either because we Giaours were glancing at
him, or because, having lost one of my shoes, I was
like a chicken in the rain—*stans in uno pede;* or,
what is likely, because with the twitter of the birds
among the pillars and in the dome, there is heard, in
strange discord, rattling into the mosque, the rat-a-tat

and rub-a-dub of the French drum, alternating with
the blare of the trumpet, which echoes and re-echoes
through the mosque! It is a reminder, perhaps, of
the '*infandum dolorem*' of other days. In fact, as
we emerge to the street, we perceive a French regiment on the march, knapsack on back, and going to
the seat of war. The conquest of Algiers is not consummated. Far down — ten days' journey to the
Desert—there has been fighting. Some few months
ago, one of the 200 desert tribes, which has never
succumbed, broke out, got thrashed, moved over into
Morocco, recruited its strength, and is (as our Consul
tells us) on Algerian soil or sand, ready to renew the
struggle. Hence this regiment is on the march!
There is a little moral here, about military rule, which
I would dilate on, but Hahmoud attracts me by
rushing up to a venerable Arab! They embrace and
kiss. He is a civil magistrate among Mussulmans—a
Kadi! By this intimacy and salute I perceive that
Hahmoud is well thought of. I am going to see the
Kadi with him.

Our next entertainment was quite in contrast with
these experiences. Towards evening we visited one
of the squares. Here was a crowd of some 300, of
all the motley costumes of this most motley city. The
Kabyles and Bistri are most in number. These are
the gentlemen from whom John Owens, the actor,
must have copied his 'make up' of Caleb Plummer—
I mean his dress, which consists of a piece of bagging
with a diamond mark, and 'Right side up! *Glass !*'
Coffee sacks are luxurious apparel—fit for the feast
of Lucullus—compared with the ragged burnous of
some of these children of the desert—the Bistri; and
of the mountain—the Kabyles. In fact, come to
Algiers if you would start a paper-mill! Rags are
plenty. Come, if you would see rags in all their

picturesque perfection and multiformity of ghastliness,
filth and variety—ragged at the top, ragged at the
bottom, ragged in the middle, ragged in fringes, and
ragged in texture ; ragged behind and before, ragged
in holes and in patches ; the old rags adorned with
new rags, and so stuck on, as to add aggravated graces
of raggedness above the reach of art ; a raggedness so
elegant in its touch as to conceal still more elegant
elements of raggedness ; raggedness which has not
merely resulted from the natural and acquired taste
and condition of the ragamuffins themselves, but is
inherited from ragged ancestors who wore and enhanced by their wearing and genius the ultra-ragged-
ness of their habits, which they have handed down to
children born with rags on them, and who have preserved, and enhanced with the lapse of time, their
precious legacy of rags ! Here we found this incarnate
raggedness, sitting and standing, in black rags, white
rags, and many-coloured rags—rags pinned, sewed and
tied, and some neither ; worn by negro and native,
by young and old ; and all happy, in beholding in the
centre of the ragged circle, an African from Timbuctoo
who had but one rag around his loins, and no other
covering save a tuft of wool on the shaven apex of his
lithe body and a serpent twining and writhing through
it, and around his neck and down his back ! This is
the Snake-Charmer, Mumbo-Jumbo, by Gumbo!

The ragged battalion give way to allow Dr. Bennet
and myself a place in front. We are careful of our
contact. There lies within this circle—on his back—
legs apart and body heaving—a passive lumpish fat
body, hid in rags. He is a part of the charm ; but as
we do not know what he is, or does, except that the
snake now and then creeps under *his* rags, we will let
him lie. The other is the performer. He plays with
the snake ; mumbles his words ; and then screams,

half-panting, his periods like a tired stump speaker in a high wind. Then he strikes his tom-tom, and with a wild dance—almost the counterpart of the plantation negro with a banjo in Alabama—he makes the circle wider as he moves. He stops, talks to his snake, twirls its shining folds into all sorts of forms, and begins again. The crowd understand his language. He calls on them to applaud, and they cheer after a strange method. Then he uses his head on a Kabyle head ; having the advantage of skill and skull, he pushes his antagonist over. The crowd laughs. He then drops his snake, picks up a large piece of limestone, calls out of the crowd some one conspicuously ragged to test its soundness. Then, with a hideous face and cabalistic talk—his face made more horrible by a long wire running out of his mouth through a hole in his cheek—he slaps his naked abdomen with the stone, and finally breaks it in twain. Then we have some juggling ; then he gouges out one of his eyes and puts it back, whereat all the rascally raggedness rages in rapture. He has an eye to his exchequer. He rests his lungs and limbs by passing round his tambourine. He obtains some sous. The least I had in my clothes was a silver half-franc. I thought of the recent decision of the Supreme Court. I knew there was an implied contract for specie. I felt that specie was the legal tender abroad, and I paid it. When I put it in, he made me quite a hero ; danced round the ring with it, and proceeded with greater activity to discourse to his snakeship, and prance about his prostrate companion in the ragged ring. What all this means, I leave to those versed in the lore of African Fetishism.

Many of the negroes here are not of our Congo-Guinea kind. They have straight noses, and other features regular and handsome. They are from Soudan, or Abyssinia. Especially is that the case with the dark servants whom we see attending the Moorish women in the streets. But most of the negroes are like those in America ; if possible, more black. I saw one yesterday in the café, where the Timbuctoo negroes congregate for coffee, who was so unctuously black that he actually shone. We have heard of objects ' dark with excess of light.' Light may blind. But some of these Timbuctoo people, fresh from the desert or beyond it, are light with excess of dark. But I do not perceive that, as a class, although they mingle with other classes, they hold relatively any better position, socially or otherwise, than the negro in America. The negro here of the Guinea type, is the American negro all over. He or she is vain, jolly, and subservient ; likes gaudy colours, banjo-music, and a diabolical, mystical sort of religion, full of emotion and superstition.

But I must leave this topic until we visit the Desert, if, indeed, the wars down South permit the journey. Not that I expect to find the southern boundary of Algeria. That is as fickle and as faintly marked as the sands of Sahara, which are moved by every wind. Even the French itineraries, while they give metes and bounds to Algeria, on north, east, and west, naïvely say : '*Au sud, elle a pour limite, bien vague encore.*' So vague in fact is this southern country, and the rule over it, that one wonders why so many French lives have been given to the fortification and colonization of French power there, until he remembers the eventful history of Algeria. Then he would not wonder. It is hard to hold water in a sieve, or to chase and capture the wind. It is equally hard to conquer such a people, or such a number of tribes, independent and nomadic, as make up Algeria. The people, besides, are so un-European in habit, and so different in religion, that it will take years, if not ages, to crystallize the alien domination, and introduce the new rule. The Mussul-

mans alone number over 2,000,000, without counting
the tribes of the Desert.   The Kabyles, whose country
on the east, in the mountains, we shall visit, number
700,000.   They were the last to be conquered, and the
reason was that they had and have a Democracy.
They have local self-government, elect by majorities,
and have a confederation.   The Arabs number 1,391,812,
and are divided into 1200 tribes, and these tribes are
broken into 10,000 among themselves.   The Sahara
has a population living a nomadic life, or cultivating
the oases, of 600,000—all Arabs, and numbering 200
tribes.   Of course a great many tribes are so errant and
belligerent, so free and so useless for commerce or
conquest, that the French do not look after them.
Still we have accounts of French officers and adventurous merchants, pushing their way over the deserts,
making treaties or trades with tribes or caravans, and
having in view the French domination.   But it will
be a long time before the Azguers, Hoggards, Airs, or,
to include them all, the extreme Berbers, or Touaregs,
acknowledge French power, or succumb to French
arms.   The French may tap the avenues of trade on
the oases, and establish trading laws and interchanges
with Soudan by caravans; and even, now and then,
with a five months' journey over the desert, reach
Timbuctoo; but there is no power, not sympathetic
with these tribes by habitude and faith, which will
ever hold them steadily in subserviency or in tribute.

The history of this country illustrates this statement.
I need not go into it at length.   Whoever first found
Algeria, or Numidia, or Barbary, Libya, or Mauritania,
or Africa, or what name or nation soever it is called in
history, and undertook to hold its tribes in thrall,
never succeeded except in a small measure, upon the
coast, until the Arabs came.   They had the nomadic
habits and the same instincts, traditions, and blood.

They brought with them a religion which captured the Oriental imagination and held Algeria. What Carthage could not hold; what the Romans, with their colonizing genius, could not retain; what the Vandals, coming hither in thousands, did not control a hundred years; what the Greeks, by their pliancy and with able generalship, held only for 150 years — the Arabs captured easily under their common sympathy and religious zealotry. 'Paradise before you! Hell behind!' cried out the followers of the Faith, and under this shibboleth they held, as they had captured, Algiers, even Morocco, and even to the heart of the desert and beyond, where the Romans never ventured. The Arabs became the masters of North Africa. Indeed they never left Europe till the year Columbus discovered America. Had America been known to them, before 1492, New York might have had her ancient mosques and San Francisco her alhambras! Pilgrims, from the farthest East and West, might have—

> " Trod with religious feet the burning sands
> Of Araby and Mecca's stony soil."

The Turks when they came to Algiers only continued the Mahomedan rule. When, in 1830, France landed her 38,000 men from her 25 ships in the Bay at Sidi Ferruch (whose black rocks we saw some ten miles west of Algiers, on the coast), and fought the battle of Staoueli, on the 19th of June, 1830—on the site of the Trappist monastery, where we spent a Sunday; and when she drove the 40,000 resisting Arabs from the field with the bayonet, and crowned the battle by a pursuit which took Fort l'Empereur and the Casbah, which commands Algiers—then and there ended the rule of the Deys, and then was begun for Algiers a new career! How long will it last, and how far will it extend? These are not for the tourist's pen, but the politician's ken.

In my next chapter, avoiding historic talk, I will venture to describe my visit to these battle-grounds, to the Trappists, to the Moorish cemeteries, and the tombs and marabouts. That chapter, at least, will be freshened by out-door country air and new experiences.

CHAPTER IX.

ALGIERS — GARDENS, TOMBS, CUSTOMS, ETC.

"I sat down under His shadow with great delight, and His fruit was
sweet to my taste."—CAN. ii. 3.

I PROMISED an out-door excursion, out
of the heat and dust of Algiers, and into
the hills and among the vegetable wonders
and beauties which surround the city. Our
first trip shall be to the Jardin d'Essai. This garden
lies upon the east side of the bay, and is reached by
a road along its shore. It is about three miles to the
garden. The railroad clips off a palmy side of the
garden next to the sea. The grounds now belong to
a company, who are compelled to keep them open to
the public, and who have turned their establishment
into a commercial adventure. It was started by the
Government, somewhat after the manner of similar
gardens at Paris, London, and Washington, for the
collection and acclimatization of all the rare grains,
plants, trees, fruits, and flowers. It has within its
domain six thousand different species, and has distributed three millions of plants. To this place we
found many omnibuses running. As the day is
Friday — the Mahomedan sabbath — these vehicles
are filled in great part by the Moors of the city, who
go out to the cemeteries and there have their picnics
among the tombs. As one of the largest cemeteries
is in this neighbourhood, the omnibuses are full. It is
not a little significant to see the French peasant in his
blue blouse, or the French woman in her plain frock,

elbow the wives of the Moors, who are dressed in their bundles of white muslin, and only their eyes visible through the lattices of white! The very names of the omnibuses are significant: 'A Mon Idée,' 'La Bien Aimée,' and 'Il Trovatore.' It looks as if Civilization, riding in an omnibus, were crowding the exclusive Orient out of its ancient customs. It seems as if the old mode of dressing here were kept up more for coquetry and affectation, than for the seclusion from vulgar eyes of those houri who are supposed to be within the folds of their ample mantles and trowsers. On our way to the garden we meet Arab horsemen dropped into *their* abundant clothes between their high saddles. They look rather soiled and dusty. These Arabs wear a high turban, banded with camel-hair ropes. Sometimes they have a gun neatly swung over their backs. If they happen to be *sheikhs* or other men of consequence, they have followers, or a cavalcade of them. It is quite romantic to see one of these devotees of Mahomet—as the engraving illustrates—fully armed and robed, suddenly stop, and turning toward Mecca, pray!

The road is a cloud of dust. The carouba trees which line the road need to be washed in a shower. They, too, are dusty. So are the dark legs of the Kabyles, trudging homeward to the mountains, on and behind their camels and donkeys. Occasionally we meet Kabyles mounted upon their donkeys. They generally sit as near the tail as possible. Having emptied their sacks of wheat, charcoal, sheepskins, or olives, they have a cute way of pushing their feet into their vacant sacks for stirrups. Some of the donkeys are smaller than the Kabyles themselves. The picture is very funny. It looks inverted; for the donkeys ought to be on the Kabyle back! Talking of the Kabyle back, I saw a curious instru-

ment which they used for scratching their backs.
Why they scratch—whether it is the itch produced
by the African sun, by the long woollen burnous or
by the fleas, or what—yet they use a wooden instrument, shaped in size and form like a wooden salad
spoon, called a '*gratte dos*;' and with it a deft Kabyle
can reach over his shoulder or around his body, and
give a good, honest 'old scratch.' This is much more
economical (if not so satisfactory) than the Duke of
Argyll's plan. He planted posts over the county
which he owned in Scotland, against which the
peasants relieved their itching bodies, and for which
they said : 'God bless the Duke of Argyll!'

On our way to the garden we cannot but remark
the French appearance of the houses. Even in the
cafés, frequented by the natives, where the cross-
legged are sitting, sipping coffee and smoking, there
is the French style. On the outside of the walls are
painted the foaming cup, and the pipes crossed and
tied with a ribbon, and some balls, indicating billiards.
We perceive that the plane tree or sycamore and the
mulberry are quite common; and lining the roads
in the meadows are orchards of bananas. In fact,
the banana is a great crop here. The leaves are a
little brown and ragged from the harshness of the
past winter. I notice that they have to be guarded,
and are, like other tropical plants, screened from the
north winds, by tall, stately rows of cypresses, especially
on the sides of the Sahel Mountain, or between it and
the sea. As far up as the beautiful temple which
crowns the mountain — a white, Moorish-looking
building, but, in fact, a theological seminary of the
Catholic faith, you may see these evidences of tropical
growth, but guarded by the cypress lines. My friend,
Dr. Bennet, who is looking after sunbeams in winter,
for his patients, and for his own satisfaction as a

savant, at once concludes that Algiers suffers from
the north-west wind. The trees thus exposed, show
by their inclination, as well as by their slow growth
and imperfect development, what the Doctor desires
to know about the Algerian climate. He observes
few orange trees and no lemons; and everywhere—
even where palms exist—the cypress is there to shelter.
He compares, or rather contrasts Algiers with the
Riviera. He notices that while his favourite health
station at Mentone is comparatively free from the
mistral and north winds, those winds pick up moisture in coming over the ocean hither. The Doctor
will find many evidences, *a posteriori*, to strengthen
his conclusions before he leaves. I refer those interested, to the new edition of his book, which will
contain incidents of his search in the regions of
Algiers after a better climate. Just now, under the
hot sun, and with all the ardour of a lover of botany
and horticulture, and with a view to perfect his garden
at Mentone, he is absorbed in this garden at Algiers.
I have already endeavoured to depict his garden. It
is hung almost in the warm air of the Riviera. It is
warmed by the geothermal as well as by the solar
caloric, and irrigated plenteously from above. The
Doctor, being thus interested, dashes into the merits
and demerits of this African Paradise in a way that
makes me wonder. Now he points out the bamboo
and cork trees; now he astounds me with the declar-
ation that those groups of flowers are the mesem-
bryanthea. Now he drops his yellow umbrella; out
comes his note-book. Eureka! he has found, not
only one, but two, three, four—a whole colonnade of
new palms — palms which to him heretofore were
habitants of the conservatory and small in stature—
lo! here are avenues of them; some native like the
draccona; some from the Isle of Bourbon; some

date-palms in fruit and flower, and some so old they
are—*without date ;* every variety, and so arranged
are they, that the tall green plumes with the golden
blossom alternate for nearly a mile of avenue in a
double row on each side—with the wide, waving leaves
of the lesser kind. 'Every variety,' says the Doctor,
'is here—proof that there is a hot summer rather than
a warm winter.' Their rough stalks and yellow flowers
—as well as the little scrub-palms which we see all
round Algiers, indigenous to this soil—are critically
examined. Here are lofty brethren from Gabon, cultivated by the side of the timid palmetto from
Carolina, the one as proud as Lucifer, the other as
meek as Moses. In fact, the palms have taken the
palm in the horticultural, 'essai.' They thrive next
to the cocoa and the banana, although trees from
Bombay, Brazil, Australia—all those trees which, like
the palm, love to have their heads hot and feet moist
—thrive prodigiously; for here is irrigation and plenty
of sun, even in April.

While the Doctor is making his observations, we
perceive some ostriches through the bamboo avenues.
They are enormous in size. They are running about
in their wiry enclosures. We approach them. They
plume themselves greatly on our admiration. They
step off, lifting their wing in a most tragical way, with
a long stride and swing, as if they had on buskins,
and had Hamlet on the brain. They are priced in
the catalogue for twelve hundred francs. A young
one can be had for fifty dollars! M. C. Riviere,
Director of the Garden, reports great success in rearing
these ostriches. Within the palisade enclosure, containing a quantity of fine sand, the female ostrich
deposits her eggs. She seems uneasy, and seeks a
suitable place. She forms a small hillock of sand,
slightly concave at the top, and lays one egg in it, to

which she afterwards adds others. She lays every two
days, for two or three months, with an interval of
repose. Incubation lasts forty days, during which
time the male and female sit alternately. On 12th
of March, before we were there, five were hatched,
and three the day after. The male, like a gentleman
as he is, takes great interest in the incubations, and
only leaves the egg when pressed by hunger, and then
the female takes the place, but not for such long
periods. An extra circulation goes on in the uncovered portions of the male's body, to generate the
heat necessary for the process.

Leaving these pretty little birdlings to their laying
of eggs, hatching of young, and digestion of brickbats
and horse-shoes, we meet another illustration of the
animal kingdom belonging to these shores—a man
with his two veiled wives and two children. We see
that the wives are young. They are painted between
the eyebrows, so as to make the brows seem one brow.
Their hands are red with henna, and their finger-
nails are black. They wear red socks. The children
have their hair stained red. We hardly have time
to admire before the Doctor bursts upon us, with a
magnolia grandiflora rotundifolia! He has found
it by the banks of the lake! We look his rotund
majesty up; and a choir of blackbirds and nightingales
sing a pleasant lyric from a grove, deep in shadow,
in praise of the wonderful glories of this tropical
garden! We find soldiers in red trousers, handling
the mattock and spade, and digging up bananas for
transplanting. The French soldier is here permitted
to do work; quite a reform, and worth considering.
Other rare plants we see; immense scarlet geraniums,
the *fleur-de-lys*, the Arabian fig, climbing plants like
the African ivy, some in flower, hanging on great
palms forty feet high, and purple in bloom; jasmine

in *trees*, all blooming; Barbary figs in profusion; and, to crown our view, an olive of monstrous proportions, literally garlanded with the white rose; then we visit the hot-houses, a dozen or more, where every exotic the vegetable world produces is to be found.

At length we wend our way along paths where water-lilies are seen in lakes, and the air is heavy with odour, to the grand entrance. Here we find under the shade of a great plane-tree across the dusty road, at the Restaurant des Platanes, several Arabs playing draughts—not for strong liquor, for the Koran forbids it, but for coffee. An Alsatian—an old soldier—with two wooden legs and two crutches, is sitting at the gate. While waiting for the Doctor we learn that he was a soldier in the Algerian wars as early as 1847, and that he lost his legs by their freezing in the snows of Africa! He had been pursuing the Kabyles, in the Atlas, and we have had demonstration to the eye that the snows are there yet, although the sun is African and 80° Fahrenheit. We sit here observing the omnibuses go by, crowded with the motley loads. It is said that the railway in India is killing caste. The omnibus is doing the same here.

I spoke of the cemetery. On our return to the city we called in to see it. At the entrance we found some vendors of Algerian delicacies. Then lying upon the grass and lazing about the tombs we see some Maltese sailors anxious, like ourselves, to see the novelties. I have a pictorial illustration, better than my pen can do, of the scene. The ladies of our party are attracted to a little Jewess of ten, her jacket broidered in gold, and her Greek cap set upon the side of her head jauntily. We go up the hill-side, among the faithful. We are met by a turbaned white-beard. He conducts us into a small mosque. We peep within. Finally, we take off our shoes and venture in. Here are families of Moors in groups, sitting cross-legged on their mats; mostly women and children. They are unveiled. As I protrude my dark sombrero within the sacred precinct, I catch the look, half reproachful, half coquettish, of a fair girl, of dark, almond-shaped eyes, whose face is stained with cerulean tints, and whose ankles and wrists are heavy with gold and silver ornaments. Her look seemed to say, 'Oh! you—you—naughty sir! You dare; *how* dare you? Tut! tut! Oh!' Of course this is an 'imaginary conversation,' to which I respond by going further. I actually go through the little court, and under the ground, where are lights aflame, and two unpleasant bodies recently buried, wrapped in shawls; the odour unpleasant, and the mourning not poignant. The families laugh, chat, giggle, eat, drink, and have what is called in New York, a jamboree over the bones of their kin. Some try modestly to put on their head-gear; some turn their faces to the wall, and with eyes askance, seem embarrassed; but, on the whole, they enjoy the intrusion. All at once a wild looking old man begins to rail at them for their shamelessness. Our conductor says that he is touched in the head. I am not so certain about it. He may be a fanatical Moslem. Religion makes people wild without being exactly crazy. He raises quite a hubbub, and as we leave we hear him still railing— a voice from the tombs—'a doleful sound.'

We breathed easier when we were again on our promenade *à la voiture*. Now we go to the Valley of the *Femme Sauvage*. The story of the valley is, that a beautiful lady was crossed in love; left Algiers to assuage her grief among the mountains; came to this vale twenty years ago; her money ran out; she cultivated lemons; she was *soured* of the world; for she never spoke to any one. She gave away lemonade, and the good people who drank it gave her something. The story is true; but the name *Sauvage*, which might imply a *wild* woman, is interpreted to mean only a *timid* creature—one who could not grapple and battle in life with its harsh experiences. The road to her former home is very charming. Her house, however, is closed. The brambles fill her little garden. She is dead! The valley is sheltered from the north wind, and Flora is abundant and Ceres advanced. The pomegranate is here, ready almost to blush, and the fig is immense in size, and far in advance of the fig of the southern shore of Europe. The nightingale sings in the groves of the lentiscus. Along our upward path, the rock-rose (so common in both its white and red dress, all through Algiers) decorates the banks and bespangles the rocks. Everywhere, above the shrubs of the country, tangled with the white hawthorn, fighting its democratic way, and even fighting its neighbour, is the prickly pear; everywhere is the blackberry. It is the bramble of all nations—as cosmopolitan as the chicken! It asserts its right to live and flourish. Why not? It is the plain, common blackberry of America, and does not care for the aristocracies and regalities of the European flora! Why not, at least here, may not the *black*berry be African. The earth has a red tinge and the rock is silicious sandstone. We pass by the jail-like Moorish country-houses, windowless, but with portholes barred, and with the same inner quadrangular court observed by us in the city houses. The mulberries as yet look bleak and untropical. Like the chestnut trees in February on the mountains of Corsica, which alone left the impression of winter—they are leafless. Strange, but true it is, that only in

the Genoese Riviera have we seen all the trees in
winter covered with the garniture of summer.   The
sunbeams were ever with them.   Not, of course, by
night, but ever by day.   My learned friend holds
that all these places so bepraised, which, like Algiers,
are celebrated for a temperature equable by night and
day, are not so good for health.   Algiers is not so far
south as Nice ; and at the latter place the nights are
cool, if the days are warm.   He argues that the earth
is turned from the sun half the time, when it is night,
and that thus nature, which does all things best, teaches
us that times of repose, times of cool, tonical rest,
are needed for recreation.   Hence he argues that, as
at Mentone and elsewhere on the Riviera, the mountains shut in the coast from the north winds, and
make the night cool, if the day be warm—there are
more health-giving influences than at Algiers ; and
that protection from the North is worth half-a-dozen
degrees of latitude towards the South, for sanitary
restoration and vegetable production.

These are results from observation thus far.   We
may, in our progress after winter sunbeams, find more
reasons for appreciating Algiers.   Certainly, wherever
the spot is guarded, there the conditions of a sanitarium appear.   We hope to find this further inland.

One thing was very striking in the Valley of the
Wild Woman.   Across the road, amidst the tangle of
vegetation, where the lemon was hid in the nooks
from the wind, there appeared on the hill-side, cut in
the rocks in the shape of a temple, supported by
Egyptian or Assyrian figures of women—a splendid
tomb !   At the top of the temple was the form of
an eagle.   The grotto beneath, in which some body
or bodies once reposed, was covered with foliage and
vines.   But there the old sculpture remains.   It will
thus stand till some earthquake—no rare occurrence

here—disturbs the tricks which art has played with nature.

In our observations about Algiers, I am at a loss whither to turn for incidents. I find enough to write about—novel to a stranger—from my window at the hotel. Glancing out upon the bay—where I perceive the divers going down in their armour to dig out the accumulated sand which chokes the harbour—I could find for a pen-picture a company of a hundred 'peculiar' people watching the performance; or, moving upon the waters of the bay, what is it I see? Not a man walking? Is the miracle repeated here? It is a marine velocipedestrian. He sits securely and works a water-wheel below him with his feet, which propels the cigar-shaped canoe, to which the wheel is fixed. This is a dim description; but as I see at a distance and from my window, I cannot do it better.

If you would know Algiers better, stray into the French quarter in the evening and visit the Café Chantant. You will hear excellent music and see good acting. You may call for coffee and cognac; or drop your sou into the basket of the half-dressed cantatrice and danseuse, who comes down from the stage among the audience for her coppers. If, however, you would prefer French experiences in Paris, and are here to study the Arabs and their melody, go with Hahmoud, as we do.

Upon a raised table, amidst a crowd of fifty coffee-drinking and cigarette-smoking natives, sitting cross-legged, are the musicians—three—two with tambourines, or drums, and one with a sort of hurdy-gurdy, with a bagpipe sound, or flageolet squeak, or something. The music is barbaric, but a Kabyle brother starts up and keeps time. He is bare of foot and shaven of head. He has two stringy handkerchiefs in his hands. He dances, and oh! heavens! what dancing! He wheels, he steps, he jumps, he cavorts; he sways his handkerchiefs, as a coquette her fan, as if to say: 'Am I not all grace?' He astounds you by an unexpected spasmodic hitch, as if somebody had stuck an awl into him; then halts, as if he were injured or astonished by being suddenly jerked out of his burnous or cuticle. Then he tries to bend at once both backward and forward, gracefully; and with a leap in the air, arms akimbo, he subsides into a quiet Kabyle in a corner, to enjoy a smoke, while another takes his place. This is Kabyle recreation.

Now for something more horrible, which I almost sicken to paint. Hahmoud insists on our going up into the old city to see some Mahomedan rites. The devil himself is an Algerine dervish. These are not the dancing, whirling dervishes of Constantinople. There are several sects of Moslems peculiar to Morocco and Algiers. I think there are seven. I believe Hahmoud belongs to this one. He did not like to confess it; but I saw him, as we entered, quietly salute the chief dervish with a peculiar embrace and kiss. To this performance the ladies went; but they had to go up stairs into one of the galleries of the inner quadrangular court of the Moorish house, among the Moorish women. Several nationalities were represented in our crowd. Expectation was on tip-toe to see and hear. It was a hot night, and the room was close and full of people. Some two dozen Moors were present. It is dark in the room; only two dim candles and a charcoal fire, which smouldered in a skillet. The object of the latter appeared to be to warm up the drums which the dervishes beat, and which, when the sheepskin got loose, they heated over the fire to make it tight. Perhaps there were some fumes in the skillet to make the dervishes devilish. While our ladies above were taking coffee, very black and sweet, in nice little china cups, in the galleries, with their Arab hostesses, we sweat downstairs, leaning expectant in the dark against the whitewashed walls or against the pillars of the court.

Now the chief begins. He chants passages of the Koran, while, standing around him semi-circularly, a half-dozen respond with whining tones. Others, with their drums, sit cross-legged in a row, before a little stand with two long, lighted wax tapers. A monotonous drum chorus begins; then a long-haired dervish bounds up like a jack-in-the-box, as if shot in the rear; and, being up, a brother unbinds his garments and spreads out his hair, and then he jumps up gently at first, keeping time to the music. His head bows as his body sways; then faster and faster, till his hair flies around wildly and his hands are swinging insanely. He is joined by another who is more staid. The last looks as if he ought to know better. The first one, exhausted, falls down in epilepsy and is carried out. No. 2 is joined by No. 3; then No. 4 appears, and, by this time, No. 1 re-appears, and the group collectively, like a brutal nondescript —are all at it. No. 1 having worked himself wild again stops a moment. The others stop. A brother appears from behind with a red hot bar of iron.

No. 1 laps it with his tongue. I see it smoke. My blood runs icily. He slaps the incandescent iron with hand and foot. Then the ministering brother offers him to eat some delicate stems or pieces of glass. He crunches and swallows them. His digestion is excellent. If it had been candy, and he had been a juvenile, he could not have relished it more! Then No. 2, the intelligent, stops and has a long wire run through his tongue and out of each cheek, protruding four inches. He snarls meanwhile like a caged hyena.

Then No. 3, who has been rather quiescent, commences to snap and bark like a hungry dog—eyes popping out, and face all savage and imbruted. *Barked?*—He howled, he growled. Finally, the ministering brother comes out with one of the thick leaves of the prickly pear, a foot long, in form of an ellipse, an inch thick, and full of thorns; all the dervishes drop down on all fours and are biting at it and into it, and crunching it.

The froth of their mad mouths hangs to the green prickles and slavers the green rind. Ugh! What more? No. 4, in an ecstacy of fanatic diabolism, swallows a scorpion. Whether they have taken out the poison, or whether the afflatus is so enormously exciting, that poison is innoxious; or .what, God knows! We summon Hahmoud in haste; beckon our ladies from above in the dark, and seek relief and breath in the narrow streets. Upon these infernal orgies we have nothing to comment. It is as near making the human a wild animal as anything can be.

It is worse than the negro performances every Wednesday, upon the seaside, at the Jardin d'Essai. Here all the blacks of Algiers come to celebrate the fête of Nissam. It begins when the beans begin to blacken. Up to that time the negroes abstain from eating beans. The 'sacred' is mixed with the profane in this festival. They celebrate Belal, a sainted black female slave, who had been in Mohammed's family. They pray and gorge with food. An ox, covered with flowers and gay foulards, is sacrificed. They dance round it seven times before they give the death stroke. As the ox dies, whether soon or not, in agony or not, so is the prognostication of good or evil. Then begins the negro dancing. Then the prophetic negresses, retiring under a tent near the sea,

are waited upon by the crowd to learn their futures.
The crowd bring chickens to the prophetess. She
wrings off their heads, and throws the body into the
waves. If the 'headless rooster' swims and struggles
—so; if otherwise—*not*. That is clear. Then begins
more dancing and chanting, and a wild sort of music,
called Derdeba. It should be called Diablerie, for
it is a jolly row. Thus these black devotees of Mohammed outstrip all the rest of their co-religionists,
except the brutal Dervishes. To beat *them*, I defy
all the powers on or under the earth! Justice, however, to the better class of white Mohammedans
demands that I should say, that they disapprove of
these mumbo-jumbo orgies. Especially have they
endeavoured to crush out these negro extrava-
ganzas.

The negroes are here pretty much as elsewhere—
not of much account. I say this not from American
prejudice; for where you see such mixtures as here,
you forget all about colour and caste. It is the testi-
mony of others. A volume I have opened says, in
speaking of the races here, that the Kabyle is a good
worker of the soil; the Biskri a fair boatman and
porter; the Mzabi is busy as a baker, butcher, grocer
and small merchant; the Laghonati is a laborious
bearer of oil; but 'le nègre blanchit les maisons, ou
exerce quelque grossière industrie.' He whitewashes
or does worse, or nothing. That is about all of him,
here.

One object worthy of note is the cast of the body
of the Christian martyr slave. Refusing to abjure his
religion, he was cast, not into a dungeon, but in
plaster; not pounded, but fixed alive in the mortar,
and thus smothered in a horrible way. Rumours of
this tragedy ran over Europe at the time; for it was
done by one of the piratical Deys of the last century.

But it was verified only the other day, when the remains were found on removing the wall of the old fort. The body was, of course, dust; but the mould was there, showing how the skin, lips, and brow, the very bones and attitude, took the form of agony under the terrible torture.

Our next outdoor venture is along the sea to the west of Algiers. We go to the battle grounds —to which I have referred—where first the French in 1830 landed and fought. We go to see the harbours where the pirates lay in wait for their prey, and a print of which I have inserted herein. We go to see the Community of La Trappe, some fifteen miles from Algiers; and we go on Sunday. Our way is past the barracks of the soldiers, through Mustafa Inferior, much improved and improving; and out of the city on a road just freshened with last night's rain, and with the ever present spray from the sounding shore. The road looks newly dressed in asphodel, convolvulus, and marigold. How the marigold worships the sun! What numbers of them. The meadows are like a cloth of gold. The great breakers thunder against the black rocks, those basaltic ribs of Atlas, which protrude to the seas and resist the thumps of Neptune! The sea dashes far into the rocks, and forms numberless cascades. All along we see the little palm scrub. Dr. Bennet says: 'They seem to spread out their fan-like hands in welcome.' I respond: 'Therefore is our hospitable hand called the palm.' Surely, it is so. There is more than one meaning in the Bible phrase, 'Ethiopia shall hold out its *hands* unto God.' Twenty fingers there are to this palm; all spread out in the most cordial, reverent way. Beyond us already looms up a mountain range, like the Estrelles, near Cannes. At its foot is the bay where the French debarked in 1835. They were four

8

weeks coming from Toulon to this point in their
ships. At that I wonder not, if such a sea prevailed
then as now. We forget to look in front at the great
'Tomb of the Christian,' arising to our sight. We
forget to note the sportsmen, with pointers, hunting
quail. We forget the acres of geranium in full flower,
acres on acres, illimitably, planted here for the essence
of its leaf. We are hardly drawn to observe the red
gladiolus, the rock rose, the arbutus, ciscus, lavender,
holly, prickly broom in yellow flower, hiding the moor-
like country for leagues, and the rosemary, so fragrant.
Is not the sea all in white, blooming in tempestuous
grandeur, and throwing its blossoms in a wealth of
luxuriance far over the dark rocks, and even into our
faces upon the highway? And can we forget to see
the hand of man, as well as of God, in history, written
in these rocky inlets and island breakwaters? Here,
upon these wild rocks, we see the ancient Moorish forts!
Under their guns once lay the Corsairs, fierce, daring
men, 'linked to one virtue and a thousand crimes.'
Behind these islands, and in the coves, lurked other
'coves'—the marine devils, whose diabolism brought
Christianity to the doors of the mosque and gave
Algiers to France.

And yet, when we think that only seventy years
ago, there were 30,000 Christian white slaves held
in and about Algiers, is it not marvellous that the
civilized nations allowed this so long? We have
seen an army of 20,000 men and an expenditure of fifty
millions by a Christian nation to rescue in Abyssinia
half a dozen silly missionaries. But never a hand
raised to rescue 30,000 Christians who toiled at the
galleys, or worked on the jetties, or in the fields and
in the hot sun, at the call of Mohammedan masters!
Well, I feel a little proud that the United States,
when an infant, had one blow at these people,

and won their respect by thoroughly thrashing them.

But we are seeking the Trappists. They have been established since 1843. They have 4000 hectares. There are two and a half hectares to an acre. The heath, like the land along the coast, changes at once. It smiles sweetly, as soon as we approach their domain. There are 112 members of the community. We drive up to the portal through a line of caroubas and mulberries. We behold over the portal an image in white, of the Mother of God, and within it, the Saviour upon the cross, and the skull and bones! Upon one side is written in French: 'Celui qui n'a pas le temps de penser à son salut, aura l'éternité pour s'en repentir.' On the other: 'All the pleasures of earth are not of as much value as one penitential tear.' All through the establishment are mottos to teach the frailty of all earthly hopes, and the overruling duties which pertain to the other world.

In one thing we are disappointed—our ladies are interdicted from the convent. A courteous Brother, in brown sack and capuchon, meets us, helps us to descend from the carriage, invites us into an ante-room, and bids us wait until the Brothers rise from their breakfast. It is 2 P.M. The Brothers begin their day at 1 A.M., and have gone for a short rest after breakfast. Our ladies must remain outside, and we go through. Our first observation is a cluster of nine palms, all growing from one stock, apparently, on a mound. They form a rare picture. In fact, we bring home a photograph of it.

In the first room we enter, we perceive a government chart of that fight, which occurred on this very ground of Staouel, 14th June, 1830, and which resulted in the success of the French. We meet numbers of the Brethren. They all bow. None, not even our conductors, are allowed to talk within these walls, sacred to meditation. We see some in white, and some in brown garments. The white are the Levites and adminster at the altar. The others contemplate. All work—all do charity. Many poor are relieved by them. All the work of the farm, the black-smithing, the cooking, the making of clothes and sandals—all is done by the Brethren. They showed us their long stables of mules, horses, cows, oxen, and pigs, and the principle of association was most graphically illustrated not only in the exuberance of the farm production, but in the very comforts of bed and board by which they sleep and live. It is popularly thought that the Trappists are engaged in digging their own graves and filling them up again. I did not see this. Nor did I see that they mortified the flesh too much. Their solemn mien and formal manner disappeared when out of doors and out of their precise duties. Hahmoud, our Moorish guide, was with us. He had been there before. How merrily twinkled the eye of Father Joseph, as he offered Hahmoud some wine. 'No, my religion for-bids,' said Hahmoud ; and then with some quiet jokes on Hahmoud's peculiar ways, the Brothers passed round the good wine—red and white—of their own vintage ; and with it the best of bread and honey, grapes, and almonds, all from their own lands. The very wax made into candles for their holy observances comes from their own hives, and is manufactured by their own hands.

Altogether, this was a joyous visit. These good monks made it a white stone in our travelling calendar. Far aloof from the world of gain and pleasure, with good libraries and good consciences, no temptations to beset them, earning their own living by the 'sweat of their brows,' given to hospitality, ever constant in prayer, they live far, afar on this African coast and make its desert places to blossom as the rose. Surely they live not in vain in time, although they profess to live only for eternity!

Yet many come here and try this life and fail signally. Only five in twenty succeed!

CHAPTER X.

AMONG THE KABYLES.

' And in the mountains he did feel his faith.'
WORDSWORTH'S ' Excursion.'

THIS chapter is penned at Tizzi Ouzi, in the Kabyle country. We are now in its very heart. We are between Algiers and Fort Napoleon. The latter place is upon one of the highest mountains of the Eastern Atlas range. It is a few hours' ride on donkeys from the snow summits of the Djurjura. We stopped last night here, where I write, and we are again here to-night, after going through the Kabyle valleys and over their mountains to Fort Napoleon. We might have pierced Atlas and come out on the little paths leading either to Bougie, on the sea, or to Tunis. We preferred, rather than go further east, to return to Algiers, and visit the southern and western parts of the country; perhaps go to the Desert before we leave for Spain.

This village, where I rest, is somewhat French. It is a prominent military post in the Kabyle plains. It has about 250 French, and nearly 2000 native population. But it is not a fair sample of the Kabyle towns. We have to-day passed on foot through three or four towns, which are samples, and into their hovels or cabins. There is much to write about these people. But, before I begin, it would be well to tell why and how we came here. There were two incitements :—

First. On our approaching the city of Algiers—while thirty miles out at sea—there appeared on the horizon, shutting it in, a magnificent mountain range. This was not on the shore. It was not the Sahel hills which environ Algiers and its bay. But tipped upon the top with silver head-gear—Kabyle style—old Atlas, the veritable, classic, patient Atlas, appeared! What child, studying his geography, has not heard the word? Its very philology is enchantment! Atlas! Its mention calls up a big, burly, muscular, round-shouldered person, with few clothes, whose brows are knit, whose biceps muscles are strained to their utmost tension, the adductors of whose thighs are steady under the superincumbent weight of a world! Old Atlas! We have seen his head silvered, not by the snows of time, but by the flakes of this past winter. We have seen his shoulders protrusive and burly, all garlanded with olive and ash. We have seen his ribs extending far out of his range till the sea cooled them by its hydropathic treatment. To-day we have been creeping, not beneath his great legs like pigmies under Colossus, for his pediments were not visible, but over his burdened shoulders and around his brawny neck! Old Atlas! What a great name the myth gave him in the heathen imagination! His range was, to the Greek and Roman, the spine of the world! Hercules came to the Riviera on the opposite coast. According to the pre-historic account, he went further, to Spain. There being killed, his army came here. Here they made an end of their conquests, and set up pillars at Gibraltar, yet called the Pillars of Hercules. Beyond these pillars was the flaming bound. There may have been rumours of the Gardens of the Hesperides beyond, but only rumours. Over it—beyond it—all was unknown; or, if known, known in the Nubian geographer's account of the *mare tenebrarum*, as the sea of gloom, beyond the horrid black beetling ramparts of the world, against which the shrieking and howling waves—such as the earlier voyagers round the Cape of Good Hope describe—lashed themselves in furious tempests with ghastly, grim, and shapeless demons! Up to this bound, and under the staunch earth to its limit, Atlas held the *orbis terrarum* —bore it up amidst the raging of the elements! Beyond it, therefore, in honour of Atlas, the unknown sea was called Atlantic!

Under the shadow of Atlas, as the sun sinks to the West, into and beyond that ocean where my heart goes —to the 'New Atlantis' of Bacon—I send to you of the Western World, upborne by giants of another mould, I send my greeting!

The Atlas range runs from Tunis to the Atlantic. Standing as we did on the mountain at Fort Napoleon, we could see lying under the flashing meridian rays of a sun about eighty degrees of Fahrenheit, the Mediterranean, as we had seen from that unstable element these very mountains robed in the blue drapery of distance, and hanging under the light, fleecy clouds of the land! So that one of our objects in coming to this interior Kabyle country is gratified.

The *second* incitement of our coming hither is to see the Kabyles. I confess to a strong predilection for them. I know they do not appear so well in history as the Arab. They do not dress so statuesquely. No sashes or cinctures bind their flowing robes like those of the Arabs. No fez cap or abundant turban hides their head. They are the common people, and therefore of uncommon interest in my eye. They do the work, raise the grain, attend the flocks, make the local law, fight the fights, and hold the religion of this country, with as earnest a soul as any class of labourers, patriots, or religionists on the earth. They tickle the ribs of Atlas till he laughs with plenty. As to their personal appearance, their heads are generally shaven, except the crown, which has a short tuft of their raven-black hair. Their dress is very primitive. They wear wool, summer and winter. Their sheep give them their Roman senatorial robe and its capuchon ornament. Linen and cotton they do not yet use or know. In sight of the telegraph, they still dress and eat, and watch their herds, just as Abraham did, or any other Oriental patriarch. They do more, and better; they raise good crops, and are not nomadic.

I have already said that they numbered 600,000. I should have said that they furnish one-fourth of what are called the 'indigenes.' Algiers is made up of nearly three millions of people. Of these 200,000 are Europeans, nearly all French. There are some 60,000 French soldiers here. The rest of the three millions are Mussulmans, 2,692,812 in number; and Israelites about 28,000. Of these indigenes, the Kabyles seem to be distinct from the Arab or Moor. The illustra-

Kabyle Man and Woman.

tions which I have introduced are intended to represent these diversities. The Moor is the descendant of the

Turk or other Mussulmans. He is distinguished from
the Arab, because he is a 'man of the house.' He
lives in towns and houses. The Arab lives under the
sky and in the tents. He sings songs about his freedom
and his out-door tents. The Arab is very dirty. The
Moor is proud of his ancestry. His fathers fought
the old wars. For seven hundred years, how they
fought the Hidalgos—has not Irving told us? The
Arabs are distributed into tribes, called after the patro-
nymic name. The tribes are divided and subdivided.
The chiefs are the supreme rulers. Their domestic
government is aristocratic. They have an hereditary
nobility, a military nobility, and a religious nobility. I
mention this only to show that, while the Arabs are
the pretentious and supercilious part of the population, they are not the most important indigenous
element here, as they are generally believed to be. The
Kabyle may or may not have the same origin and
religion as the Arab; though he has many of the same
habits as the Arab, still he has a distinct and independent
polity. It is democratic, on the American model: or
rather we are, or should be, democratic on the Kabyle
model. For is he not older than even the nomads of
Job's era? He goes back to the twilight of antiquity.
He is considered, at least in Algiers, as aboriginal. Some
hold that he is a melange of many nations. I do not
think so; but we shall see. His face is brown, not
black, but varying from a light to a deep sunburnt
brown. His features are regular. The women are,
when young, at least, not ill-looking. They are smaller
in proportion than the men. The latter are fine-looking,
walk erect and gracefully, are courteous, hospitable,
good-natured, and, as the French found, they are
patriotic and brave. They were the last to yield to
French power. From my window I can see the plains,
hills, and mountains, which were ensanguined with their

blood within the past five years. These mountains were lined with their braves, in the last great fight before they yielded to McMahon. The struggle was terrific. They are hardly yet subdued. They have only recently paid their tribute in cash at Fort Napoleon. It was collected before last year, with the bayonet.

Their polity I said was democratic. The tribes live in villages. You may count the villages by the hundred from any elevation. These villages are subdivided into communes, or *decheras*. Each dechera has as many *karoubas* as there are distinct families. The members of the karoubas elect a local legislator, or *dahman*, who represents the interest of his commune, in the *djemaa*. This is the local congress or parliament. The president of this body is an Amin. He is mayor of the village, and possesses judicial and military powers. He is elected by the Assembly and has to be re-elected every year. I asked one of the French officers of Napoleon how the election was conducted by the villagers? He said that they meet *en masse*, and vote *viva voce*. The majority rule. The Amin is the intermediary between the French and the people, and as such he is held responsible. I saw a dozen of these mayors of villages, sitting serenely cross-legged in the court at Fort Napoleon, giving their advice as to local matters to the officer, while several hundred anxious suitors were waiting for the decisions.

Again, the Amins of the tribe name an Amin of Amins. He is the political chief or president of the united tribes. The French supervise or control this election. When the various local legislatures or *djemaas* assemble, it is a sort of federate congress. It is called the Soff. It is this system which has made the Kabyles potential. The French officer at Fort Napoleon confessed to me that the Kabyles were industrious. He took us through the machine-shops, where they were learning to make furniture, &c., and had turning-lathes. He praised their ingenuity, but he said they were quarrelsome, litigious—always claiming and defending their rights or objecting to wrongs. He said a Kabyle was then in prison for shooting another who *had stepped on his land* after being forbidden. Well, I rather like this, for it shows a sort of individual independence, which accounts for the fact that the Kabyles of these mountains and valleys were never conquered by the Romans or Turks, and never by the French until 1857. Marshal Randon promised to conserve all their domestic and ancient institutions, and then and then only, they agreed to the French domination. So you see that this part of Algiers has some local self-government; that the authorities conciliate and mitigate the vigour of the military rule for the general tranquillity. As an instance of the acquiescence of the French in these ancient and native customs, I mention what was told me by the judge at Fort Napoleon: A Kabyle woman was maltreated by her husband; she left him; but alas! with another man. She was overhauled at Bougie *en route* to France or Tunis, and brought back to the mountains. Word was sent to the husband. The officer said that he must deliver her to the husband, though he was sure that the husband would assassinate her. The Kabyle is very jealous. He holds his wife with a tyrannical grip. I said, ' But why do you turn her over to certain murder?' He said, ' If we do not, we raise turbulence and trouble. We agreed to respect the Kabyle customs, and this is one of them. If, however, he assassinates his wife, we shall capture him and set him to work.' To my understanding this was very foolish or very—French.

The Kabyle belongs to the mountains. His little houses, made of cane and covered with the same and straw, or in some places made of stone and covered with a rude red tile, indicate considerable social advancement. His women are not so sedulously hid and veiled. We had no trouble in seeing them and going into their houses, both in the plains, on the mountain sides, and even perched upon the tops of the Atlas spurs.

If I were to speculate about the Kabyles, and with the valuable work of John D. Baldwin before me, I should say, first, that races are seldom found pure; secondly, that Africa, even in its interior, is not inhabited by savage blacks, like the Guinea negroes; thirdly, an opinion based on conversation with Dr. Beke and other explorers, that the African proper, if not white, is a 'red race,' *i.e.* brown or olive coloured, like the Kabyles; fourthly, that in northern Africa, although there is a great intermixture of black and white—growing out of conquests of Phœnician, Greek, Roman, Goth, Turk, and French—yet, so far as that portion of the continent is concerned, the Berbers, or Barbarians, now supposed to be the Tauarigs, or Touaricks, are of the prehistoric, primordial stock, from which the Kabyles are doubtless an offshoot. These Tauarigs are of the Desert and not the people to acknowledge the relationship; they are proud and reluctant to recognise any power but their own. Even their camels are said to be more aristocratic than the beasts of other tribes.

But whether these people, who live alone and in a great degree untainted by commixture, are aboriginal and pre-historic; whether they are coeval with the first forming of the Mississippi delta, only 100,000 years ago; or the Florida coral reefs, still thirty-five thousand years older; whether they are Cushite, Semite, or Aryan; whether out of Arabia, Egypt, or India; whether they are the second birth of a race,

aroused to self-consciousness by some new physical
developments;—one thing is as certain as any other
connected with these nebulosities of history, viz.
that the Kabyle is very like this same pre-historic
Berber. The Kabyle is not black; neither is the
Berber. Their colour comes alone from solar exposure; it is not organic. So of the Berber. Mr.
Baldwin, in his book, in describing the Berber, uncon-
sciously describes the Kabyle, as I have seen him. They
have towns, and an excellent condition of agriculture.
They are very skilful in the cultivation of fruit. Their
method of political organization is democratic, somewhat after the fashion of the old Cushite municipali-
ties. To quote Mr. Richardson, he affirms (what I
allege of my Kabyle friends) that in the Touarick
countries all the people govern; that the woman is not
the woman of the Moors and Mussulmans; that she
has much liberty, walks about unveiled, and takes an
active part in the affairs and transactions of life. I
shall have occasion especially to verify these latter
statements. That Berber and Kabyle each has a con-
federacy, has had a literature, that they once had extensive supremacy in the ancient world, that the whole
continent was once controlled by them from this
northern coast,—all these facts help us to account for
the wonderful relics of civilization which exploration has
disclosed to the sunbeams of the African coast. But
if other proof were wanting to establish the relationship of Kabyle and Berber, the similarity of language
does it. For the discussion of the question, I refer to
General Daumas' book, 'La Grande Kabylie.' One
thing is beyond discussion, their language is from the
Orient. Its history, grandeur, and glory, and that of
its cognate languages, are well described in the following extract from the volume of J. R. Morrell :—
' With reverence we approach the ancient and venerable

tongues of Northern Africa, but mostly the Semitic, of yore the speech of angels, and the vehicle of the Almighty Himself, when He walked with man and spake unto the fathers. The accents of tenderness and love transcending the heart of man, the utterance of a sweetness emanating from higher harmonies, flowed in the soft Syriac stream from Immanuel's lips; and that mysterious writing on the wall, the warning of the despots once again startling the vision of the New World, was traced in the primeval Ninivean characters affiliated with the great Aramæan family; and lastly, the glowing yet sublime language of the Koran must ever command the respect and admiration of Christian charity.'

The reader being somewhat prepared for the inspection of the Kabyle communities, and of their portraiture, as it appears in these pages, I go back to Algiers and travel with him to this point, in the Kabyle land.

As to the mode of coming hither: We left Algiers early in the morning, in a carriage. Dr. Henry Bennet and two ladies were of the company. Hahmoud was upon the seat with the driver. He is our Arab conductor, and bears a great name here. We had three horses, with bells on them, and a relay of horses sent on the evening before. We soon pass eastward out of the City of Algiers. We drive along the sea, under the brow of Sahel, and reach the broad, long, fruitful Plain of Mitidji. This plain was once the granary of the Roman world. It is yet, or could be, as fruitful as of old—*potens armis atque ubere globæ*. For forty miles east, and from the sea side to the swelling hills and rising mountains, which are the steps to the summits of Atlas, there were many shifting scenes, but they all presented views of cultivation and evidence of labour. Of course here and there was to be seen some shrub-covered land. It is like parts of

Corsica or the heaths of Scotland. It has been left to the pelican and the brushwood. Over it the sheep and goats brouse and the cattle and camels feed under the care of shepherds and drivers. The road is good, but not yet as complete as French roads in Corsica. Some fine bridges are made of iron, but they are not to be used till the 15th August next, the centennial anniversary of the great Napoleon's birth.

We invariably start for travel in Algiers before breakfast; or, taking coffee, travel towards breakfast. To an unaccustomed tourist, almost the first object on the route to Fort Napoleon, is the cactus. It is so common and so high. It is the hedge or dividing line between farms. It was in use here before the French came. It is not only good for its fruit—the prickly pear—but it would 'turn' any animal except a crazy dervish. Lions, tigers, hyenas, and jackals, yet in plenty here, hardly dare attack a cactus fence. We perceive, also, some bamboo hedges. Upon the road we meet a team which smacks of the Yankee—a waggon drawn by two oxen and a horse in the lead. It is driven by a Kabyle sometimes, but generally by a French colonist. We meet heavy waggons with ten horses or mules. The animals are feeding out of baskets as they go. Now and then a big bird—the heron—dashes by us and lights in a distant marsh, or presents a good shot from the top of a hay-stack. Sheep and goats are seen. Charcoal in loads, in waggons, and on donkeys, passes by. In the early morn the blue smoke of the charcoal-burners is seen curling through the mist of the distant mountain. The Atlas range seems far upon our right and front. Great shadows hang down its sides, like the furrowed folds of a garment. The clouds to the south, towards Blidah and Milianah and the desert, lead us in our imagination thither, though our course lies towards

the Kabyle land and breakfast. We pass by houses—
long, one-story houses—walled in. In fact, they are
known as ' fortified farms.' In the recent wars, and in
the interest of military colonization, like that of the
Roman and French, these farms were a part of a
system. In the fields of the Mitidji Valley, over which
we are rolling, we perceive the ploughs at work. Far
up into the mountains, five, six, and sometimes ten
ploughs are going on one farm. The plough is a rude
wooden one, with a tongue or pole, and another pole
at right angles with the tongue, to which, far apart,
the oxen are rudely yoked. They pull slow, and
slower, and *seem* to stop. This plough has been used
in the Orient some thousands of years. Here the
natives, when they have been offered the shining share
of civilized people, let it rust, and plough in the old
soil a few inches deep in the same old way. You
perceive in yonder field, walking with dignity, holding
one handle—the only one—some Kabyle Cincinnatus,
robed in his flowing bournous, every inch a Roman,
clad in his toga. Indeed, the Kabyles claim to have
been mixed with the old Romans, and to have their
customs and manners. Perhaps so. The Romans
never conquered them. That is true. But it was the
Roman custom to colonize by settling down the soldiers
and marrying them to the maidens of the land. If
they could not conquer them by arms, they did by arts,
or by a combination of both.

There is something very beautiful in this grand
plain of the Mitidji. Not only its fields of waving
oats, barley, and wheat, just ripening; not only its flax
fields, in bluest bloom ; not alone its flowers and
shrubs, two out of every three of which we have seen
in Corsica or in the Riviera; not alone its yellow
genista, a flower of Gascony, and from which the
Plantagenets took their name—for they were Gascons,

like the flowers; not alone the ferula, so common here, resembling in its green the asparagus, and in its blossom like the aloe. This is the same ferule, known to the Roman schoolboys and celebrated by Roman poets; and not unknown, though not philologically, as the same ruler by which Young England, Ireland and America have been seasoned, *a posteriori*, by bellicose pedagogues. It is the same plant as that, known to the classics, within whose cavity Prometheus—that 'thief of fire from heaven'—brought down surreptitiously, for the benefit of mankind, 'Sunbeams' and fire. This plant is therefore—notwithstanding juvenile associations derived from the time when I was flogged along the flowery path of knowledge—a favourite with me: and I, therefore, recall its Greek equivalent— νάρθηξ.

Yonder, in the plain, are a dozen camels and as many donkeys, feeding, some of the former upon their knees, and some of the latter saying their grace with the most hideous outcry. These remind us of what the inner monitor confirms, that we, too, need breakfast. In a little village we find it all ready, for we had sent our horses on, and word with them. The name of the village I cannot recall, but the inn was the Hôtel de Col Bernaycha, and its sign was two pipes crossed. The village has cane huts with straw roofs, and one little stone, white-washed house. It was French, and there we were ushered through the hotel into a summer-house, amidst some chickens, dogs, and little pet piglings, of the wild boar tribe, and served with five courses.

We are not long in despatching the meal. Soon we are on the move. We meet French soldiers guarding prisoners. The latter trudge along sadly. We cross some little streams, having on their banks fringes of oleander. We arrive at Isser. Here we meet some half-dozen camels and drivers—Kabyles. Hahmoud knows the head man. I desire to ride a camel, never having essayed it, since I 'swung round the circle' of a tan-bark ring, when life was young and Africa *terra incognita*. The camel drivers consent. The French population gather about our carriage and watch proceedings. The complaisant head of this caravan descends from his donkey. He calls one camel, uses his stick upon him, says, 'scht!' 'scht!' with plentiful sibilation, and the camel prepares to come down. This is a part of his early education. His front knees first touch the earth. He groans at the indignity. Then, after more sibilation, he gathers in his two hinder knees, and gives another groan. He looks mad, chews the cud vigorously, observes my motions as I proceed to mount, and not seeing a turban, and not knowing me for one of the 'long robe,' he exhibits a strong disinclination to being 'backed like a camel.' There is a coarse, wooden skeleton of a saddle on his hump. I lay my hands on, preparing for a spring. 'No, no!' all cry. 'Not that one. Try another!' 'Why not?' 'Oh!' as Hahmoud translates the Arabic, 'He is too *méchant*,' —*i.e.*, wicked. Others are lowered on their knees. At last I mount; the Doctor takes another. We are so successful in our camel ride that we persuade one of the ladies of our company. She mounts, with Hahmoud's aid, a grey one, but her ride is cut short. Her camel, too, is cross. The Arabs say that the beast is afraid of the black dress of the lady, being accustomed only to the light woollen burnous. It was hard for her to mount, and being mounted it was a lively trot for my lady. She clung to her saddle with a death grip. The camel started off. Far up in the air, swaying with the motions and feeling uncomfortable at the sinister looks of the camel, she calls a halt, and is at last rescued from her lofty, ambitious and precarious position. I am sure that the artist has but faintly pictured this interesting ride.

Soon we proceed, and having reached the confluence of the Sebaou and Oued-Aissi—two principal streams in this part of Africa—we find the bridges not yet made. We perceive the convict soldiers at work driving piles for the bridge. But we have to cross. The rivers are not full, but when full are very wide, and the bed, nearly a half-mile broad, is full of boulders. Of course the crossing-place changes with each day's current, and is therefore perilous. Our driver summons some Arabs to ford in advance. It is lucky that it is afternoon, for we learn that in the morning the current is strong and deep. The melted snows of the day before reach, by soaking through the earth, the middle of the plain by morning. The Arabs fold the drapery of their clothes high about them, and—wade in ! We follow in our carriage. A few sous in recompense, and we move on to Tizi Ouzou, where we are to rest over night.

From it we can, by a glance through the dim air, perceive Fort Napoleon, far up and beyond in the mountains. Tizi Ouzou has much celebrity, not alone for the vigour of its fighting against the French, but for its great fertility. The mountains above the village are crowned by a marabout tomb. The marabouts are the priesthood most sacred in the Moslem faith. They are the repositories of literature. Their homes are sacred, and their tombs preserved. Upon every mountain where the Arab population is gathered you may see a small, square building, with a dome. This building is white, and renewed by the reverence of its devotees. Many a marabout, by his songs, sermons, wit, and wisdom, has given the French the trouble of suppressing fresh insurrections. The 'sacred war,' as it is called, which Abd-el-Kader waged from 1837 to 1848, was preached as the Crusades were, by priests. The marabouts rouse the people by a secret, emotional religious power. They will yet cause the French much trouble. Fanaticism is very unruly.

The valley of Tizi Ouzou, where I write, is cultivated. The green fields climb up above to the summits, 4000 feet. The flocks are abundant. There is no separation—not yet—of sheep and goats. The French houses are low and white-washed. We had hardly descended from the carriage and into our little inn here, before two Kabyles—Hahmoud's friends—invited us to walk across the valley, and see their town. One of them, in good French, promised to show us the inside of his house, and to introduce us to the women-folk. We pursued our path up the mountain side, and there, behind the high, prickly-pear walls, lay the irregular, streetless, flat-roofed town. The mosque is the most considerable building, in front of which sit, cross-legged, some score or more of Kabyles, who salute us. A dozen of their children are playing corner ball—our old American play,—only this, that those who are 'in' ride on the backs of the 'outs.' It is a political lesson. We visit the fountains. Then we enter one of the huts. A little yard, roughly paved, and shut in by the prickly-pear, is the antepast of the domestic entertainment within the low door of the cane hut. The hut is dark, and smutched with smoke and soot. It has no chimney. All of them are thus, even those of stone. These were not even lighted, as others which we saw were, by a little charcoal fire on a brazier, in the middle of the hut, or in a hole in the ground. The smoke got out, any how. It got out of the door just as the mother of our conductor drew our ladies in. You should have seen the old woman of these premises! What hilarious grunts and giggles of satisfaction she displayed. After I struck a match, she pointed out her properties and advantages. She pointed out her jars for olive oil, saying: 'Umph!' With a gurgle of delectation, she showed us a pan for cooking her bread. She lit some rags, as the match went out, to show us her mat, which she slept on, or rather matting to be placed on the stone bed. How easily are the unsophisticated comforted! When one of my lady friends gave her some coin she was made happy—could not let her go. The children—and there are many in the Kabyle country (Algiers is prolific, even the girls put on the veil at nine)—the children were watching for our exit. While the boys were boisterous and boasting, and displayed their morsel of French, learned in the schools, the girls were as timid as gazelles. Coaxing would hardly bring, even to one of their own sex, foreign to them, the trembling forms of the bluely-tinted and tattooed girls who followed in the rear of the boys for the sous! As we marched down the mountain, there were not less than a hundred children dancing in our wake. Some were bold enough to talk. Some had charge of their kids and lambs, and dropped them to say 'Buono.' One of our ladies to one of the youngsters of ten years, thus :—' What will you do when you grow up, my boy?' 'Oh! I shall be married.' 'Will you make your wife work?' 'No! I will do the work; she may stay at home.' 'But what will *she* do?' 'Oh! make bread—and—my burnous.' 'Do you wash your own burnous?' 'Oh, yes.' 'How?' 'With my feet.' 'Why not with your hands?' 'Too heavy.' So you may see, from these chance talks of the shaven-headed youngsters, something of their elders. It is true the Kabyles treat their wives pretty well. They seldom, unless sheiks,

have more than one. We see them in the streams
washing their own burnous. The women carry water
in great jars to the huts from the fountains as they
carry their babies—on their backs—with hands behind
to guard. This seems to be their principal employment.

The French have had some of the Tizi Ouzou
children at school. They have shown great pro-
ficiency. We overheard some of the children who
followed us say of the others: ' How proud these
schoolboys are ! ' The schoolboys say, *'Quelle bêtise !'*
and trot along after us, ' proud ' of their civilized
tongue and the attention we bestow.

I have been rather particular in these frivolities of
the children, for they are indices of the government
established. Is it not an interesting problem, even for
Americans to solve: How may subjugation by the
military be made at once tolerable to the conquered
race, and elevating to a higher refinement ?

But we are soon at our hotel, leaving political and
social problems under the beauteous, pink veil which
Nature is arranging for the evening party, whereat
Old Atlas will gallant his family to the upper star-lit
halls.

We, too, pass within the veil, and sleep till morn.
Then we start again. More rivers to cross, and, still
worse, they are dangerous; for, if the horses should
halt, the carriage would sink in the sand of the stream.
We wind up, up, and around the groves of ilex, ash,
and olives, in a spiral path, decorated all the way up
with the little red pink silena and the pheasant's eye.
We are overlooking valleys, rich in vegetable life;
and like a splendid piece of mosaic, far below, like
a picture from an Alpine height,— you may see
the square fields of newly-ploughed ground, alternating
with light and dark green fields, as far as the human

eye can penetrate either with or without a glass. We
perceive the Kabyle villages. We go through more
and more of them. They grow better on the mountains; because built there of stone, and tiled. We are
treated courteously; invited in; our watches and ornaments are gazed at with unsophisticated curiosity.
The Peruvians did not look at Pizarro's feather or
Almagro's horse with more simplicity and eagerness
than the Kabyle girls examined the dress of my lady
companions. The work on hand was the pressing of
the olive—dark and oleaginous. The old press used
in the time of Daniel—a simple wooden screw, which
has been whittled out for a common press—is all the
mechanical power they have. We see the women
kneading bread. We notice many Kabyles lying
around and doing nothing. We ask why? They are
waiting for harvest next month. But generally they
are working people. Their fig-trees are loaded; their
olives are thrifty; their wheat looks well; their cattle,
camels, donkeys, and horses, seem the best. Their
houses, beds, and dress, alone show the lack of civilization and comfort.

But there are many drawbacks to their prosperity.
Everything now looks well. But the locust may be
on his devouring path from the desert, as in 1867.
The fogs may come from the sea, and all the summer
harvest may be lost! The sirocco may come in June,
and blast by its hot breath the flower of the wheat.
This has been, and may be. But the Kabyle still
labours on and ever. He is, to-day, what he was when
Rome found him.

We finally wind up to Fort Napoleon. It is like
all the other places where battles have been fought in
Algiers—a walled town. It has a good hotel, but the
roguish landlord, anxious for custom, offers Hahmoud
a handsome bribe to hold us over night. Hahmoud

is honest and refuses. We have a good lunch of wild boar's meat, and between the dishes we amuse ourselves by observing the pictures of the auberge. Here is Napoleon III. in red pants, sash over his shoulders, five medals on his breast, a cocked hat, and as large as life! On the other walls, the English cockney is ridiculed in a pictorial series of Parisian caricatures, very imaginative! Above us is the skin of a leopard, like that which we saw at Tizi Ouzou. The leopards are nearly obsolete here. So are the lions. A few now and then appear. We have seen none yet; in fact, no wild beasts, except a jackal. But of the wild beasts of Algeria I must write at length. My porcupine quill, presented to me by one who plucked it, has much to tell.

Fort Napoleon overlooks 170 villages. It has but 500 soldiers, but it is strongly fortified. We obtain a splendid view from its highest terrace; from its centre 'all round to the sea.' The long, broken, snow range of Atlas, which has followed us hither, seems but a short distance from us. The village is improving since the war has ceased. Some officers pointed out the two Kabyle towns last to yield to France. They had their local Joan of Arc. Her name was Fatima, and she inspired an intense hatred of French rule, and helped the Kabyles to fight. We see here what we have seen before—a grand market of the natives. Calves, cows, kids, sheep, goats, horses, donkeys, figs, and charcoal, are here traded off and on. Some 1000 Kabyles are gossiping and buzzing like so many bulls and bears in a New York gold market or a Paris Bourse.

We were kindly treated by the authorities at the fort, and before night, we dashed, under the crack of the whip on the flanks of our three horses, and followed by crowds of Kabyle children, to the base of the mountain. In

9

passing one village a handsome young gentleman—a
Kabyle Alcibiades—in a very clean robe, accosted us
in French. He was astonished that we had come seven
thousand kilometres to see him ! He had a very vague
idea of America, but an enthusiastic admiration for the
Italian girl, who is one of our companions.　He ran
after our carriage several miles in an ecstasy of love
at first sight.　The scapegrace !　He had two wives
already.　He said that he could afford another, as his
last was an orphan and cost only three hundred francs !
I said, ' How much do you propose for mademoiselle ? '
' A thousand francs, and if you wait here I will go up
the mountain for the money ! '　We did not wait, but
dashed on, and the Kabyle after us.　I was reminded
by his flowing robe and naked leg, of the verse of
old Purchas (before Chaucer) in his ' Musical Pilgrim '
—describing my Kabyle, *tunc pro nunc*, as a—

> ' Man with doublettez full schert
> Bare legget and light to stert.'

What time we made, or he made, for many miles !
How he performed that journey ; with what strides
and with what hopes ; how the Arab horses glanced
round now and then at the airy bournous of this
swain,—is it not more than written,—graphically
sketched on enduring copper, whose impressions I
present on the neighbouring page ?

The Kabyle men are shaven, but the women wear
long hair and have girdles, which the men have not.
They all have glistening white teeth and perfect.
Good food and digestion,' says the Doctor.　' No wine,'
says Hahmoud.　It is true.　Mohammed forbade
wine to the faithful.　Hence, they say, white teeth
and good digestion.　We see no rows, no drunken-
ness.　The only intemperance is too much marrying !

Hahmoud says : ' If they drank, what with jea-
lousy of their women and their independence and
guns, their troubles would multiply and their prosperity
decrease.' In fact, as we were told by the officers at
the fort, one of the great troubles among the Kabyles,
not only between the towns and tribes, but individuals,
springs from the jealous feeling as to their wives. *La
Vendetta* is almost as rancorous and persistent here
now, as it ever was in Corsica. We are told that to-
day, while at the fort, a man with a pistol had been
walking and watching for an enemy.

There has been much fighting here, not alone on
domestic, but political matters. Every field has been
a camp ; and every little mountain town has run red
with blood. The Arab makes more display ; but the
Kabyle effects more. I do not admire the Arab way
of treating the women, by shutting them in houses and
tents, or worse in manifolds of linen drapery. Even
when they travel, the Arabs, like some we met, build
on a camel's back a harem, in which to hide their
women under shawls. It makes an interesting picture
of the Orient, but has no good sense to recommend
it ; and more than that, it does not improve the temper
or ensure the chastity of the Arab women. This mode
of isolating the women is as unnatural as it is tyrannical
and devilish. No wonder the Arab women are reputed
cunning and loose. The Kabyle women are otherwise. The men respect them. I have seen the Arab
men riding and the wives walking. Not so the Kabyle.
If he rides, his wife is before or behind ; and the
Kabyle man carries—-the baby ! On our pathway to
this place we observe many signs of the ' Grande Halte
du Maréchal,' from which I infer that here the French
troops bivouacked on their pursuit after the natives
in this Kabyle land.

If I could give in a few words my observation of
the Kabyle, I would say that he is industrious and
ingenious; the Yankee of the Mohamedans; democratic in polity, frank in intercourse, and independent
in character; a mountaineer and a farmer; a man of
bravery and of intelligence, only his religion enthrals
his energy.

CHAPTER XI.

BLIDAH AND MILIANAH—THE ARABS.

'And he will be a wild man; his hand shall be against every man, and every man's hand against him; and he shall dwell in the presence of all his brethren.'—GEN. xvi. 12.

'The Arab is the hero of romantic history; little is known of him but by glimpses; he sets statistics at defiance, and the political economist has no share in him; for who can tell where the Arab dwelleth, or who has marked out the boundaries of his people.'—*The Crescent and the Cross.*

WE return to the city of Algiers from Kabyle land, and from thence we move to the west and south. The desert is before us, and Oran is beyond. A splendid stopping point is the city of Milianah, which is reached *via* Blidah. We have thirty miles of railroad out of Algiers to Blidah. We start at noon and remain there all night. The railroad runs eastward for a space, along the sea. A strange phenomenon appears on the waves, for half a mile out. The sea is 'incarnadine,' with a vegetable fucus. The effect is peculiar, tropical, and striking. Soon we turn to the south, and pass over the plain of Mitidji. It is not so well cultivated as its rich soil would demand. The fact is, it is malarious, and although it has buried a generation, it has yet to bury more before the old fruitfulness comes out of its soil. Blidah is celebrated for its earthquakes and its oranges. It has been badly used by the former. In 1825 it was shaken up horribly. Seven thousand people perished in the ruins caused by the earthquake. Another destructive earthquake occurred in 1867. Every house made of the boulders and mortar, as the poorer houses here are constructed, was shaken down. There was great loss of life and property, and great suffering was the consequence. Our driver told us that he had been shaken, and when he awoke he found the walls of his house down, and his head out of doors. He showed me a scar on his forehead from the falling of his tiles. But Blidah makes up soon for these disasters. She has twenty-two gardens of delicious oranges. They are large and fine. No such oranges grow anywhere else that I know of. The Sicilian, Nice, Mentone, Spanish, or West Indian oranges cannot compare with them. Why are these oranges so very rich and sweet? Is it the sun? But the sun is the same at other places. If you would know why winter sunbeams ripen the orange with so palatable a saccharine juice, you must go into the arcana of nature. I ask my friend, the Doctor, sometimes such puzzling questions. He does not answer me as the peasant in Corsica did, when I questioned him, why one mountain was all green and the other all bare? The peasant says: 'It is the caprice of the Eternal Father.' The Doctor would say, 'It is owing to the exposure —northern or southern. If under too scorching a sun, the land will be bare. If under too harsh a northern blast, the same. If it has no water, no clouds, no irrigation, or other conditions of vegetable life, then that life will not appear. Rocks are the bones of nature. They come out of the skin to show that the patient is not so well, and will not grow.' But you ask: Why is the orange so sweet, and the lemon so sour, all under the same sun, and from the same soil? Why do the dates ripen after being plucked, and other fruit not? There you are trying to force the arcana. Only thus much will the oracle respond: Grains or fruits, even when unripe, have starch in large quantities (amylum). It is partly changed to sugar in ripening, whether the process takes place before or after gathering. This sugar by fermentation becomes spirit. That any vigilant whiskey inspector knows. In all three conditions — starch, sugar, and spirit—the chemical components are the same : carbon, hydrogen, and oxygen. Good! It begins to be clear why Milianah has the sourest lemons, and Blidah the sweetest oranges; for is not the combination of these components, at the two places, in different proportions? And does not this constitute the difference? But why should the rocks grow cedars of Lebanon, and the fat plains have nothing but grass and flowers? The oracle is dumb. Let us eat our oranges, and drink of the spirit of wine, and be glad. Allah is Allah! and Mahomet is his prophet! Let us be content to look at the orange in its bloom, and its golden orb of fruit, and be thankful, without further inquisition. The size of the orange-trees at Blidah indicates that they are guarded from the north wind, and not harmed by the sirocco. I measured in one garden an orange-tree whose trunk was six feet round! The Doctor confesses that no such trees grow on the Riviera. There are some 40,000 trees growing and bearing here. Of course, this makes Blidah quite lively. It has a population of 8000, more than two-thirds European, and among them many Spaniards. The cactus, or prickly-pear, is much grown here. It is a sort of hedge or protection to the other fruit. The malaria once prevailed, but drainage has made it an infrequent visitor. Blidah is 600 feet above the sea-level. Generally, the malaria stops at 300 feet, as in Corsica. But the plain is so enormous between Blidah and the sea, and the mountains — Sahel range — along the sea, so enclose the streams from the Atlas south—at whose feet Blidah reposes like a young bride in her orange blooms,—

that it required much labour and ditching to make
Blidah habitable and healthy. Above Blidah is the
Atlas, and, as usual here, behind a misty shroud;
because the air from the north comes saturated from
the moist plain and the sea. But the mountains do
not look less lovely because they are enshrouded, and
their deep shadows are very beautiful upon their
northern flanks.

I spoke of the orange orchards. We visited the
largest one. Through avenues of plane-trees we come
to its gate. We perceive at the gate a fine dog on
watch, and over his kennel some facetious person has
written: '*Parlez au concierge, nommé Turc!*' I
cultivated the good humour of our Cerberus, and he
allowed us to go in. This garden is walled thickly,
and guarded also by cypresses.

But there is a more delightful resort than this for
the inhabitants of Algiers and Blidah. It is the Gorge
of Chiffa. After being fixed in our quarters—for,
the hotel being full, we had to be lodged the best we
could over a confectionery establishment—we proceed
to the Gorge. It is a two hours' ride. As we go out
we perceive upon the plain the Chasseurs d'Afrique,
practising in the sun. The flash of their swords, the
words of command, and the white horses dashing
about so picturesquely mounted, make it a lively
scene. We are happy in meeting on the way mine
host of the Gorge, who turns about to prepare our
dinner. While dinner is preparing, we pass on to
penetrate the mountain still further. The Gorge is
celebrated for four things—its cascades, dashing down
from mountain heights, 4000 feet; its monkeys, after
which the inn is named; its having been visited by
Louis Napoleon in 1860; and its gardens of quinine
and tea-plants, and winding paths up the mountain.
On our return to the hostelry we are invited to walk

up the paths. The torrent from the mountain—
which runs at right angles with the Gorge of the
Chiffa, and empties its seething waters into the Chiffa
—has made such a wild, romantic valley, that the
hand of Art has seized upon it to beautify it with
paths and plants, flowers and fountains; while Science
has made its quiet nooks, unvisited by harsh winds,
a conservatory for experiments in quinine and tea-
plants. 'Will the party please walk up to the summer-house where the Emperor took breakfast? Your
dinner will be served there.' We will. We did. It
is a fairy spot. The birds sing all through it, far up
some thousand feet, whither the walks tend and wind
among rocks, trees, and flowers. Here moss of every
colour and age grows, made beautiful near the grottos
of fern, and both fed by the dampness of the torrent.
The African ivy hugs about the rocks, which hang
imminently over our heads, or lie where they have
been tumbled into the midst of the torrent. What
with the song of the torrent never ceasing, the carol
of the birds—a whole choir attuning at once—the
ba-ha-ing of the goats above and sheep around, the
croak of crapeau, and the chattering of the monkeys,
who are wont to come out of a warm afternoon from
the rocks above to eat and talk, old monkeys and
young ones—the latter on the backs of their mothers
—but all little monkeys—these are the beings whose
noises salute our ear. But to the eye, what with the
willow, the micoulier of Provence, the castor-oil tree,
the quinine, and tea—'Ah! is it not pleasant,' says the
Doctor, 'to see Nature doing her best, as she does
here?' Human nature must do likewise; and so we
go to dinner. Upon the side of the hostelry some
genius has painted a race of hounds after a wild boar.
The dogs are mounted by monkeys; even the gobbler
which saluted us as we passed up the valley, is depicted

mounted by a monkey and in the race. These illustrations furnish us amusement, as we wait for the trout, quail, and other dishes of which the dinner is composed.

The tops of the mountains begin to lose the last radiance of the day. We start for Blidah. The moon comes out to give us the Gorge under new lights and shades. Verily, it is grand. The lights are on one side and the shades on the other, and they reach far up into the sky, from which the cascades leap and play in the moonlight musically and fantastically.

We slept at Blidah, and had no earthquake. We awoke to find the starlings, which fill all these villages of Algiers, in full song. We are soon in our carriage, out of the walls and gate of the place, and on our way to Milianah. We pass an Arab market, which is held every Friday over the country. Thousands are chaffering and bargaining. We begin to find some Morocco people among the native population: they are darker and wilder looking. We pass some gipsies from Spain. We go by a vale known as the Valley of Robbers. It used to be quite a haunt in the good old days before the French gensd'armes came.

We are half-way to Milianah before we know it. Our driver has been bragging a good deal about the Arab horses, their endurance and speed. We listened incredulously. Now we begin to have faith. He tells us that one of his horses, a white one (and most of the horses here are white or dark gray), can go—has gone—126 miles in twenty-two hours! Now I believe it. We have travelled behind these horses for a fortnight. We have been promised, for example, to reach our destination in nine hours; we were content; but we reach it in *six!* We are delighted. Every time yet have we been more than delighted with the performances of these horses. It is either a cunning way the French have of promising far less than they perform, or else when under way, their horses beget a more than ordinary momentum. We were promised forty-eight hours from Marseilles to Algiers; we did it by the steamer in forty! So that it looks like a French rule, and it is a good one. As to the Arab horses, I have seen the best of them. Nothing can excel the elegance of their bearing and the speed and hardiness of their thoroughbred training and work. The best stallions of Algiers are in the hands of the French officers. The cute Arab will risk a flogging, and something worse, in order to steal into the precincts of the stallion with his blooded mare, for the probability of a thoroughbred colt. I do not see, however, evidence of that attachment which it is said the Arab bears to his horse, and which the Arab songs lead us to imply. The spur he uses on his horse is worse than torture. It is a sharp spike, six inches long. I wondered that these Arab horses were so plenty and so cheap. One of the finest was priced at 500 francs, or 100 dollars. The Doctor says such horses would bring in London £150. I think in New York they would bring 500 dollars. But I wondered no longer when I found that the Government prohibits their exportation. If it were allowed to-morrow, Algiers would be full of horse-traders from Marseilles and Paris. Why is their export prohibited? It is impossible to do the work here, under the summer sun, either of the army or the diligence, with any other kind of horses. So it is alleged. The reason given by a French officer for these fine qualities of the Arab horse, is, that they are not closely stabled. The French follow the Arab custom, and give them, as the Doctor gives his consumptive patients, plenty of fresh air and sunbeams. Their innate good qualities having been improved by many generations of this careless care, we find the Arab horse the best type to-day of his kind!

As we approach Milianah we find the flora changing, for we are rising. The clematis and elder bushes appear, and everywhere on the hills and in the fields, the prickly broom, giving to the very air a golden hue. Blidah was 600 feet above the sea; Milianah is 2700; so that we ought to be prepared for great vegetable changes. As we approach near, the tall, fresh, green poplars stand up like sentinels about its walls. Gardens of vines, weeping willows, lemons and figs in wonderful abundance appear. Algiers city is not so far advanced in vegetation as Milianah, and the latter is so far up in the air! Wherefore? The solution is easy, Milianah is sheltered beneath Mount Zakkar—6000 feet high, white with marble and snow — but a complete protection from the northern winds; and her foliage is exposed to the south and its balm of solar radiance. Hot it is, no doubt, in summer. Here against the rock on which it is built the genuine sun of Africa pours its vertical rays, and doubtless burns and bleaches. But Milianah is the Damascus of Africa. I was ready to say here, as poor Buckle said, on the last of May, 1862, at Damascus, where he died: 'This indeed is worth all the toil and danger to come here!' Milianah is not only beautiful for her vegetable grandeur, but, like Damascus, because of the fountains and streams by which it is caused. She not only turns a dozen or more mills by her water-power, but irrigates the city and silvers the perpendicular rocks on which her ramparts are erected, with cascades, which leap from the terraced sides of the mountain and flow through many conduits throughout the plain below for miles. The Moors, who once made Milianah the seat of their power in Africa, knew more of irrigation than any other nation. Southern

Spain, the driest part of Europe—Murcia, Valencia, Andalusia, all the lands which they held, from Bagdad to Gibraltar — were made to blossom under their system. Spain still preserves the system. We shall see much of it when we arrive there.

I wish that I could give you a photograph of Mili-anah, warmed somewhat by the colours of the flowers which make it so fragrant. Make to your mind the imagery of a plain, out of which, rising through several miles of gardens, there winds as it rises, the road, up to the gate in the rear of the city; and before you get there, picture the limestone rocks grottoed, honeycombed, and irregular at places, but all decorated with vine and leaf and cascade, and surrounded by a staunch wall, within whose fortified escarpments a luxuriance of vegetation seems to surround a city of elegant proportions, with tower of church and dome of mosque, and all flashing white and clean as one of its own cascades under the African sun—then you have Mili-anah! It is the glory of Algiers! Enter within its gates! Walk around its plaza! Here we find em-bowered in foliage, in the centre of the large square, a Venetian Campanella. It stands alone and sounds the hour for Moslem and Christian. Go down the wide avenue to the south side of the city, and you find yourself looking from the precipitous walls, upon the grand views beneath and afar! You see no frowning beetled brow of rocky fort, fortified by art and nature. That is here, but it is visible only from below. You gaze down amidst the wild bryony, creeping about the rocky sides, making hanging gardens of these walls, creeping about where the cactus, the rocks, the pomegranates and the fountains, the figs and the waterfalls in promiscuous luxuriance form a fore-ground. While at the end of the long plain, more than twenty miles distant, the mountains stand, one

range above the other, and the second above the third,
long intervals between, for seventy miles and more,
until the eye from Milianah seizes, as upon its last
outpost of the vision, the mountain range from which
the beginnings of the Desert appear! Our way lies
there!

What a leap from Milianah to yonder wall of Atlas!
Yet we must partially go over it. Not, however, until
we exhaust Milianah—having visited its market, where
the vendors stand with their donkeys loaded with charcoal; visited its plaza by evening, where we saw the
fat woman, weighing 400 pounds, painted on the
booth larger than life, and heard her speak in French
of the immensity of her obesity, and showing her
jambes and arms, prove to Arab and European that she
was all their fancy and the artist had painted her;
visited the public garden, where we gathered cromatella
roses as big as your hat—a small round hat; talked
among the booths with intelligent Jews, of whom some
had been to England, and one old man had a son
in America; seen the Jewesses decked out in gay
colours, and the Jews in their dark dresses—admired
the easy air of the latter, and the beautiful eyes of the
former, for 'hath not a Jew eyes?'—and a Jewess too!

What more, then, hath Milianah? I could fill a
chapter with its olden renown as a Moorish city, and
its military glory as a French fortress. Here were
once twenty-five mosques! Now there is but one!
Here once lived the great Emir, Abd-el-Kader. His
house is now occupied by a gunsmith, who works his
machinery by the water which once fructified the
Emir's gardens. Lemons and roses once in Moslem
days, now fusils, and revolvers. Civilization marches.
Here, long before the Turk and Christian, and their
historic vicissitudes, the Romans made this the head
of a colony. The French say: 'We only resume,

after some unpleasant years of interruption, what the Romans began.' Here, in 1830, the Emperor of Morocco ruled. He departed soon under pressure. In 1837 Abd-el-Kader made his brother Bey of Milianah. He did not last long. In 1840 the French took and held it against the multitudinous and daring attacks of the Emir. For twelve months 1200 French soldiers, under the brave Colonel d'Illens, held this place. At the end of that time 700 were dead, 400 were in the hospital, and 100 weak men still held it. They had determined to blow up the magazine and perish, rather than surrender. General Changarnier rescued them. They live in history. Poems have celebrated their heroic resolution.

It makes this, and other places which we have seen, interesting, to know that Canrobert, Bertheneze, Desmichels, Clausel, Bugeaud, the Duke of Orleans (whose effigy, in bronze, ornaments the square of Algiers), Valle, Pelissier, Randon, McMahon, Niel, and others, whose names figure in the wars here, and in the Crimea and Italy—the heroes of Sevastopol and Magenta— and some of whom, like Changarnier, rose above the law of the sword, into the elemental law of liberty for France—here, on this ground, made their first efforts, and won and wore their first laurels.

Algiers has been the training ground for French heroes. It is so still. It is objected to the present government of the colony that it fosters the sword, and imperils French civism and liberty at home. We are at a loss to know how the latter is in jeopardy, inasmuch as it is not, and may not—unless Napoleon becomes wise,—be in existence. Whether Algiers is helped by the military rule, is not so hard a problem as whether France is hurt. Of this, however, when it comes properly under my eye. That organ is just now full of Milianah. True, I see the French soldier as we dash out of the gates; for is he not there to salute? I see the Arab move around this beautiful city, subject and discontented; but is there any hope of his being rescued? Only one; let him do as Abd-el-Kader did—go off to the Orient, where many of the best Moorish families have gone, and in Syria, under the 'Sick man,' get better! There his religion is held so sacred that he may refuse his wine and, without Christian interdiction, multiply his wives.

CHAPTER XII.

PLAIN OF SHELLIF.—TENIET-EL-HAAD.— CEDARS—DESERT.

'Yea, the fir-trees rejoice at thee, and the cedars of Lebanon, saying : Since thou art laid down, no feller is come up against thee.'

ISAIAH xiv. 8.

WRITE from Teniet-el-Haad. It is the last fortified town held by the French this side of the Desert. In the last chapter, we had started for this place out of the beautiful walled City of Milianah. Milianah might have been better described; its associations and surroundings are attractive. It is not far south of the old Roman City of Cherchel, which is seen sleeping under the sea! Earthquakes make strange bed-fellows! Milianah is the last city of refinement to be found before we move toward the Desert. It is so attractive both for vegetation, waters, and sky, that it requires an effort to leave. But once started, it requires an effort to stop. Our horses gallop through crowds of donkeys and men about the gates, and then down, down, we go—winding off our miles, like thread from a spool, until we drop 1000 feet as easily as ever a player, in an Irish sensational drama, leaped from a fictitious crag into an imaginary lake, upon a painted island in an illusory scene,—upon a feather bed. By the Doctor's barometer we fall easily in an hour over 1000 feet! We are in the plain. The pampa grass grows along the road. We meet the heavy laden teams, with their many horses and many bells, tintinnabulating along the dusty route; a strange team, as patient as the Pennsylvania team of thirty years ago, only they have no dog or tar-bucket under the waggon, and yet, unlike the Conestoga, they have three horses at the wheel and six tandem. They are the avatars of civilization. The donkey and camel must get out of their way. They, too, must soon go out before the railroad. The Arab tent, so exquisite in fancy, and so dirty in fact, must give way before the Occident and its steam. Even here, on a level 1300 feet above Algiers, according to the barometer, the railroad is in progress. We see the Arab men and women working at it. What hands enterprise employs! The Pacific Railroad once worked squaws. Within two years the Desert will be within a few days of the railroad! Then there will be something fresh to draw the gambler from Monaco, and the epicure from Nice.

As we approach Teniet-el-Haad, the thermometer opens mildly at 76°, and we ride over the great plain of Shellif, between Milianah and the mountains without turning a hair of our Arab horses. The plain is well-cultivated. Like the Metidja, it is full of grain, almost ripe for the harvest. Where the grain is not, there is the poppy, wild and red; and the marigold, all yellow,—spangles the green garment of Atlas, which here sweeps down smoothly over the prairie. We stop at a stone well, round-walled, and worn with the chain. The Arabs are thick about it. It is the same kind of well, according to the pictures, at which Jacob met and wooed Rachel. We come to the region of small palms, and here in these fields, which seem to be claimed by no one, the smutched and dirty tents of the Arabs are spread. Around them

we see the goats, sheep, and donkeys. We approach
the mountains, if not the Desert. The signs betoken
this. Is not the vulture circling above us far? When
a camel drops in the dusty path, does he not appear
at once as a speck on the horizon in one instant, and
in the next is he not in the carcase?

As we advance the day gets hot. The wind blows
from the south. The air grows close and stifling.
We recall Byron's line—'Death rides on the sul-
phury siroc;' and at once consult science. The
thermometer says 92°. We can stand that, for we
are on the rise, and Teniet-el-Haad will be ours
before night. Teniet is 4000 feet, and surely it will
be cold enough there. Over mountains and plains
till evening, we work our way, and finally come upon
a walled town of a few hundred people. Soldiers
appear, cavalry and infantry; the gun sounds for
'sunset,' and the music plays. We are in reach of
European civilization again. We find a lodgment
at the Teniet Military 'Cercle.' The proprietor finds
us some rooms in the rear of his inn. They are
situated upon the 'Rue Mexico'!

Luckily we had purchased some powder with which
to kill fleas. It is a sort of dust. As we were in a
military hotel, we found that the powder was effective
on the light infantry. Many a flea bit the dust. But
the powder had no effect upon another troop; I will
call them the heavy artillery. I will not mention
their familiar name. I will only designate them under
their Latin appellation of *cimices*. But the Rue
Mexico will be remembered by us, not alone for its
name, but for its conflicts. I wish I could describe
it. We reach this rueful rue through the kitchen of
the 'cercle,' thence through a back yard into a little
alley ten feet wide. As we turn into the alley from

the yard we see painted on a one-story, whitewashed, stone-house,—

RUE MEXICO.

Whether the authorities, whose drum sounds here as it did on the docks at Vera Cruz, and in the Plaza at Mexico, intended to honour the French triumph in America by this designation, or whether the word *rue* was a playful *double entendre* on the forlorn path which French imperialism followed in Mexico—I know not. I only know that we lived in that street, nearly two days, and made desultory efforts to sleep there two nights. The Rue consists of four houses on one side and five on the other, one being a stable. I have seen an Arab gallanting his donkey down our boulevard! These houses are covered with a red tile. At the west and fashionable end of this street are seen the tents of the Turcos; and above them, upon the hill, is the fort, in yellow stone. The houses are all whitewashed. I perceive a young Arab, minus his clothes, approach. He whistles a French tune. He is progressing. Three turbaned people and a dog are examining a string-halt mule for a trade. A tall, dilapidated African wench, in tatters, makes up the *tout ensemble* of our street. But the swallows do not disdain to sing there. The dogs sleep in it, regardless of fleas. The people have not yet heard of Maximilian's failure, and the name of Mexico is still glorious. Does it not run parallel with the Rue Napoleon? and where so near the Great Desert can you find a better street than the Rue Napoleon?

The houses in Rue Napoleon are numbered. Some have 'insurance' signs on them. Respectable denizens have their sheep in the houses. An African, of choicest ebony, has a wheelbarrow and is wheeling whitewash through that street! Chickens and goats are there! One end of the rue runs into an elegant open stable. Above the stable, in fine perspective, is a mountain, and a road winds up it, and a camel is on the road— and the African landscape is thus complete! Three thousand soldiers are here in Teniet. It is the outpost of the French occupation. From hence, as a military base of operations, the fights were made. Ammunition, provisions, and guns are here kept for emergencies. An emergency arises, to wit:

When we arose in the morning and looked up and down the Rue Mexico—lo! there is a hurrying to and fro of brave men! The trumpets are sounding, the drums are beating, and something is up! Is there an insurrection? There is. Where? among the Indigenes? It is. Where? in the desert tribes? Aye, marry is it! Does it mean destruction? It does— dire, direst. On with your armour, men of mettle! Mount your barbs, ye chasseurs! Pack on your backs your knapsacks, O Turcos! The Locusts are upon ye! As an enemy they are worse to Algiers, by far, than fire and sword of fanatic Moslem! The news is hurried into Teniet, that the army of locusts which ate up every green thing three years ago, is on the march! Already their videttes have been seen by us, not knowing what it meant, far up into the alluvial plains, near Milianah. The van of the locust army is approaching; nay, is already here and beyond Teniet-el-Haad. Five hundred men—before I can understand the situation—are already on the march to meet the host before it reaches the fertile plains. The locusts come from the desert, and with instinct equal to reason, they are making for the road to Milianah. They do not travel over fields and mountains, but on the highway! At night, when they are tired and torpid, the soldiers gather them in heaps and throw lime on them. By day they fight them back with branches of trees and noises—guns, drums, trumpets, blunderbusses, and thunder. In this way they may save the country at the north.

It was a terrible devastation, that by the locusts of 1867. It was the more so, inasmuch as the natives had made no provision for famine and loss of crops as in former years. Before the wars with the French it was the custom, when crops were gathered, to hide the surplus in the ground for an exigency. During the wars, this surplus was taken by the French. The natives, Arab and Kabyle, since then have sold their surplus. Hence, when the crops were destroyed by the locusts in 1867, the natives were in the power of the brokers and hucksters, and found no relief. The famine was, therefore, terrible upon them. Over one hundred thousand people perished of hunger.

I remember to have read, in some notes to a poem of Southey, that the Arabs of the Desert rejoiced at the advent of the locusts, because they devastated the rich plains of Barbary, and thus afforded them the opportunity safely to push through the Atlas gates, and pitch their tents in the desolated plains. Where such terrible consequences follow, it is worth while to ascertain not only how to avert (and in this instance, unhappily, the effort was too late, and, therefore, fruitless) but to investigate the cause of these insect plagues. They are one of the results of the intense sun of the desert— for the beams of which, I was *not* in search. Scientific men, in investigating the maximum degree of vital manifestation under the sun's rays, have concluded that it attains its highest point, as well as its greatest variety, richness of hue, and sometimes venom, where the solar beam is most intense and the luminary most nearly vertical. Hence, life, especially insect life, increases as you go from the Pole to the Equator. Humboldt has written of its horrors in the South American swamps. The beetles and birds of Brazil are described by Agassiz. The pyramidal ants of Africa, the white ants of India, the parasol ants of Trinidad, these are the schoolboy's wonder! The scorpion valley we have ourselves found in Algiers; but the locust phenomena outrank them all, either as marvels or as scourges.

The masses of locusts not only darken the sun, but their migration is conducted on a plan so remarkable that human reason can hardly out-march, out-flank, or out-general them. They have been known, in their short lives, to do more damage than the armies of men. Even after death, when great masses have been thrown into the sea, they have, when thrown back on the shore, poisoned the air by their decomposition. In 1858 they moved from Barbary on England. They have been known to cross from the Continent to Madagascar. They are the same enlightened insect which providence used in Egypt nearly 4000 years ago, of which it is recorded that 'the locusts went up all over the land of Egypt, and rested in all the coasts of Egypt; very grievous were they. . . . . They covered the face of the whole earth, so that the land was darkened; and they did eat every herb of the land.'

If any one will explain why England in August was full of ladybugs, and how they reached that fast-anchored isle, I will explain the locust flights. With feeble wing—a leap rather than a flight—these insects, born of the sun, have come to England to eat the vermin which infest the hops, with a view to beer and ale. Surely there is a special providence in these miraculous flights!

But our pathway will cross that of the locusts, if they push on their columns. We are to move on toward the Desert. We are to see the great forests of ilex and cedar, south of Teniet, upon the Atlas, from which the present chapter is penned.

The forest is a day's hard ride and many hours' walk from Teniet-el-Haad. More; it is a good two hours' walk from the end of the road, called by the French '*Le Rond Point*,' or turning. It is so called, because it is the only point within many miles among these mountains where a carriage may turn to go back. It is the point to which the French officers at Teniet-el-Haad often ride for a day's recreation and pic-nic among the mountains. Indeed, we left below us, at the foot of this mountain, a considerable company of them. They are bivouacking in the woods, near the hut of a lumberman, and under the wide-spreading umbrage of the cedars of Lebanon, which help to make the forest here.

I write where I sit, upon the topmost and most southern of the range of Atlas, into whose heart we penetrated at Fort Napoleon, in East Algiers. Here, in Southern Algiers, we have gone through this range of Atlas, and have now an uninterrupted look to the far south —so far that nothing intervenes between us and the limit of our vision, as far as the eye can reach from a point 6000 feet above the sea. That is, we have an eye-grasp of objects over 150 miles distant! What we have seen, when so far north from here—as if in cloud-land, or in arctic-land, or in dream-land—is here and now real and near ; for we are not only amidst, but have surmounted, the mountains which gave such a glory to distance.

Apart from the gratification of the eye, there is something very attractive to me in the mountains. I naturally go toward and into them. These African

mountains have a spell about them; they hide the
mysterious. Beyond their walls what is there—not?
The unknown is ever wonderful. They form a part
of that range which makes Italy. They are classic
enough to help Scylla and Charybdis into their olden
bad fame. They make Sicily possible. Crossing
under the sea, from Sicily to Tunis, they are only
1500 feet below its blue surface. They are as plainly
marked to the eye of Captain Maury, or the philosophic geographer, as if they were above, all clad in
snow and dressed in greenery. East of Tunis, and as
far as Egypt, there is a level desert; west, these mountains move in majesty towards the Atlantic. Were
there 1500 feet more upon the ridge which binds
Sicily to Italy and to Africa, we should have one
continent less! Let him who would abolish Africa
reflect on this. We have seen the glories of this
range in its most conspicuously interesting aspect, and
from its grandest positions.

I find it best, as a saving of hand-labour, if not as a
matter of interest to reader and writer, to take my
shots at scenery 'on the wing.' If I wait till next day,
or till I return to my hotel, or have more leisure—
when every day is crowded with fresh incident and
new phases—something of the interest and all of the
freshness of description are lost. Therefore, I take my
ink in pocket, and my 'pen in hand,' and open my
eyes and write. Whether seated on a crag or in the
grass; upon a fallen pine, or ensconced amidst the
broad arms of the olive—whether in a Kabyle cabin
or an Arab tent—the best way to reproduce the object
to the eye at home is to catch it before it 'lights.' If
I could concentrate into one focus the eyes of your
mind, and fix them here on this pinnacle of grandeur,
and then, inspired by one lofty mountain thought,
turn it round till it sweeps the horizon, you would

10

have a panorama entirely unique and sublime—an
endless chain of eminent 'royal highnesses,' each
worthy to wear the crown of Atlas, or the diadem of
snow wherewith the Alps are honoured. It seems as
if the mountains here were once mobile billows,
and had been stayed as they stand, by the Living
Word! Turning directly to the south, you perceive
mountain beyond and above mountain, until the
vegetation, here so gigantic, gradually seems to die
out, and the hills begin with a few patches of isolated
green to make their halting and timid march towards
sterility. Still on, and on, until we perceive where
they end, and, by the aid of a glass, where the yellow
line of sand begins. There, at last, between us and
the horizon—just before our eye, no illusory mirage—
there, is the first footstep of the Inscrutable in the
sands of the Great Desert of Sahara. It may not be
the desert itself, but it is enough for a sample. It
is the Algerine desert; almost as formidable as Sahara.
We know that Sahara begins there to be what it
becomes further on, in its consummate desolation.
Beyond this line of sand two or three dark green spots
appear. Are they oases? In the midst we see a town,
by the aid of a glass—called Chelala—white, oriental,
but dim—a resting place for travellers over the great
sea of sand. Then beyond this — most uncertain,
waving, and vapoury—a hundred and fifty miles from
our lofty vision ground, is a line of mountain where
my pre-historic Tauarigs, or Berbers live and levy
tribute of the caravans, or plunder. To the left, on
the west, are three surpassing peaks of mountains.
They are a part of this range of Atlas, rugged and
glorious. As we look out from our mountain ground
of vantage, they seem awful and mysterious, swimming
like clouds in the upper ether, or standing like weird
sentinels at the gateways of the Desert. The heart

beats as we gaze—beats tumultuously—and the hand trembles to record its throbbings. Turning about and looking north, the excitement is not lessened; for our vision reaches Milianah, which is over ninety kilomètres (five kilos to three miles) from Teniet. Over this distance we have come—over mountain and vale, through cool blast and sirocco heat—not without fatigue; but the fatigue is compensated by this one magnificent view, for ever burned by the 'Sunbeams' into our memories.

We are so far out of the track of travel, even for French people—and I was about to say, being on this untravelled mountain, for Arabs—that it is worth while to give the reader the *modus operandi* of coming hither. Algiers, to English and Americans, is only known as a conquered colony, and the battle-ground of Abd-el-Kader. The French themselves know only a few of its places by personal observation. I mean the French tourists. A French writer, in a pamphlet which I have picked up (being a plea for a more liberal policy for Algiers), says that Algiers is as far off as China, as the French look at it only through the wrong end of the lorgnette. There is much truth in the remark, not alone for its political metaphor, but for its literal meaning. We astounded the French travellers we have met, when we told them that we were bound for Teniet-el-Haad. Indeed, but for the fact that two of our party were invalids, unequal to 120° Fahrenheit, and the sirocco which blows at times to that point of the thermometer, we should now be moving over the Desert itself. As it is, we must be content with a Mosaic vision from the mountain of promise. We are not permitted to enter in upon the land itself. It puzzles me almost to know, much more to tell, how we attained this grand eminence. Teniet is itself 1200 feet above Milianah, and Milianah is 2700 feet above the sea. We undertake by a carriage road—as rough a road as any country can show (made at first for military, and then for forest purposes)—to reach the Great Forest. We pass through every variety of scenery and climate to reach it. We find the flowers, trees, and birds of the temperate zone as we rise up into the mountains. The hyacinth, the blue and yellow orchis; the blue-bird, jay, and cuckoo; the oak, ash, and cedar,—all salute us. What oaks these are! Some are deciduous, just drawing over their naked wintry fingers the green gloves of spring. Some are, like the evergreen oaks of Florida, hanging with sad, grey pendants of moss. As we rise upon Atlas' sides we see, far off below, our road from Milianah, over peak after peak, and into plain after plain. We are going west rather than south, and before us are the mountains of snow and cedars, and behind us the green wheat and barley fields of the Teniet environs. I count six ranges of mountains in the east and north, and as we move still on and up, the few red roofs of Teniet seem like specks in the distance. At first we see a few young cedars. They are conical in form, but how unlike the mature cedar which we soon meet—the limbs of which are covered with green, and are as thin and flat as a table. Some are gigantic. It is hardly the same graceful, little tree we first perceived. *Quantum mutatus.* After fighting the north wind and the sirocco for a thousand years its trunk is, as it ought to be, immense, and its limbs and foliage beaten down at right angles with its trunk. As you look down on these forests, it seems as if you could walk over their level floor of frosted green. These are the veritable cedars of Lebanon.

I know that efforts have been made to depreciate the 'glory of Lebanon.' It is said that Lamartine is responsible, by his grandiose description, for the poetic aggrandisement of the cedars of Lebanon; for they are, he says, grand and impressive; they tower above the centuries, they know history better than history knows itself; they astonish the people of Lebanon. Evidently, they did not astonish Madame Olympe Audouard, when she visited Lebanon; but M. Alphonse Lamartine did. She found the trees dwarfed and ugly, and Lamartine imaginative in more senses than one. 'Shall I carve your name under M. Lamartine's, Madame?' said her guide. She asked if he had been with the poet when he carved his name. 'Not at all,' was the remarkable reply, 'for he never came here; but, like a wise gentleman, remained in Beyrout, and sent me here to cut his name.'

Such at least is the story as it goes the round. Whatever of truth there may be in Madame's representation, these cedars of Lebanon deserve all the eulogy which Lamartine has bestowed on them. Dean Stanley, in his exhaustive and elegant volume on Palestine 'recognized the sacred recess of the present cedars of Lebanon.' He proceeds to describe the scene and the impressions; from which we learn that Lamartine did not exaggerate. Above the moraines of ancient glaciers, and even above the semicircle of the snowy range of the summit of Lebanon, is a single dark massive clump—the sole spot of vegetation that marks the mountain wilderness. This is the Cedar Grove. The outskirts are clothed with the younger trees, whose light, feathery branches veil the more venerable patriarchs in the interior of the grove. There are twelve old trees remaining, called by the Maronites the twelve apostles! Their massive branches are clothed with a scaly texture, and contorted with all the multiform irregularities of age. From these David had received his grand impressions. The shiver-

ing of their rock-like stems by the thunderbolt is to
him like the shaking of the solid mountain itself. Dean
Stanley further remarks upon the peculiar grace of
the long sweeping branches, feathering down to the
ground—as we have the transplanted cedar in Europe,
and which, he says, is unknown to the cedars of Lebanon.   He pictures the latter as we saw them in
the Atlas; the young trees holding up the old, and
the elder holding up the younger trees.   He speaks
of their height and breadth, and does not forget, what
we saw all through the forest of the Atlas, that it
was full of birds of gay plumage and clear note.   It
was out of these ancient trees that the Temple of
Solomon was made; so that it was called the house of
the forest of Lebanon.   Tyre and Sidon built their
ships out of these cedars, and their fame went with
ancient commerce to the ends of the then known earth.
Sennacherib could find no image so suitable to the
expression of his power as this: 'By the multitude of
my chariots am I come to the heights of the mountains, and to the sides of Lebanon, to the height of
his cedars, and the beauty of his cypresses.'

Our experience with the chariot was limited, compared to that of Sennacherib.   We ascended to the
height of his border, and the forest of his park, on
foot.

The cedars of Judea, aye, even those described by
the Psalmist, cannot be, as we believed them to be
before the enchantment of observation, larger or more
beautiful than those of the Atlas.   Where do they not
grow here?   There are two forests of them, many
miles square in area.   Some of the trees burst through
rocks; some grow out of peaked mountain tops;
some look so natty that it seems as if they were a
company of Parisian ladies, exaggerated a thousand
fold in size, and carrying their parasols with a genteel

crook of the elbow and the latest Grecian bend. The
colour of the cedar foliage is that of tea green, with
a sort of whitish frost work. Some of my readers
may have seen the cedars of Lebanon. Some at least
have seen, in the Jardin des Plantes in Paris, the first
one ever brought into Europe, in 1740. Its counterpart is here reflected in a thousand forms and much
larger and multiplied infinitely.

Among these living cedars, above us and below us,
we find many blasted and many charred by fire; but
even in their desolation, they seem sacred and sublime.
The live oaks about them, covered with brown, gray,
and green mosses, would of themselves repay us for
this forest trip, had we seen no cedars. We missed
our trip to the larch forests of Corsica, owing to the
snows of February. Those larches were 170 feet high,
and immense in diameter. These cedars are not quite
so high, but as great in girth. They are broad and wide,
fit emblems of Atlas—upon whose head they grow.
I measured one of the two under which we lunched;
and in a fair way, three feet from the ground. It
measured thirty feet around; not equal to the 'big
trees' of California; but big enough!

Nature exaggerates her growth here to compensate
for the vegetable lack in the Great Desert so near.
Nature does nothing in a hurry. Having reached the
climax of vegetable glory in these oaks and cedars, she
gradually moves down the ravines and mountains,
until there is not a blade of grass left where the Desert
may be said to begin its empire; for we are outside of
the Tell, or cultivable region. And even in these
forests, before we see the Desert, the mind is prepared
for Desolation. One-fourth of the great trees are
blasted; but their skeletons remain. The bones of
their gigantic bodies are here; and in all postures.
Some are erect, as if they died proudly facing the

blast; some devotional, upon their knees, as if warned
of death and praying to be delivered; some grotesquely
on all fours, or flat on their back, big enough even
when down and dead, to frighten the ordinary woodman. But I must record the truth. The woodman
has come hither, and he does not spare these cedars of
Lebanon! The Scripture is fulfilled; 'Howl fir-tree,
for the cedar is fallen!' The Government have conceded to the railroad company the privilege of cutting
these trees as sleepers. They are being slaughtered:
but it will be years yet before the glory of Lebanon
departs. It is so difficult to reach them among these
mountains, and so far to haul them. It is a pity the
workmen could not clear up the forests by taking out
those which are half-alive and half-dead; or, if you
please, remove those unhealthy cedars with large pro-
tuberances, called gout stones, or those which resemble
the description of the old Indian orator—'dead at
the top.'

But we have not yet reached the heart of the forest.
The barometer registers only 4500 feet. The vegetation varies as we rise. Here is the schoolmaster's
ferula again, and the blackberry pushing its way to the
very edge of the Desert, and inviting comparison with
the old oaks. Flowers appear in great abundance.
The wild pea, the butter-cup and the daisy, deck the
road-side! How beautiful the wild roses look—the
French eglantine—draping the old rocks! How came
Narcissus here? Where is his mirror of tranquil
water, in which to dress his vanity? Here is the
purple, velvety pansy, as large and as sweet as that of
England! It lies in clusters, as if native to the Atlas,
and used to the breath of 5000 feet elevation. Here,
too, is the hawthorn; not numerous, but here! How
came it in Africa? Is it indigenous, or is this May in
England, or April in America? Where is the solution?

Listen! You hear it in the voice of the birds. The cuckoo has had an indigestion, or the swallow, or jay, or that other unknown bird of passage, whose note is familiar to English ears. They have brought the undigested seeds here, just as the birds in Corsica have sowed that island with the wild olive. What that other bird is—even the learned Doctor, our companion—does not know. It seems to say, as it sings: 'Come! come! sit here—sit here!' The Doctor answers : 'I must kill one of you, my beauties—when I get to Surrey, in order to ascertain the bird which helps us to plant the pansy and the hawthorn on the ridges of Atlas!' But the kind-hearted Doctor will not kill; he would rather read Audubon.

Still we travel up—through the zones. The road becomes nearly impassable. We walk mile after mile, around the awful edges of precipices, looking down sheer thousands of feet. We startle the owl, as we startled the pelican in the plains of the Kabyle. 'The owl of the desert and the pelican of the wilderness.' The oriental imagery of the Bible is here and thus illustrated, for the owl flies towards Sahara, as the pelican flew for the waste marshes of the plain. Still we move up and out of the Temperate toward the Arctic zone. Snow appears, and near it, some green spots of grass. The Doctor, who is ever alive to botanic utilities as well as beauties, has found here the maritime squills. The natives call it a wild onion ! Squills! My mouth expectorates in the pronunciation! It covers all the Algerian land, high and low. It has followed us east and west from Algiers. He has found something else. It is the *hedera mauritanica !* 'Good!' I say; its associations are more agreeable. It is the African ivy—as beautiful as that which varnishes with its glistening green the Gothic glories of the English and Irish abbeys and minsters. Now we approach the rocky eminences, looking down from which we see our turning place, the Rond Point.

It is a green plateau; down below us, and across the wide valley over there, we hear the French officers enjoying their picnic. We give the Indian war-whoop. It is responded to by an Arab shout. We finally approach the spot. Horses are found tethered under the trees. Bottles of wine are opening, and game disappearing from the table. We are met with hospitality. The officers say that we are about 5000 feet above the level of the sea. We select two immense cedars of Lebanon, having a circumference of about thirty feet each, as our tent! Under it we picnic. An Arab boy, the same who killed the porcupine here last evening and to-day gave us the quills, is on the alert to bring us a jar of mineral water, all cool, from the famous springs near. The Doctor analyzes it. No need of his charcoal filterer for this water. It has a name and fame as old as that of the Roman days.

Finishing our lunch, we feel as if our object were not accomplished. There are mountains still above us. We are on the last ridge of this range of the Atlas; but we have no view of the Desert. Shall we ascend? It is an untried path, but it looks feasible, except this, that we have two ladies, and we two men are supposed to be invalids. The Doctor has a cavity in his lungs, and as for me, no matter! We resolve. The Doctor proposes in a methodical way to rise with the occasion. His aneroid barometer will measure his upward path. One hundred feet and then five minutes rest; another hundred and then ten minutes; another hundred, fifteen, &c. Thus he will save his breath and his lungs. At least, he will try the ascent. I suggest, as the mountain is very steep, and time is of the essence of the operation, that we had better begin at once, as we have 1000 feet to climb. We could not go much higher, as our aneroid barometer marks only 6000 feet, and it would blow up if we attempted more. Allons! First rest, by all hands, on a cedar tree, hewn and ready to be made into ties for the railroad at Milianah! Second rest, two hundred feet, Doctor in a rocky, curule chair, cushioned with venerable moss; ladies at his feet, near a charred cedar, hollow, but decorated with the honeysuckle; the other invalid in advance, prospecting the ravine for easy paths. Third rest; Doctor rouses, hits Atlas about the jugular, and falls in the ring; wind still good! Fourth rest; he is able to start ten minutes sooner; on and up we go—on and up, until the barometer indicates that we have risen 800 feet. The Doctor forgets his methodicity and his lungs. The writer is in advance, but runs against a perpendicular rock 200 feet high, and reposes in despair near a snow-bank; he makes a battery of snow-balls, assaults the party below by way of recreation, and is assaulted in turn. He retreats gracefully before numbers, makes a detour of the rocks, rises to the top of the ridge, and lo! disappointment and perspiration! another mountain beyond, and another valley below, and no view and no Desert yet! The Doctor assumes command and directs operations. I am scout. We turn to the west, follow the ridge. The barometer is near the bursting point; the ladies resolute; the Doctor still sound. At last, at last, after refreshing our lips with *African snow*, under the cry of 'Excelsior,' we reach the summit! We are rewarded. We come upon the wild, wonderful spot, where I write this chapter. True, we are not so high as other mountain tops. We are not quite as high as Mount Washington. We are more than three times as high as the Torc mountains of Killarney. We are higher than the Catskill Mountain, but we have a view such as I have never before had, and which I have I fear in vain endeavoured to describe.

Around is snow and grass; and our table for writing is a rock covered with a greenish dark moss. Again and again we gaze off into the distant south. We see no caravans winding their way to or from the Desert, but we see the mountains of the wild tribes, who levy their tribute and defy the French. The level, herbless plain grows more yellow, almost red, as the sun sinks; more like 'a thirsty land where no water is.' The air far away seems hot to look at; yet we look at it from a cool, snow-surrounded mountain. We play at snow-balls here, and within view, the ostrich hides her eggs for hatching in the burning sand.

As we gaze in deep amazement at the view, clouds begin to gather on the west and north. They are full of moisture from Labrador, says the Doctor, and are trying to do something for Sahara! If these mountains were only larger, there would be glaciers and rivers, and then Sahara would not be—sand! She would be all through as fruitful as one of her own oases. These rocks require only pulverization and water to be—food. Give them water and the sunbeams will make them fertile. This is one of the Doctor's thoughts. He sees camelias in muck, dates in clouds, wine in running brooks, and good in everything. He is, in fact, an optimist. He finds utility in fleas. They tend to make people cleanly. He even found some excellence in the scorpion I killed. He did not tell me what. I suppose because it furnished food for —Dervishes! But in the economy of nature, he does not exaggerate the influence of clouds and mountains. Many of his climatic and sanitary conclusions about Algiers are based on those very phenomena of wind and rain and mountains; for here the clouds are drawn by gravitation to the mountains; and but for these clouds, the oases would be sand, and irrigation would lose its fertilizing power for the plains. Without these clouds no one could sing the missionary hymn about 'Afric's sunny fountains,' much less its 'golden sand.'

But for these clouds, or for the cause which draws them hither, what might not the opposite coast be? If Sahara—a furnace—sending to the upper air its heat, and sucking the north winds with their clouds and rains from the coasts of Spain and France, leaving them dry, and refreshing with copious rains the plains of Algiers; if Sahara were not as she is—a sucker—Northern Africa would not be so cool and damp, nor would Spain and France be so warm and dry! These are not paradoxes. But we have no time for reflection. The storm comes. We are far away from habitation or succour, in case of danger. We retreat in disorder—a little damp, but all safe. Teniet-el-Haad we reach, and we sleep on the borders of barbarism and sterility, in the 'Rue Mexico.' We sleep; unconscious of the terrible fact, which we afterwards ascertained,—that we had been in that part of Atlas where lions and tigers are common. We sleep; for we are very weary, and the fleas have lost their power to disturb us. Is the Spanish proverb true which says, ' *Quien duerme bien no le pican las pulgas* '? He who sleeps well cares not for fleas!

I close this chapter in Milianah, to which on the next day we retrace our steps. We pass through the camp of scorpions—where I killed an ugly, venomous specimen. It was reckless on our part to pic-nic in this neighbourhood. Not because of the colony of convicts—some of them sneaking about in the brush; but on account of the scorpions. But our pic-nic was a delight. We almost filled the ideal of the worthy Fuller,

in describing the early monks, who left the city for
the wilderness. As for their food, the grass was their
cloth; the ground their table; fruits and berries their
dainties; hunger their sauce; their nails their knives;
their hands their cups; the next well their wine-cellar,
and what they lacked in the cheer of their bill of fare,
they had in grace.

Lunch being over, we pass over streams lined with
oleander in such profusion, that when it comes out in
June 'Afric's sunny fountains' will be all aflame. We
follow the meanderings of a stream whose foam is not
amber and whose gravel is not gold; but we find its
bed incrusted with—salt! On inquiring of our drivers,
we find that we are near 'Salt River.' I had supposed
that stream, the synonym for the Limbo of departed
politicians,—was somewhere in America, although I
knew Africa had something to do with it. My companion, who is slightly tinged with the fanatical,
suggested playfully that I ought to go up to the
sources of Salt River. 'I should find some friends;'
but as a few choice friends still remained absent, I
refused.

One peculiarity this Salt River has. It is illustrated in the sketch. Fifty Arab women go there
daily, and after the water has run over the rock and
the sun has evaporated it, they scrape the rocks, and
thus 'earn their salt.' We met four Arab women
and one child on their road home to their tents from
the salt region. They had kettles full of salt, and as
they are samples of the Arab women, I will photograph
and anatomize them. They were dressed in a—
chemise, which had two loops to let the arms through
at the humerus! These openings are unnecessarily
continued down to the ileum, thereby allowing the
curious profane to obtain a vision of the—female form
divine and dirty! Their feet are encircled with anklets

of steel or some dark metal, or ebony. We did not handle them. The feet presented no useless conventionalities of sandal or shoon. Their outspread phalanges took firm, yet graceful grasp of the earth. No fear of scorpion did they show. In the absence of their male protector—for they were all wives of one lord—they showed no fear of us Giaours. Their eyes sparkled as they saluted us. One of the number tried to say : 'Bon jour ;' but her guttural Arabic made it sound like—'Bad job !' A handkerchief confined their raven hair, over which there was a turban of enormous altitude, requiring the Doctor's barometer to measure it. They did not look beautiful; nor do their smoky tents or filthy surroundings look enticing. They do not wash much. The Arabs reproach the Moors for living in houses. The latter reproach the Arabs for not making their ablutions, which is a part of Mohammedanism. Both female and male Arab seem to be varnished over with layers of dirt. The nitrogenous elements, decaying in their bodies and going out of the skin, produce an odour very unlike that of jasmine or attar of roses!

We meet, however, when nearing Milianah, many families of Moors, and some Arabs, too, very unlike the females from Salt River. I present one of these ladies pictorially, as in marked contrast to the Salt River dames! These ladies of Milianah were riding, and were covered with white veils. All but one eye was hidden, and sometimes even that; but they will peep a little. They sometimes walk, bearing their baby behind on their backs, the husband a rod or so in advance. Slaves follow them, and they all follow their full-turbaned master. He still lives, or rather lingers, near Milianah, and praises the days past and sighs to be afar from French rule and Hebrew liberty, afar off in Syria, with his old neighbour and brave chief,

Abd-el-Kader! While going up, we meet a score
of families coming down the mountain from Milianah. Among them is a couple on horse-back—
Jacob behind on the crupper, and Rachel astride in
front. All these people look proud and cleanly, and

Lady of Milianah, full dress.

no wonder. They have been to Milianah—city of
fountains—to the baths. They are tending home-
wards. As we move up through the leafy and floral
paradise to the walled city gates,—the waters plash
down the garlanded rocks, the clouds begin to grow
orange and red over Atlas as the sun sinks, and the
warm breath comes from the south, showing that the

north-west storm of yesterday has been smothered by the sirocco. As we ascend and look below us, and follow the white thread of the spiral road we have come by, and thence glance over the plains and mountains to the distant range of the Atlas, where two days ago we stood, we are not sorry that a kind Providence has permitted us to see so many of His terrestrial wonders, and to return to the vicinity of comfort in this beautiful city of Milianah, upon this most Delectable Mountain!

CHAPTER XIII.

CONFLICT OF CIVILIZATIONS.—FAREWELL TO AFRICA.

'The crescent glimmers on the hill,
The mosque's high lamps are quivering still,
——but who, and what art thou
Of foreign garb?'

BYRON's ' Bride of Abydos.'

THE reader will perceive that I have travelled over five degrees of longitude, and very considerably inland toward the 35th degree of latitude; that is, from Fort Napoleon on the borders of Constantine on the East, to Oran, where I write, not far from the borders of Morocco. He will understand that I have travelled not hurriedly in railroad cars, but in carriage, and on foot; and have thus had opportunities to correct first impressions, and to know the country for what it is. He will also perceive that I have indulged in details, seemingly trifling, but with a view to elucidate the questions growing out of the conflict of antagonistic systems and civilizations. There is no country like Algiers in this regard. Here is a Mohammedan people under Christian rule; the religion of the ruled tolerated—almost fostered in a way—by the ruler, and the ruler doing all in his power to attract the affection and loyalty of the subject, and that failing, holding the people pinned to the throne by the bayonet. There have seldom been less than 100,000 French soldiers in Algiers; yet I do know, from conversation with leading Mussulmans, that their

hatred of the French is inveterate and irreconcilable.
The Mohammedan religion and customs are, however,
decaying before the French. But it is still a problem
whether France had not better do with Algiers, as
the conscript did who caught the Algerine tigress,—
let her 'go.' I do not see that she intends to do
that. Her hold is rather growing tighter. True, she
does not realize much,—in fact, pays more than she
receives in revenue. But she is building bridges, walls,
docks, forts, turnpikes, and railroads, and inducing
immigration by offers of land—rich land—at cheap
rates. But the immigration is not increasing very
fast. The land is untaken. There is protection
enough given now by the army against native outbreak. There is no trouble in reaping, if you plant,
provided there be no untimely fog, or sirocco, or
locust raid. The land is very rich, well watered, well
rained upon, and labour is cheap. It is, however,
complained, with great emphasis, that the home Government discriminates against Algerian produce; not
only by taxes here, but by heavy duties, even against
the grain here raised and imported into France.
Algiers is treated as a foreign country, and that too
by a nation which claims M. Chevalier as its political
economist, and free trade as its policy. Because the
farmers of Southern France howl for protection against
the importation of Algerine wool, wheat, barley, and
horses, the Government yields. The advantages of
Algiers as an agricultural colony—once acknowledged,
and so great in grain that Rome was fed from it—are
dissipated under the insane clamour of 'protection to
home industry.' Even the Arab horse is 'protected'
against visiting the world out of Algiers!

I have already incidentally shown why Algiers is so
fruitful. The desert below, by its heat, draws the rain
clouds from the north. Their cisterns are sucked over

the sea, which keeps them full, and they are emptied on the mountains and plains of Algiers. Hence, when the soil of Southern Spain and France is cracked and seamed with heat, Africa is damp, misty, cool, and fertile. Her fields and mountain sides are carpeted with all the hues of Flora. Her streams are fringed with oleander and tamarisk, and her rocks are draped and tasselled with the Mauritanian ivy. Her very moors (I do not mean the Moors) are mosaics of every dye; her road-sides are odorous with roses and jasmine.

From the soil, mountains, harbours, climate, waters, flora, and the history of Roman successes here, as also from the proximity of this colony to France, I infer that France will hold Algiers, and in time make it an exception to her general system of colonization, by making it a success. France has had Algiers about thirty-six years; but never so as to control it in tranquillity until recently. Her experiment, therefore, has not been fairly tested. Had not England been jealous of French power in Africa, or in the Orient, France would have had control of the three pashalics of Morocco, Tunis, and Algiers as long ago as 1827. A treaty was made for that purpose with the Turkish power, which had held them from 1516, but the intervention of England with the Porte prevented its consummation.

French blood has flowed freely here. Half a million have perished to hold Algiers. The war with Abd-el-Kader, whose plume cut the French like a sword, for when he was in the saddle all Moslemdom were his retainers—lasted from 1837 to 1848. The French did not keep faith with him on his surrender. They imprisoned him in the island of Marguerite for four years—a beautiful isle near Cannes, which I have described, out of whose barred castle-windows

he could look out upon the sea toward Algiers—a sea
as unstable as his own vicissitudes of fortune. Perfidy
and blood, confiscation and plunder, these are the
penalties exacted when the strong war with the weak.
By some overruling law of political gravitation, which
attracts the minor States to the greater; or which
compels the less civilized people to yield to those
of superior civilization, Algiers has become absorbed
in France. The word is, that God has commissioned
France to redeem North Africa! France accepts!
The military and civil administration of France, with
its system of magistrates and prefects, its division of
military and civil territory, departments, arrondissements, and communes, combining a central, provincial,
and municipal government, leaving much of domestic
matters to the native people, especially as to religion,
marriage, and indigenous customs, remitting much of
the administration of justice to the Kadis, or Mussulman judges, all these are features of the policy prevailing here to-day. But there is no representation
as yet of Algiers in the Chamber of Deputies, and
great complaint is made that the interests of the
people are neglected by the irresponsible pro-consular
system.

Still, Algiers does show prosperity. Her agriculture
flourishes, but her market is restricted. Oats, beans,
sorghum, and all the cereals are readily raised, and the
production seems to be augmented with every year.
The forests of cedar, cypress, and oak have been
noticed in our visit to the South; but the fruits of
the olive, the palm, and the orange almost rival the
grains of the soil for their production. The tobacco
is held to be next to that of Cuba; and its production
and manufacture is a large business. Everybody
smokes in Algiers; cigarette is the favourite style with
all. The cotton culture has been quickened by recent

events in America. The quality raised is that of the
long staple, and of the species familiar to the Carolina
coast. A volume might be written of the mineral
springs and mineral resources of Algeria. Every
coloured and veined marble—blue, red, white, and
black—is found here. Porphyry is to be had in Constantine. Wages are not high; they vary in the
different provinces. A good carpenter will make from
three to five francs a day, and a gardener or common
labourer about two francs. But it costs little to live.
Beef and mutton are common and good; milk of cow,
goat, and mare is ever at hand. One thing may be
said, that Algiers is prolific in births. Doctors unite
about this. There is something about Africa peculiar in
this respect. The births far exceed the deaths. Commerce increases, despite restrictions. But the United
States have no part in this commerce. I find no
record of any American vessels at the ports. Our
consular duties are restricted to rescuing naturalized
citizens from the French army.

The principal drag upon the prosperity of Algiers is
the Mohammedan faith and polygamy. I am not
illiberal toward the Moslem. He has much of interest
in his religion, even for a Christian. The Mohammedan is not so intolerant as he was; nor has he been
so intolerant to the Christian as to the Jew. The
Koran itself refers to the Saviour 'as one who came to
save from sin; as one conceived without corruption in
the body of a virgin, tempted of Satan, created of the
Holy Spirit; as one who established an Evangel which
Mohammed confessed!' But it offers to the faithful
a heaven of sensuality: nay, seven of them—one of
silver, of gold, of precious stones, of emeralds, of
crystal, of fire-colour; and the seventh heaven—a
delicious garden whose fountains and rivers are milk
and honey, whose trees are perennially green, whose

fruit is so beautiful and delicious that a drop of it in the sea would change its brackish taste into sweetness. The mansions of this heaven are filled with all the imagination desires, and the believers espouse there the most wonderful of lovely houris—*for ever young and for ever virgin !* Thus you see that the highest heaven of the Mohammedan smacks of his earthly home where the senses are gratified, and where there is no limit upon his loves.   Of course so long as this is his religion, and when, too, wives are purchased for money at pleasure, the family, which is the base of the social pyramid, cannot be said to be a blessing, but a curse. The author of the ' Crescent and the Cross ' thus hits the nail on the head :—' In Paradise he finds the extreme of sensual enjoyment, as a reward for the mortification of the senses in this life ; so that his self-denial on earth is only an enlargement of the heroic abstinence of an alderman from luncheon on the day of a city feast. His heavenly hareem consists of 300 houris, all perfect in loveliness. What chance has his poor wife of being required under such circumstances !— it is *supposed* she has a heaven of her own, in some place or other, but as to *her* substitute for houris the Koran is discreetly silent.   In Paradise is to be found every luxury of every appetite, with every concomitant, except satiety and indigestion.'   Hence, the Mohammedan has for-ever in his appetite and faith a canker to his prosperity. He must give way.   France, not over scrupulous in her own domestic ways, is nevertheless reforming even Algiers in this particular.

The influence of the Crimean war upon the Orient reaches even to Algiers. The Turkish empire was not so much shaken as it was propped by that war ; for the real power of Government to-day in Constan-tinople is not in the Sultan, who sits cross-legged, sipping his coffee, and smoking his chibouque on the

Bosphorus. His favourite occupation is to feed his
chickens and ducks—of which he has a poultry yard
full; while Russia, France, and England, from their
ambassadorial palaces at Pera, dominate as well over
the roofs of Stamboul, as over the various nationalities
which make up the East.

The harem has been invaded. The chief wives, or
the wives of the chief Turks, seek European society—
or the society of European women. They import the
tawdry ornaments of Vienna and Paris, and their
dresses are no longer the velvet jacket and trousers,
but they have the stays, the gaiters, the long trains,
and chignons of their fashionable sisters of the West.
They are even learning the piano!

All these domestic reforms are traceable to the
influence of the Crimean War, which opened, with
much authority, the secluded portals of Oriental life.
France is doing the same just as effectually in Algiers,
because she wears the velvet glove over her mailed
hand. For instance, she not only permits, but aids
the Moslem schools. Therein the children of the
faithful are taught, as of yore, by Moslem teachers.
Many of the institutions, about Algiers especially, are
under the care of French matrons. The girls of
Moslem families come to these institutions to learn
the arts of domestic life, including, as I have explained,
the refinements of embroidery, toward which their
delicate, henna-tinted fingers seem to have an instinc-
tive tendency. While at these schools, they are
guarded from Moslem eyes of the opposite sex, but
not from those of the Christian; and, when they return
at evening to their homes, they are muffled up in their
awkward mantles and head-gear, and conducted by
some one who is approved by the parents. My wife
visited one of these establishments. She tried to tell
me how the girls put on their long, winding, won-

drous involutions of dress for the street promenade. But I cannot repeat her description.

Again, I think the French influence, while it tole-rates the existence of Mohammedanism, has its effect upon polygamy; not to abolish but to mitigate. It will in time abolish it. As we used to hear it said about American slavery, that we might repose in the hope of its ultimate extinction, so we prophesy that much about polygamy. This remaining 'relic of barbarism' will, perhaps, be found last on the soil of my native land. Proud thought! For example, you rarely find a Mohammedan in contact with the Christian community, and having relations of business or otherwise with it, who has more than one wife. Not but that he can afford more. Out on the plains, where the Arabs roam, or up in the mountains, where the Kabyles farm, there you may find the sheikhs, or chief men, who have means, also having several wives. The people generally adhere to the Christian practice. The wives appreciate it. They are very much more docile and dutiful, when they are alone the mistress of the household. The other day, while at Algiers, my wife was invited to visit a Mussulman family—that of Mustapha Rayato—a merchant of Algiers. The lady of the household—and there was only one—gave a sparkling answer to the American lady, when the latter inquired after the *other* wives. If you will allow me, I will extract the scene from the journal of my wife, not alone for the colloquy about polygamy, but as a better description of the Moorish domesticity than any *man* can give. Thus the journal :—

'Our visit had been previously arranged, and a promise made that we should see all the trinkets, jewellery, &c., of a Moorish household. As Mustapha Rayato led the way to his house, he enlarged upon

11

the narrow streets through which we were passing.
He said that once he, too, could afford to live in
an elegant house in better quarters, but the French
came, and all was changed.   He had sold his house,
taken one much inferior, and kept his shop like one
of the common people.   We did not see the degradation of that, but out of respect to our host, we for-
bore comment.

'We enter the ordinary Moorish house by a large,
double, common, wooden door.   It opens into a small
square vestibule.   A neatly whitewashed stairway is in
front, and a door at the side opens to the inner court.
The double piazza incloses the court, and the lower
rooms are devoted to the servants.   The stairs were
of slate; and I noticed their extreme cleanliness was
not in the least disturbed by the boots of the gentleman, since our host had quietly dropped his shoes at
the door, and encased his feet in another and clean
pair, ready for him at the foot of the stairway.
[Mem.—A good idea for the lords of creation of
other nations.]   The floors of the piazzas and rooms
were of porcelain tiles.   We were ushered into the
salon de réception; one of the four upper rooms reserved for Mustapha's family.   A low, cushioned
divan ran the length of the room, and in front, scattered over the floor were carpets, rugs, and silken
cushions.   Madame Rayato rose gracefully to receive
us, and gave us the Arab salutations.   Then, we chose
our seats as best suited us: I, on the divan near our
host, my Italian companion on the carpet in front, and a
charming little French madame whom we had encoun-
tered on our travels (for the gentlemen were excluded),
upon the cushion by our side; Madame Rayato on the
other side of her liege lord, with her group of three
very pretty daughters; while number four, a pretty
child of about that number of years, nestled between

her father's knees, and alternately bestowed and received caresses from her handsome papa. Indeed, it
seemed much like a Christian household in this respect; a beautiful domestic picture. Mustapha's face
radiated with pleasure, "as he floated down the calm
current of domestic bliss." [This last remark is not
in the journal. It is that of the author.] I introduce, not the madame, in my engraving of the
"Moorish woman of the period," but a type of the

Moorish Lady of the Period.

well-dressed Moorish lady, dressed like her. She had
decidedly Spanish features and complexion, as we
thought; we afterwards found that she was descended

from a very wealthy and noble Spanish family—*i. e.*, a Moorish family once celebrated in Spain before the Moors were driven thence. Her hair was dazzling in its blackness. It was cut short at the neck, and covered with a silken foulard (handkerchief) whose embroidered and fringed ends hung jauntily upon one side in the form of a heavy tassel. A diamond sword fastened it on the forehead; an agraffe, with pendants, ornamented the point at the tie; and these were matched with earrings of pendant solitaires. She wore a crimson velvet basque, cut nearly in the Pompadour style, with lace chemisette, and confined at the waist by a girdle of the same, the girdle and basque embroidered in gold. Large, full muslin "pantalons," over those of a thicker material, completed the toilet of the madame. A dozen strings of pearls were around the neck, and several pairs of bracelets, in unique designs, of gold and diamonds, and massive gold anklets, were the ornaments of arms and feet. Of course, she had put on the additional number of jewels, to redeem the promise made to us, that she would display all her bijouterie. The diamonds were set in silver, so that we were able to bear their brilliancy without envy. Had the setting been gold, I would not speak for our integrity. Afterwards, jackets of gold cloth and "pantalons" to match, were shown us, for fête occasions. The children were attired like the mother, only with less jewelry, and instead of the foulard head tie, they wore the little fancy Greek cap of gold coins, tied coquettishly at one side.

'Salutations over, and jewelry examined, we explained our different nationalities, we speaking in French and Mustapha Rayato translating into Arabic for his wife. We inquire if they have French masters for their children, since their country is becoming essentially French? "No. Our religion does not

allow instructions in any other than our own language."
"Did you take your wife with you to the Paris Exposi-
tion?" "Oh, no! Moorish ladies seldom travel."
"They do not always stay at home, surely?" "No.
We go to our country house for the summer," said
Madame, "and we ladies go to the cemetery on
Friday. Besides, we visit among each other." "But
in the evening, in the city? Do you not drive out as
we do, to take the fresh air?" "Oh, no. Never!"
"Why should they?" says M. Rayato. "They have
the court here, and the freedom of the house." "Your
daughters marry so young—when do you expect to
give up your pretty charge there?" pointing to his
oldest of fourteen, in size and precocity, already twenty!
"Oh, in two years most probably," at which the young
lady coyly concealed her blushes behind her mamma.
M. Rayato remarks that no dowry is needed on the
part of the girl, but the would-be husband must bestow
a certain sum. "You must be happy to know that
you can add to that sum, however, for your daughter's
comfort?" "Oh, yes! Allah be thanked! the French
have not taken my all." Our French "madame"
appeared to enjoy all these diatribes against her nation
as much as we did.

"But, M. Rayato, how is it? You have but
one wife?" Here Madame R., curious to know
the question, opened her great black eyes half suspi-
ciously, and said, on the instant the translation was
given: "If he had another, I would strangle her!" and,
with equal quickness, turned on us, saying: "But why
is it that Madame has but one husband?" To this
we could make but one reply, and that the one previously given by her, in answer to our numerous other
questions: "It is our law—it is our religion!" Here
a neat, bright-eyed mulatto girl, who proved to be our
acquaintance of the night before at the religious rites,

appears, and brings in a low table or stool. She follows it with a silver salver, on which were china cups in silver filigree holders, filled with odorous coffee. Napkins of the length of a towel, with embroidered golden ends, were unfolded for each, and laid upon our laps ; then the sugared coffee, without cream or spoons, was passed, followed with blocks of fig paste, handed to us in spoons.   Both were very good, and our hosts were kindly urgent that we should par-take again ; but glancing at the time, I found, to my horror, that we had but twenty minutes to reach the hotel, pack our trunks, and leave for Blidah ! I will not answer for our hasty adieu, or the impression left on our hosts by the hurried manner of our exit, but in a few words of thanks for their great kindness, and good wishes for the future of Mademoiselle the elder, we descended the stairway and literally rushed for hotel and railroad.'   Thus endeth the journal.

From this chance conversation thus reported, one may perceive what does not appear on the surface, that there is a good deal of inflammable material in Algiers growing out of French domination and these hetero-geneous elements of society.   A spark may make a tre-mendous explosion some day, even in the midst of Algiers. Let me illustrate. Last year a beautiful French girl, a child of eight, Rabel her name, and daughter of an engineer on the railroad, was found with her throat cut and evidences of attempted outrage, lying upon the basin or quay under the Rue Impératrice, leading to the railway.   The child had been sent at seven in the evening with an umbrella to the depot for her father. It was raining. Having missed her father and returning alone, she was thus assassinated, after an ineffectual attempt at rape. The indignation was wild. It was accounted the beginning of a plot among the indigenes to murder the children of the Europeans.

Algiers was on fire for a month. At length, after the police had in vain endeavoured to find out the villain, a Mohammedan, who had been educated in the French Moslem school, denounced his brother-in-law, one Ahmed ben Mustapha, called Sordo, an irreclaimable scoundrel. He had returned to his home, the evening of the murder, with the umbrella, and bloody. He had been seen to burn the umbrella and throw the debris into a cistern. There it was found, with two buttons of the child's dress. His sister saw them, and telling her husband, the conscience of the latter compelled him to make known the guilty one. Sordo was tried. I have read the pamphlet of his trial. It was before a jury, and after the French method. The Judge questioned the accused, and so implicated him in the toils of his own prevarications, that he was at once convicted. He suffered death. The reputation of the indigenes was not only saved, but it grew in favour, because of the conduct of the brother-in-law.

France does all she can to mitigate the prevailing prejudices; but more in a social than in a political and economical way. A story is told that, shortly after the French took Algiers city, the general in command gave a great ball and invited all the native inhabitants. Ices and wines were passed around. Directly dishes, cups, and glasses became scarce; and it was found that the natives, not familiar with the customs of the French, had 'put away' those articles under their sashes and clothes. Explanations followed, and the articles appeared!

The French appear to sneer at the Arab; but they are, nevertheless, utilizing the native elements here, as best they may. I have seen Arab women, under the direction of a zouave, breaking stones for the road. I have seen the Morocco people in this province of Oran, pumping water and making mortar for the

bridges. I have seen as many native soldiers as French.
I have seen the Arab tending sheep for the French
colonist — those large black-faced Southdowns, were
they not? At any rate, they are heavy with mutton
chops, and rich with a golden fleece. I have seen the
railroad track for many, many miles, lined with native
workmen. An Arab does the work among the horses
of the diligence. We see him as we ride along the
road for leagues toiling amidst the wheat and barley
all but ripe, or standing sentinel against the birds! As
this is novel to us, may I picture it? Indeed, I have
it engraved. At first I could not understand why some
half-dozen turbaned individuals were standing like
statues of ancient Romans, and all at once began hallooing at each other across fields of wheat. Then we
saw them with slings, swinging them with rare handicraft, round their heads, till—crack! and off went the
stone a half mile into a flock of hungry birds who
were in the wheat. The birds, affrighted, rise. The
Arab sounds his warning to his fellows—and, crack!
another sling from another quarter! Thus they attend
their fields, and in spite of the Koran, take a sling! I
understand now how young David became such a pro-
ficient, and why the giant of Gath fell—beneath a sling.
David, we learn, attended flocks; and no doubt was
employed to frighten crows from the corn. David, like
Samuel, when young was a good boy! He degenerated.

There is something so oriental, biblical, and interesting in this sleight of the sling, that I requested the
performer to perform on my account, rather than on
the bird business. He complied. I endeavour, in the
absence of a library, to recall the history of this
peculiar weapon of war, and, as I perceive, of industry.
But having no library—only my Bible—I must be
content with researches therein. It is a favourite
weapon of the Syrian shepherds; was adopted by the

fighting Jews; and, judging by its quality, it belongs to the light brigade. I learn from 'Judges' and 'Chronicles,' that the tribe of Benjamin, like my Arab friends, were experts in its use; and from 'Kings,' that it was used in attacking and defending towns, and in light skirmishing—as in this delicate warfare on the birds. From 'Maccabees,' I have the apocryphal account of its use by the Syrians, with some refinement in its manufacture. The simple sling, which I made when a boy, is the same which these Arabs have, viz., two strings to a leathern receptacle for the stone in the centre, termed the pan, or *caph*. At least Samuel the prophet says so; and Samuel ought to know. According to that good son of Hannah— whose name I have the honour to bear—the sling stones (1 Samuel xvii. 40) were selected for their smoothness. A bag was carried round the neck, as a sort of shot-pouch, or the pebbles were heaped up at the feet of the slinger. We know what sort of a stone David took for Goliath, and how it hurt. The old rude sculptures of Egypt show some of these slingers at their art; but I never understood how neatly they could be aimed, and how swiftly shot, until these Arabs taught us. Their skill rivals the classic fame of the dwellers in the Balearic isles.

Our ride from Milianah to Orleansville, runs through a valley, very wide and perfectly cultivated. The birds were busy at the grain and the Arabs at the birds. In England, some years ago, the farmers were in a rage at the little winged thieves, for destroying their crops, and commenced to destroy them. They soon found out their mistake. The birds were really conservative. The Algerines are on the same track. They are after the birds now; but when the worm comes, they will call for the birds. What a lesson here for—political and social sages!

There is nothing more beautiful than fields of grain
waving in the zephyrs, and just transmuting or trans-
muted into gold.   There are vast areas of grain here,
and where grain is not upon our Western route, the
Abyssinian jasmine, the white flox, the flaring red
poppy, the yellow, prickly broom, and opulent marigold, fill the meadows.   On all the route to Orleansville, and thence to Relizane — where the railroad
begins to run west to Oran—we perceive why Africa
is so fragrant and efflorescent, why her sands are
'golden' in another and better sense than those of
Australia, and why her fields are beautiful with
bountiful harvests.   We see, too, every few miles,
that the capital and enterprise of France is making
a railroad track along and parallel with this coast.
Every few miles, already erected and waiting, are
good substantial stone houses for the railway officers,
the only houses to be seen out of the little villages,
not counting the Arab huts and tents as such.
These railroad edifices are the nuclei of towns hereafter to be called into existence.   The land around
them is offered at about six dollars per acre ; and the
wonder is that so few seem attracted to these alluvial
plains.   Labour is plenty.   The Arab *will* work.   I
have already told how I have seen him at it.   Notwithstanding he is so berated by the French, he is used
and useful.   Our driver could not speak ill enough of
him.   I confess that, compared with the Kabyles, the
Arabs are not so industrious, but I find them much
better than they are painted.   Our driver talked of them
as Western people in American territories talk of the
Indians.   They are vermin, robbers, murderers, useless
cumberers of the earth, only fit to be exterminated.
But one must make allowances.   The Arab has lost
his pasturage.   His tenure of land, for his flocks,
never very strong, has been rendered very uncertain

He wanders very much now—still nomadic in his wits and about his fortune. He is in perpetual unrest as to his right of property. He may be lazy and may think that 'property is robbery,' and try, on this principle, to increase his store. He may grow tired of his nomadic flock breeding,

——'fold his tent, like an Arab,
And quietly *steal* away.'

Our driver approves of that verse in its most literal significance. He thinks the poet has travelled through Algiers. But I rather distrust that driver. He told me some monstrous stories yesterday about lions and tigers here. Perhaps he took me for a credulous Cockney. He did not frighten me; no, sir. I have seen only one jackal, and that in the Gorge of Chiffa; one porcupine, and one scorpion. Not a lion have I met 'in the way.' One leopard I saw at Fort Napoleon, but he was dead and skinned. In fact, it was the skin I saw. I had read Gerard, and how he killed one hundred and thirty lions here, and kept after them till he found honourable sepulture in the belly of the royal beast. I have read of a native lion-killer, Mustafa Somebody, who always killed his lions when they were gorged and dormant. He was a mighty hunter. That would be my style with a lion. But I was not prepared to hear our driver say, that along this route, not long since, while driving the diligence in the night, he saw what he thought was a cart and two lanterns before him, at which his horses refused to go on. He got out, went ahead; found two enormous lions in the road. Their eyes were the lanterns. He struck them with his whip, when they sneaked off! I asked the driver, gently, 'if he were native to Algiers?' 'No, Monsieur.' 'Are you then from France?' 'Oh! oui, Monsieur.' 'From Gascony?' His exploit was accounted for—he was from Gas—con—y!

But it is not long since lions did ravage these parts. The very cactus hedges and fortified farms show that protection from them to flocks and people was sought for.   Laws were made to help in eradicating the royal beast.   Bounties are yet paid for his death.   Near the Morocco line, lions and hyenas, panthers and jackals are common.   The Government pays eight dollars for a lion or panther scalp ; for a young lion, or young panther, three dollars ; for a hyena, one dollar, and for a young hyena or jackal, about thirty cents.   There are adventurous Arabs who make the hyena business quite lucrative.   I have seen no return of the bounties paid since 1863, when there were 1578 animals paid for.   There is no better sign than this in a new country.   I can remember, and I am not old, when Ohio paid bounties for wolf scalps.   But we have seen no wild beasts here of any account.   We have breakfasted on wild boar, and have seen the little wild piggies, all brown and striped, and very pretty for hogs.   They are easily tamed, and (like the Arab here and there) are found following the French about like a pet dog.

When Southern Algiers was fairly opening fifteen years ago, and the French soldiers were working their way to the inland, where the wild beasts did most abound, I remember to have seen in the Paris ' *Charivari*' some funny caricatures of the experiences of the green French conscript.   One retains its place in my memory.   It was that of the unsophisticated young soldier just arrived, who, seeing a lion asleep, ran up, and in a glee of satisfaction caught him by the scruff of the neck, and called out to his companion with great delight that he had him !   I do not remember what was the result.

In these letters I have said little about Roman or other remains, preferring to photograph the living present. But there is much of interest to the archæologist in Algiers. The museums of Spain, France, and Great Britain are full of Roman coins, arms, tombs and monuments. But in no part of the world out of Italy, are there such attractive evidences of the great power of the ancient world as here. Carthage leaves its impress both here and upon the opposite coast; but Rome outlives and outshines all nationalities, in her aqueducts, roads, amphitheatres, temples, and cities. Cherchell, shaken by an earthquake, is still visible under the waves of the sea, whose salt preserves the town, as ashes preserved Pompeii, and rock preserves Herculaneum. But there is one monument in Algiers, which not to see or mention would be like going to Egypt and forgetting the

Tomb of the Christian.

pyramids. I may be excused for presenting an engraving of it. It is called the 'Tomb of the Chris-

tian.' We have seen it from several points : on our
trip along the sea to the Trappists ; and to the Gorge
of Chiffa ; upon our visit to Blidah, and afterwards on
the way to Milianah. It is upon one of the lofty
mountain points, and cuts the sky so clear and clean
that it is ever beckoning your vision from distant
points. It is said by some to be the sepulchre of the
Mauritanian kings ; by others, to have been erected in
memory of a Vandal princess, who was converted to
Catholicism. Hence its name. It is 136 feet high.
Its base is polygonal, and is 195 feet in diameter.
Ionic columns support the tomb, which rises in circular
steps, in the form of a truncated cone. Various are
the stories told of this monument ; several romances
are woven about it, but they are all reduced to this,
by some scientific and learned researches : In the year
22 before Christ it was erected by Juba II., King of
this region. He had been brought up in Rome, had
visited Greece, and had travelled to Egypt. Hence
the base of the monument is Ionic, and the dome
like that of Egypt. It was erected in honour of his
wife, who was an Egyptian, and brought her religion
with her to this land. These revelations are the result
of the researches of Dr. McCarthy, an Irish savant
employed here by the Emperor for tracing and veri-
fying Cæsar's campaigns in Africa. But whoever
made this tomb, and for what purpose soever—
affection, vanity, or ambition—it is the pervading
presence of Algiers. It seems to follow you. It is
the genius of the country. It is like St. Peter's to
Rome, and Vesuvius to Naples. You cannot think of
Algiers, without recalling the vision of the Tomb of
the Christian.

There is another tomb in the province of Constantine very similar to it. It is no doubt a burial
monument of the Mauritanian kings. It is called the

Tombeau Madressen. When I presented it to the artist who has undertaken to illustrate this volume, Mr. Simpson, he at once recognized old friends in these monuments. He had never been in Algiers, but being skilled in such subjects, from his researches as the artist of the 'Illustrated London News,' in India, in the Crimea, in the Abyssinian war, and the East generally, he made me a note bearing on the matter. In it he writes that the Tomb of the Christian, and the *Tombeau Madressen* are so unlike any other architectural remains in the western parts of the Old World, that they have given considerable trouble to archæologists to explain them. Their identity with the Buddhist Topes, or Dagopas of ancient India, Ceylon, and similar erections of the present day in Ladak, Thibet, and other parts of the Himalayas, makes their explanation still more difficult. These Buddhist buildings may be described as 'round pyramids,' and this description exactly describes the tombs of Algeria. Drawings of these Buddhist Topes were in Mr. Simpson's Exhibition of Indian Drawings in London. The Buddhist Topes were tombs and temples combined.

I was surprised at the appearance of Orleansville. Although it had but a thousand people, nearly all were European. The hotel was Parisian. The country about it is hot, and the land denuded of trees. The water is not drinkable, but it serves to irrigate very well the plants within the walls. The River Chelif runs by the town. Its valley bursts through the mountains here towards the sea. Our Kabyle friends live in the mountains on the north. Many a good fight they have waged with the French hereabouts. The place is the old Castle Tingitii of the Romans. It has had much to do with the French and native wars. It is thoroughly defended by a wall and a fosse. It is like all the Algerine towns. They resemble, with the walls, towers, and gates, more a city of the Middle Ages than one of the ninteenth century. In 1843, when Marshal Bugeaud was rebuilding the town of Orleansville, the ancient basilica of St. Reparatus was discovered. This Christian church was built here under Roman rule in the third century. It has some rude mosaics, red, black, and white, and is ornamented with five inscriptions, of which two form a species of *abracadabra*. One of these is upon the words, '*Sancta Ecclesia*,' spelt *Eclesia* (with one *c*). It is a square, covered with letters. The letter S occupies the intersection of the two diagonals, or the centre of the seventh line. Starting from thence, you may read in every direction—'*Sancta Eclesia*'—repeated a great number of times. Here it is—

```
A I S E L C E C L E S I A
I S E L C E A E C L E S I
S E L C E A T A E C L E S
E L C E A T C T A E C L E
L C E A T C N C T A E C L
C E A T C N A N C T A E C
E A T C N A S A N C T A E
C E A T C N A N C T A E C
L C E A T C N C T A E C L
E L C E A T C T A E C L E
S E L C E A T A E C L E S
I S E L C E A E C L E S I
A I S E L C E C L E S I A
```

We found the most delightful accommodations at Orleansville; and took our dinner, as Plato advised his ideal republicans to dine, with sweet music—harp, violin, flute, and voice! At this feast every dish was at hand. The spirits of the old Roman epicures, who once visited the baths and waters of Algiers, seemed to have managed this menu for us. Among the dishes was the *kouskousou*, made of flour stirred up with sour milk, and garnished with beans. This is the national dish. It was tolerably good. Hunger made it seem so, at least.

But I have no complaint to make of the French cuisine. We have been nowhere yet that we could not get three or four good meals a day. While in England—at any English village inn, or even in the cities—you have only the mutton chop or beefsteak, either overdone and charred or underdone and raw; in France and its dependencies you can always have something—ragouts, soups, and meats, fruits, salads, and dessert—fresh, fine, and savoury. The native wine of this country, especially at Milianah, is good. It is like a light sherry, stiffly alcoholic, but very palatable. It smacks of the sun; it will bear improvement in the making. Unlike the full-bodied wines of Corsica, or the unfortified port of southern Spain, it is the very wine to make a breakfast, a lunch, or a dinner sparkle! The white wines of Algiers have a fine future, and great efforts are making for their perfection.

Nine more hours of carriage-riding from Orleansville, and we are at the railroad terminus. We changed our horses at a fortified auberge, half way. While waiting for the change we visited some Arab tents. I said that they were not enchanting. In fact, we were disenchanted by too close an observation. The tent fabric is much smoked. It may look better in the engraving. It is like a filthy old rag carpet, awkwardly stretched on poles. The entrance is very low—so is the tent. The women were present, and invited us to enter. There were four tents enclosed by a sort of fence of dried wild joujouba, a thorny bush common here on the moors. Inside of this fence were some goats, and sheep, and one donkey.

The latter was chasing the kids and lambs with the playful jocoseness of a kitten. There were three women—all wives—of different ages; one like a child. They were ornamented with immense steel necklaces and ear-rings, having little red coral charms pendent. On their foreheads were similar decorations—very coarse, and not unlike those of our squaws. The tent curtain was so low that we had almost to go on our knees to enter. We saw some matting, rolled up now, but reserved for night service. It was their bed. Two kids, tied together, occupied the centre of the tent. A woman was milking a goat. Charcoal embers were alive under an iron pot in one corner, and scattered around was an immense amount of rubbish. This was the tent whose romantic beauty fills the fancy of the Occident. One of the wives held upon her back her baby, its feet somehow supported in a belt at the back. The little one had to hang on to the mother while the mother milked the goat. I do not yet understand how that baby held on to that mother. The women seemed very proud of their ornaments. One of them who had the double ear-rings, at top and bottom of the ear, six inches round, was eager to take off her handkerchief and head-gear, and display her distorted ears and tawdry decorations. They have little use for fire, as they live on goats' milk. What cooking they do is done either with charcoal or the joujouba thorn-bush. By the way, as there were no hawthorns or wild briars in Syria, and as this thorny joujouba is native there, it must have been this to which the Psalmist referred, when he spoke of the laughter of a certain class being 'like the crackling of thorns under a pot.' David had been a nomad! He had, while watching his sheep, cooked an improvised mutton chop, over the crackling of the joujouba.

We are soon in the cars and on the way to Oran. The track is lined with the castor-oil plant. The plains are very level. As we leave for Spain to-night, we watch for the sea. We hope it is not covered with white caps. Alas! before we reach it we see the wind-mills about Oran swinging their long arms madly. We pass into Oran—past a fine cemetery full of dead Mohammedans and live cypresses.

The mountains about Oran are whitish with limestone, and seem almost chalky. The fields are decked to the last in colours. The city is splendidly fortified by Santa Cruz and Santa Gregory. Its rocky mountains are topped with castles, the scenes of many a fight between Spain and Algiers. The anchorage is fine. The jetty is made, like that at Marseilles, of artificial stone. Oran has 23,000 people —all busy. The Spaniards are most noticeable. Their black velvet hats everywhere appear. Some 3000 have emigrated hither within a few months, owing to the home taxes and the doubts they have of a stable government at home. In 1509, Cardinal Ximenes himself led the fight here against the Mo-hammedans. He conquered Oran. Hither the dis-graced Spanish nobles used to be exiled; and they had a saturnalia while here. Oran was called a little court. It was so lively when the Spaniards used it for a prison or exile, that it was sought for by those in search of enjoyment. Since the French have had it, it has been very gay. The population appear on the dash. Business is brisk. An earthquake now and then does something. Oran is an Algerian Chicago—in little. We see, as it were in a picture, military people of every uniform and grade, from the chasseur with his blue cap and red pants to the spahis with their red, flowing robes. Here are the Jews in sombre black and Jewesses in damask, gold, or silk; the Spaniards, from

the huertas of Andalusia, with their light shawls gracefully folded over their shoulders and a semi-turban
about their heads betraying the Moorish vicinage;
and finally, amidst a tableau of outside natives—on
donkeys and off, moving, noisy, and curious to look
at—we find the Moors themselves, careless, easy, fasti-
dious, stoical, and any other adjective to show how
utterly indifferent they seem to the active European
life around them.   The whole makes up a picture
quite equal to that at Malta or Algiers.   It would
require a Dutch artist of the old school to depict its
variety and detail.

Thus endeth my description of Algiers and the
strange vicissitudes and contrasts which make up its
life, scenery, history, and people.   I cannot close
without remarking upon the courtesy everywhere
extended to us by French and native, by officials and
peoples, by Kabyle and Arab.   Especially would I
remember the honest, faithful, and accomplished
master of the Hôtel d'Orient at Algiers.   Not because
he sent after me, through the country, to return
some money I had deposited, and which I had forgotten to draw; although that was very handsome
and indispensable.   But his hotel is a model.   If I
were to build one, I would send an architect to Algiers
to study its conveniences and proportions.   It rises up
before me, like a dream of the Orient.   Its court, so
airy and sweet with flowers; and its figures of Mahmoud the Great, Schamyl of Circassia, and Abd-el-
Kader—wrought in stone, within the enclosure—
these, if we had no other souvenirs, would preserve
Algiers in the amber of our memory.

Let not the traveller who crosses the ocean and
desires to see Oriental life rush off to Syria or Egypt
before he tries Algiers.   It is at the door of France.
Fifty hours from Marseilles in good weather will bring

him to Algiers, and less than half that will take him
from Malaga or Carthagena to Oran. I would advise
him not to go in winter. Some friends who were in
Algiers in February were nearly four weeks weather-
bound. The sea is uncertain. If you put out,
you may have to go back or run for refuge into
another port. So at Corsica in winter. But in April
there is no such risk. Take little or no baggage with
you. Be prepared, if you can, to go in a company of
four or five, to make up a carriage-load, and thus save
expense and the too rigorous travel of a diligence.
Besides, in these countries of the sun, the diligence runs
generally at night; and you miss so much. Arm yourself with plenty of flea powder! Do not be afraid of
being under an umbrella! The sun is very perpendicular and warm in its attentions. The turban has
an object. If you lunch out of doors (as we did
often) and upon the ground—find out first if the
ground has a good or a bad reputation for—scorpions; and if you can secure a companion—as accomplished as Dr. Bennet in the sciences, botanical and
otherwise, and as social in the amenities of life—do so.
But you will have to look long and far to find him.

We depart in the vessel for Spain at 4 o'clock. We
are to see the smoke of the silver and lead mines of
Carthagena at breakfast to-morrow. The little steamer
rides very lightly on the waves within the mole at
Oran. We have inspected her. The sea looks squally.
What will the vessel do in the open sea? We are
attended, and so is our baggage, to the boat by an
indigent indigene—Mahmoud, *not* Hahmoud. We
were compelled to leave the latter at Algiers. Hahmoud wears a French sack-coat and Turkish baggy
breeches. He is the incarnation of the conflict
between the two civilizations. His coat is of the
Occident, and it looks bright and new. His breeches

are of the Orient, and are seedy, patched, and ready
for the rag-basket, the paper-mill, or as manure for
the olive! Thus passeth away the glory of the Deys,
the Pirates, the Moors, and the Arabs before France!
Adieu to the African Orient! Hail to Spain! old
Spain no more, but *Hispania rediviva,* under the
revolution!

CHAPTER XIV.

SPAIN— VEGETABLE SURPRISES AND DISAP-POINTMENTS.

' Once more—once more! in dust and gore to ruin must thou reel!
In vain—in vain thou tearest the sand with furious heel—
In vain—in vain, thou noble beast! I see, I see thee stagger,
Now keen and cold thy neck must hold the stern Alcaydé's dagger!'

LOCKHART'S 'Spanish Ballads.'

WE steamed from Oran, in Algiers, into the port of Carthagena through a heavy sea. The Spanish coast might as well have been in the moon, for it was utterly bleak, wood-less, and leafless. Dry, white limestone mountains, like those we left at Oran, stand along the coast, and make it so forbidding that we wonder the Cartha-ginians ever thought it so inviting, or that for its conquest the Romans under Scipio waged such ter-rific wars.

The port presents a narrow gateway between the lofty rocks on either side, which are impregnably fortified. As we go between them, an officer boards our vessel, and we sway on the waves while we pass through his ordeal. He examines us for our health. As we steam up the bay, we perceive before us the great old fort built by the Carthaginians. It was captured by Scipio Africanus, 210 B.C. At that time, Carthagena, or New Carthage, was one of the richest cities of the world. Not for its glass, cordage, and fisheries—which now furnish employment to most of the population—but for its silver and lead mines, from which, as we sail

up the bay, the smoke streams. It comes from pipes
or chimneys, out of the bare mountains, under which
these old and long tenantless mines are reworked. The
Romans used the ancestors of the proud Hidalgos as
slaves in the mines, and sent their convicts hither.
Owing to Gothic destruction and the decay of European silver mining, incident to the discovery of the
richer Americas in the fifteenth century, these mines
lay dormant until 1839. Then joint-stock companies
began to work them, and to be enriched by the experiment. I was told by the French Consul, who
was on our steamer, that fabulous sums are made out
of these mines. The mining has given rise to a considerable commerce with England. Three hundred
collier ships come here every year with coal to be used
in the explorations.

When we landed, we found a crowd ready to receive us. They fought for the honour and the price
of taking our baggage to the hotel. The solitary
hotel—and that very poor—received us after many
hours of travail, on being delivered from the custom-
house. Ours was a French steamer, but we found that
the new Spanish Government had relieved it and all
other foreign craft of port dues and other restrictions.
We did not wait long for our permit. But the time
is not lost. We make observations. The people
look odd and picturesque. Every one has a touch
of the Moor: not the turban, but a coloured handkerchief in its stead tied about their head; not the
burnous, but the mantle, cross-barred, like Scotch
plaid, or red and flaming, and always worn with grace;
their feet, not entirely shoeless, but a sort of Arab
sandal, made with hemp, and tied with strings over the
bare feet, generally all the toes out of doors but the
big one! The pretentious sash is inevitable, but
how can I describe that peculiar, black, rusty velvet

hat? You may, to see it with your mind, imagine a round platter, turned up three inches or more on the rim, and upon it a conical-shaped bowl, turned upside down! If this description is imperfect, some of my illustrations will suffice. We perceive, occasionally, a dark rosette upon the hat. We notice, also, that we are in the land of the crazy Don Quixote; for windmills decorate all the breezy eminences. It begins to rain. No sunbeams here. The streets are full of people, and not of the best class. Beggars abound. Walls shut in the city. We learn that the gates are locked at night. The town is on a plain, surrounded by sierras. This plain is in process of rescue from a marsh, which used to breed malaria. Livy records that there was, in his day, round a part of these walls, a lake which had been there when Scipio took the city. This, doubtless, once covered the present marsh land.

Strange to say, in one of the churches into which we ventured, the modern Carthagenians were preparing to celebrate the deeds of those gallant Spaniards who fell in the recent war with Peru! Thus patriotic and high-spirited, yet they are actually demolishing the magnificent Carthaginian fort upon the mountain above the town—the most commanding object of interest around or in the city. It is already half torn down to make houses withal! All the world cried out when, years ago, Mehemet Ali *talked* of using the Pyramids for building purposes; but here, in this century, the most prominent monument of Carthaginian power and commercial affluence, and of Roman prowess and civilized sway, is already half destroyed. What are Rome and Carthage to these degenerate sons of the proud discoverers and conquerors of the New World? I will not be too hasty about answering this question.

We go out of Carthagena on a railroad; we wait at

12

the depôt two hours before the train is ready. Here
the beggars congregate, and, indeed, all other classes
of the inhabitants. The baths of Caracalla were not
more the resort of Roman quidnuncs than is the depôt
for the Carthagenians! We at length leave the city.
We are in the country. We look about for the
promised Paradise. The gardens of Spain are here!
So we have somehow heard. We look in vain. Are
we in the moon? Is this a land of ashes and scoriæ
of extinct volcanoes? Where are the orange groves,
the vines, the pomegranates? Have we become inverted? This is Africa, with her proverbial aridity;
and what seemed Africa to us, with her glorious
luxuriance of growth, was Spain! Well, both the
Doctor and myself were puzzled. We expected to find
arid, white, yellow, bare deserts in Algiers; but they
are here! Not deserts of sand, but deserts of rock,
lime, or clay, dazzling to the eye and relieved by no
green. It is as if all this country for thirty miles, with
a few exceptional spots—oases in a desert of dry, baked
lime or clay—was an extensive, old, used-up brick-
yard! Yet, still stranger, the vast area between the
bleak mountains is ploughed ground. It is ploughed
and planted. What does this signify? Are the peasants
waiting for crops from the sterile soil? They will wait
in vain. It is May, and the harvest-time is nearly
here. Are they yet to sow other seed? Surely not;
for the blister of summer will soon be on the breast
of the earth. By no drawing, squeezing or sucking,
can milk be pressed or stripped from these dry
earthly udders. What does it mean? The Doctor
suggests that he has read that rice is here raised. Very
well; but I suggest that there is no provision here for
water, and that rice requires water! Then we are
silent—more and more puzzled. We ask some people
in the cars. They do not know. Here is an immense

country, geometrically divided into lots, and subdivided into smaller lots, all as if worked industriously and as if for a good crop, and not a petal of flower, leaf of clover, spear of wheat, or any other green and growing thing, between that dry white soil and the bright, blue, blazing sky! We look for the horizon and its verdant woods; nothing there but volcanic rocks and ashes. We look at our feet for something of vegetable life; not a blade is drawn in defence of the 'garden of Europe'! Not until we reached Murcia did we solve our sphinx riddle. Our Œdipus was an old peasant, who told us that every year they ploughed and planted this whole plain, from Carthagena to Murcia; ploughed it all carefully and planted it religiously with grain of all kinds; hoping against hope, that a few days of rain might—possibly would— come to gladden their hearts and gratify their labour; that every once in a while, say in three or four years, there *was* rain enough to make something *shoot* besides soldiers; and a little to them was so much. They laboured and waited for the possibilities of results. Farmers of Pennsylvania and Kansas, or of rock-ribbed New England! What a commentary is here! Never, *never* repine against your 'most blessed condition.' The wonder of this peculiar land grows on me, as I move on toward Murcia.

I was not more surprised at certain phases of African life and scenery than I am at the appearance of this portion of Spain. Whether I have been so ignorant, or whether the writers on Spain have skipped these parts, or, having seen them, set them down as uninteresting—whatever is the reason, I was utterly unprepared for what I have observed. Indeed, if Buckle could arise and see what is going on in Spain, in a social and political way, he would rewrite a large part of his second volume, even as I have had

to unlearn and then relearn much about Spain.
I refer to the Provinces of Murcia and Valencia,
or rather to the country from Carthagena, through
Murcia and Alicante to Valencia, on the eastern
shore of the peninsula, and looking forth towards
the Balearic Islands. We have gone through these
regions nearly the whole route by carriage. The
latter here is very little slower than the rail, and much
better as a point of observation. We come hither
fresh from Algiers. We come to the old land of the
Moors from the new land of the Moors; and, although
three and a half centuries have elapsed since the
Spaniards drove the Moors hence, time has not erased
the eminent marks of Moorish manners and civilization. If this be the case here, what shall we not find
in Andalusia?

Still travelling with my former companion, Dr.
Henry Bennet, I have had my eye and mind directed
to the soil and its productions, as well for themselves
as for the proof they give of the climate most fit for
human health. We have continued our search for
'winter sunbeams' into the month of May and into
the land of Spain. The great desideratum which
the Doctor seeks is a winter station for consumptive
patients, or others affected with pulmonary and throat
troubles. He would, if possible, find something better
than the Riviera. These cities of Spain are favourites
with many medical men and invalids; and we visit
them to understand why. The great object of our
trip is to know—as Milton wrote it, in 'Paradise Lost'
—where we can find a Paradise regained, far from the
harsh regions of the North—

> ' By what means to shun
> The inclement seasons, rain, ice, hail, and snow;'

and at the same time that we ascertain the facts, profit

by the experience in our own bodies—enjoy in this purer clime, and in this chivalric land, the delights of external beauty and historic memories.

Coming to Southern Spain, one may be sure, even in winter, to find Apollo with a quiver full of silver arrows, shooting them from a clear blue sky at our mother earth. The air is dry and light. The sea of this coast is so calm that the sailors say, it is for women to navigate. The rain, if it comes at all, comes at long intervals; but the hot mountains and the demands of vegetation make water so scarce and valuable that it seems hidden, like precious jewels, from ordinary eyes.

I have said that, instead of Africa in Algiers, we found imaginary Africa here; and instead of Eden in Spain, we found it in Africa, and that this was a mystery. In reading Washington Irving over again I see that the same thoughts impressed him. I am, therefore, in good company, if I have been ignorant. In his journey to Grenada, he says : ' Many are apt to picture Spain to their imaginations as a soft Southern region, decked out with all the luxuriant charms of voluptuous Italy. On the contrary, though there are exceptions in some of the maritime provinces, yet for the greater part it is a stern, melancholy country, with rugged mountains and long sweeping plains, destitute of trees, and indescribably silent and lonesome, partaking of the *savage and solitary character of Africa !*' Had Irving visited Algiers where we have been—at least before the army of desert locusts came to devastate its luxuriant fields—he would not have spoken of Africa, except as I have, in bold contrast to the melancholy destitution of Spain. He speaks of the absence of singing birds in Spain as a consequence of the want of groves and hedges ; but Algiers is full of songsters, because so full of shrubs and trees. He

sees the vulture and eagle wheeling around the cliffs
of Spain; we saw these birds of prey only near the
desert. He sees in Spain, in lieu of the softer charms
of ornamental cultivation, a noble severity of scenery,
which makes the boundless wastes as solemn as the
ocean. All this and more we saw in the provinces of
Murcia and Valencia—as Washington Irving saw
them in Andalusia. But we remembered as we rode
that irrigation had made the waste places blossom as
the rose, and we waited for the vision!

As we go up towards Murcia, and rise into or upon
the plateau (all Spain is a plateau, ridged with sierras)
two or three thousand feet above the level of the sea,
we begin to perceive a palm or so, and find a fig or
so, until all at once the City of Orihuela appears. It
is surrounded by rocks and mountains; but there is a
river near. It looks Oriental—Moorish. There are
20,000 people here, and their plain laps up the River
Segura like a wild, thirsty animal. The dwellings are
low, and all marked with a cross, though of Moorish
style; but the Gothic cathedral towers above all.
Here we find what water is. It makes the pomegranate blush and it ripens the fig. The olive grows
darker in shade and leaf. The orange looks large,
but, as we soon find, its size is due to the thick rind,
not the large rich inside—for, unlike what it is at
Blidah, the orange here is more for commerce than
for eating. Almonds are nearly ripe; merino sheep
appear; palms grow more numerous and stately; the
donkey is larger and white; the horse is stout and
elegant; the women look brighter, and their dark
optics dance; the costumes grow very peculiar and
gay; the plaintive songs we have heard sung among
the Arabs we hear repeated—how strangely—even
here after three hundred and fifty years; and these
evidences of prosperity, contentment, and joyance

follow us, more or less, till we stand within the
charmed circle of emerald which environs the proud
old city of Murcia! Not that the bleak, bare mountains, and white, arid plains are not still our companions all along the route from Carthagena. But, as
a relief, this terrible sterility is beginning, under the
system of hydraulics, pursued here yet, and which the
Moors began, to give way to something like vegetable
vitality; but no grand garden of Spain yet—at least,
none such as we fancied, about these realms of the sun.

Why all this is, I have shown already in my chapters
from Algiers. The rain from east and north is here
dissipated by these heated mountains or sucked down
south by the desert to fall upon Algiers. Africa
becomes an Eden and Spain becomes a desert. Only
irrigation rescues the latter. Here water is a creative
power. We in America do not know it, except as a
motive power. We never feel the lack of water, not
even in our whisky; and we do not know the want
of rivers. Let not the harsh North repine. The land of
'winter sunbeams,' which gives the almond and vine,
orange and date, is not conquered by man, except
through labour. The sun may do much, water and
soil much, but it is man who combines and produces.
Paul may plant; his plants will wilt and die, because
there is no root. Call in Apollos and much water,
and then God will give the increase. Tempes do not
come spontaneously to the surface; but man makes
them with water, out of the dry ribs of the torpid
earth. Tempe and the Peneus are inseparable.

I am not sorry that we halted for Sunday at
Murcia. This splendid city is seldom visited by
tourists. It is an exclusive and lordly old Spanish
place of 60,000 inhabitants. Our posada, or hotel,
was once a Moorish alcazar, or palace. We could perceive the tracery of the Byzantine architecture and the

gilt of former splendour. Murcia is the Murgi of the Romans. It was once the capital of an ancient kingdom of the same name. The Moors built the city from the Roman ruins. There has been much fighting here between Spanish and Moor and Spanish and French. The city is even now full of soldiers. In the revolution of last September, not a soldier would raise a voice, much less a musket, for the Queen. She died out, not because the people here were republican or revolutionist, but by the general disgust entertained for her character. The houses of Murcia are painted in yellow or pink. Every window has a balcony, every balcony has a beauty, and every beauty has a bouquet; and, when night comes, the city is all beatitude and all a-twang with guitars. Serenades sound from every quarter and in every street. But most, and above all, as if it were an elegy and requiem of the dead Moors, there is that same sad song—drawling, mournful, everywhere heard, from palace and hut—which we heard the Arab Dervishes sing when they chanted the Koran!

I confess to various essays in regard to this song—first, to understand it; next, to resolve it into music; and then to reconcile it with what I heard in Algiers among the Moors. I failed in each essay. Picking up Irving, however, I read from him what it meant; but I do not believe that he heard it in Murcia or Valencia. It has more significance than he gives to it, as the simple music of muleteer, bandit, and contra-bandista. He describes it, or rather, he describes all Spanish song, as rude and simple with but few in-flexions. These the singer chaunts forth with a loud voice, and long, drawling cadence. The couplets are romances about Moors, or some saintly legend or love ditty, or some ballad about a bold contrabandista or hardy bandolero. The mule bell or the guitar is the ac-

companiment. This very nearly describes what I desire.
But I affirm that the music which I heard all through
Murcia—city and province—and in Valencia from
every girl or boy, from sunburnt poverty or grandiose
elegance, was the same music, if not the same words,
we heard from the diabolical dervishes who, in their
religious ecstasies, swallowed scorpions in Algiers!
What then? I am not to be led away by the click of
the castanet or the sound of the guitar to other and
more Spanish airs—not until I sift this song of songs,
this sadness of all song. I am satisfied that this universal music is the unchanged Arabic *Gaunia*. It is very
like the Corsican *vocero*, which I have already noticed.
It begins and ends with an 'ay !' or a sigh. It is all love.
If not for a lady-love, for something beloved: home
or horse, country or kin, or a mule, perhaps. Our
driver and his 'mozo' (carriage imp) sang it into
Alicante, and improvised their affectionate souvenirs
of that place into the national music. There is a
chorus at the end of the verse, sung with a long,
dilatory, tearful plaint, that spontaneously opens the
lachrymal duct, without regard to the sentiment sung.
Generally this lament is sung to the guitar. As a
fierce and inexorable economist, I find the guitar to
be a nuisance. The harsh north wind, the constant
drought, the hot sun, the bull fight, each and all may
be counted enemies to Spanish prosperity. Buckle
has, after the manner of his philosophy, gathered in-
ductively a collection of historical and scientific facts,
and has concluded, deductively, that the history of
Spain has been a failure, by reason of her violation of
the laws of national improvement. He has shown that
physical phenomena, by inflaming the imagination and
preventing analysis, have operated on Spain as upon
other tropical, volcanic, and epidemical lands, to its
disadvantage and decay. He evidences the heat and

dryness of the soil, the deepness of the river beds, which forbid irrigation, and the instability of the pastoral life, as reasons for the improgressiveness of Spain; but he has never mentioned the guitar. Yet is the guitar the especial devil of idleness;—for does it not beget singing and dancing? Once set it a strumming, and all unpin their eyelids at night and open their ears by day. Labour, duty, and patriotism, are all forgotten in its music. The guitar is an old deceiver. Don Juan was not the first performer, nor the Spanish señoritas the first to lean from balconies to hear. It is found on the Egyptian tomb of four thousand years ago—ribboned in marble over the enduring neck of some petrified, love-sick, young Pharaoh or Ramesis. It is the kinoor of the desert and Orient; the Greek kithara; the guitarre, githome, guitume, and banjo of the universal minstrel! It is not always the magical guitar described by Shelley in his poem; which was taught to reply to Love's questions, to whisper enamoured tones, and sweet oracles of woods and dells; or the harmonies of plains and skies, forests and mountains, fountains and echoes, the notes of rills, melodies of birds and bees, murmuring of seas, and rain, and, finer still, the soft airs of evening dew; but the Spanish guitar had one of the faculties which Shelley found in his enchanted instrument:—

> 'It talks according to the wit
> Of its companions,'

and its tinkling talk consists in making the castanet, feet, voice, body, spirit, and soul of its company harmonize with its service!

All Murcia' was musical with this minstrelsy. I know that it has been called the dullest city in Spain. It is not! Go into its great cathedral. Do not stop to examine whether it be Corinthian or Composite, or both. Never mind the carvings or the relics. Fear not because the earthquake has made cracks in the tower and façade. Go up into the belfry —rising, compartment up and out of compartment, like a telescope drawn out, and crowned, like all the cathedrals of this part of Spain, with a blue dome, shining beauteous in the sun. Then look about you. See what water will do for the dead earth; palms, standing up and aloof from the other vegetable glories, only made more beautiful because surrounded by such fire-tossed and twisted rocks as those which bound the view of the horizon, and bind the volume of Nature here presented.

Is Murcia dull? Come with us on Sunday to see the bull fight. It is said that the Irish make bulls, the Spaniards kill them, and the English eat them. Not to see the butchery, is not to see Spain. Murcia is alive with the occasion. Do you call it dull to see several thousand people streaming down the avenues to the great amphitheatre, following the soldiers and the bands, and all intent on the 'blood of bulls'? Come in : admission is only about ten cents apiece. Look about : a clean arena; a red flag on a pole in the centre. The bands strike up 'circus tunes,' in strange discord with the sweet and holy chimes of the Cathedral bells. Then the Marseillaise Hymn is played. A young Murcian soldier tells me that it is only lately that music of that kind has been permitted. I surmised as much. Then enter five persons out of the gates below the seats, dressed in tights, red and blue, spangled, like circus-actors. One is on a horse and has a pike. He wears an ostrich feather, and yellow breeches of leather. Look around the theatre : it is a gay scene! The sun glorifies it; the costumes are very pleasing and various. Directly the bull comes out. He is not quite what we expected; rather small and young. He is welcomed by shouts. Young Spain in the lower tiers is very demonstrative. Three dignitaries enter a box, which is draped in red. It is the mayor and the judges. Old women are selling and crying 'water'; and boys, American peanuts. The bull trots about rather surprised. The five fighters on foot begin to run at him, shake their red inflammatory mantles at him; finally thrust little barbed sticks into his hide. These poles are ornamented with coloured filigree paper, but have a sharp nail, so as to make Taurus bleed and dance merrily. It is very pretty sport—very. Directly Taurus gets mad. He makes a moan. He bellows. He dashes at the men right and left. They scrape sand lively to jump over the red boards of the arena. He clears the ring; and then, as if thoroughly ashamed of the business, and mortified at the tricks of rational beings, he trots round to the door, now closed, where he came in, and bellows and begs to go out! This goes on for some time, with several bulls. Several times Mr. Matador came near taking—a horn. I should not have cared. I was for the bull. Then in marched the five and set up four posts, and within the posts the chief stood. The four men had to guard the chief. That was the little game. The bull knocked the posts about their ears, whereat the crowd roared. Then, in came another bull, very lively and gamey. The matadors worry him with their garments; then with their pointed poles; then the 'man on horseback,' the picador, with a pike appears. His horse trembles. It is a poor, black, half-blinded, Rosinante; and no bloody spurring can bring him within reach of Taurus. At last the judges give the signal to kill the bull. The crowd cry out, 'No! no!' He has fought so well. I begin to like the crowd for that. Then begins more teasing, until, the bull being thoroughly aroused, panting and glaring, with threads of froth hanging from his mouth, and his neck all bleeding with the pricking of the barbed sticks, they turn in on him half-a-dozen of white and brown bull dogs. Amid roars of fun—it was very funny, oh very! for the bull,—the dogs hang on to his flanks, throat, horns, ears, tongue, and nose. He flings them about in the air and under him. It was such sport! A horn comes off, all bloody, then an ear. The foaming tongue hangs out. The bull is not down yet. The dogs cannot do it. Taurus has won his life. He wants to go out. He again thinks he can go out as he came in. He does not. His reasoning is in vain. Amid shouts, the order is given to kill, and, after five thrusts, the sword at last goes into the lungs, and the Bovine Gladiator, butchered to make a Murcian holiday, falls dead under the repeated stroke of the sword, to which is added, as a coup-de-grace, some dashes of the knife. This is sport! During the last of it the ring is opened, and all the youngsters of Murcia rush into the arena hallooing, and following with their yells the staggering, dying brute. The child thus trained to cruelty, is he not the father of the brutal man?

Mr. Buckle might have laid a little more emphasis on this 'Aspect' of Spanish nature, in his conclusions as to its character. Mr. Cobden has said, that so long as this continues to be the national sport of high and low, so long will Spaniards be indifferent to human life, and have their civil contests marked with displays of cruelty which make men shudder.

Sick, *ad nauseam*, of this rational and national sport, we left. We entered a Catholic Church next door, where the preacher was doing his best to speak of the gentle Saviour. His periods had been pointed and rounded with the shouts of the bull ring. Truly, it was hard thus to preach, and it seemed as if the preacher laboured. His congregation were sitting

on matting on the floor, and in the dim religious
light of the church. They were nearly all women,
and, as is the custom, were dressed in black. They
seemed like people of another world. Certainly they
did not belong to the world of Spain, as it outwardly
keeps its Sabbath!

Since writing the foregoing, I have seen *a real,
bloody bull fight!* I am compelled to interpolate a
description of it, for the reason that the fight at .
Murcia gives no idea of the sport. We were told
at Murcia that bull-fighting there was child's play
compared to that at Madrid. So it proved. We
now have had enough of it. A more brutal, barbarous, horrible thing never was conceived or executed
under God's blue sky. But I had to see it, though I
could not sit it out; and the ladies, who were with
me, retired on the death of the third horse and second
bull—retired as colourless as a white handkerchief.
There were no horses killed at Murcia, and there were
no bull-dogs at Madrid; but at Madrid there were
superadded the agonies and death of a score of innocent
horses and six brave bulls.

The arena at Madrid is much larger than at
Murcia. It is the Roman theatre over again—the
Coliseum in resurrection. It was packed with 15,000
people. The Queen's box, over which floated the
yellow and red flag of Spain, was empty. Spain does
not know it, but she is a republic—provisionally.
The empty box showed it. The first bull let into the
ring was a splendid one. The Murcian was a miserable
mouse beside the Madrid monster. The latter was of
the Andalusian kind—brown, big-necked, diabolically
belligerent, just such as Irving described as having
seen, in his trip to Grenada from Seville, among the
Andalusian mountains, when, in 'winding through
the narrow valleys, he was startled by a hoarse

bellowing, and beheld above him, on some green fold of the mountain side, a herd of fierce Andalusian bulls, destined for the combat of the arena.' Irving felt an '*agreeable horror* in thus contemplating near at hand these terrific animals, clothed with tremendous strength, and ranging their native pastures in untamed wildness, strangers almost to the face of man, and knowing no one but the solitary herdsman who attends upon them, and even he, at times, not daring to venture to approach them.' I confess that the same horror surprised me when the terrific animal appeared in the Madrid arena; but it was not 'agreeable.' By being introduced all at once to 15,000 people—all hallooing like savages—the menacing aspect of the Andalusian monster was not relieved. When the picadores thrust into him their pikes from their horses, you should have seen how he tossed the horses upon his horns, upturning them and goring them into a speedy death. The horsemen themselves were tumbled under bull and horse, but, as they were padded or in mail, they escaped unhurt, and were lifted again into the saddle, while others called off the bull by all the arts known to the ring. These horses are blindfolded, at least on one side; and are pushed up, with spur and lash, to the bull. Sometimes they show fight, with their heels; but generally die without a sign. One of the horses survived the loss of his insides for some time, yet the impatient crowd shouted for fresh horses. Fresh horses were forthcoming; but they cost little. They are but the frames—the gristle and bone of the horse; not horse-flesh and spirit. Such horses may be seen in a London cab or New York omnibus, kept up by the shafts. Now and then one may show a little breeding.

The ribboned daggers were inserted in the bleeding neck of the bull, to goad him on to fury. How he

pawed the dust and bellowed; how he raved madly
after his tormentors; how he shook his neck to loosen
the fangs of the ribboned daggers; how, at last, the
trumpet sounded for the espada, or swordsman, to come
forth and kill; how the crowd hallooed as the bull chased
the spangled men in the rings; how they saluted those
showing the white feather—calling them 'blackguards,'
'thieves,' 'rascals,' 'dogs;' how at last the espada,
dressed in his black velvet, embroidered pants and
jacket, his head queued and chignoned—Tatto, by
name, and famous for his skill—came forth and made
his speech to the judges, promising to kill the bull in
the name of the New Constitution; and how the bull
fought him for half-an-hour, till at last Tatto (since
wounded and nearly killed), under his *red flag* hiding
his sword, drew the bull's eye to the flag, and, quickly
drawing the sword from beneath the flag, struck him to
the lung vitally; how the bull died—first falling on his
knee, then down, then up at his enemies, and at last
tumbled into the dust, his magnificent strength, derived from the Andalusian mountains, all gone; and
how he was dragged out by six mules dressed in red
livery, amidst the lash of the dozen drivers and the
shouts of the Spaniards;—all these have been often
told, as they are often repeated here, to the delight of
the native and to the horror and disgust of the foreign
part of the population.

　　Six great bulls honoured God's day by their martyr-
dom. The blood, made so familiar to Spanish eyes,
may return to plague some one. There is a terrible
lesson to be read in these fights. I am a Puritan—
when it comes to a bull fight. If this people—who
are illustrating the lessons here taught upon the plaza
at Cuba, where the garotte does its bloody, cruel work
upon its victim—expect to have that well-regulated
liberty which will last, they had better abolish this

bloody diabolism; and spend the money given for bulls and horses intended for immolation to educate the masses of the people and prepare for that consummate freedom which I trust is yet to dawn on Spain!

I confess it was a relief to me when, after writing the foregoing, I opened the volume of a friend, Mr. Henry Blackburn, and found that the spirit of Cervantes was not dead in Spain; that if not in print, yet in acts, there was humour or satire enough to 'take off' the abominable practices of the bull ring. He pictures so well the bull fights in Seville, which burlesque the scenes I have described, that, as a relief from the horrors of the reality, and with his consent, I copy his description of the caricature. As a true artist, he pictures the intrepid señorita, or tauro-maniac who was advertised to face the bull in her bloomer costume, with cap and red spangled tunic. Having done that, he comes to what I may call the 'Tale of the Tub.' After the señorita's grace to the president and audience, who receive her grandly, she is placed in a big tub. There she stands, up to her armpits, waving her barbed darts till the bull is let in. The animal lowers his head, and after some hesitation and skirmishing, rushes at the tub. The señorita curls or coils herself inside unhurt. 'The bull,' says the author, 'soon began to get angry, at last caught up the barrel on his horns, and rushed bellowing round the ring. It looked serious for the tenant. There was a general rush of "banderillos" and "chulos" to the rescue, but some minutes elapsed before they could surround the bull and release the performer from her perilous position. When extricated she was smuggled ignominiously out of the arena, and we saw the brave señorita no more; the bull was not killed, but "bundled" out of the ring.

'The next act was "Skittles." Nine negroes ("Bedouins of the Desert"), dressed grotesquely, stood up like "ninepins," within a few feet of each other, and a frisky *novillado*, or young bull, was let in to knock them over. The bull struck out right and left, and soon overturned them all. They then sat in rows in chairs, and were again bowled over, to the delight of the assembly. This was great fun, and was repeated several times; the bull liked it, the "ninepins" *seemed* to like it, and the people gloried in it.

'The third act was a burlesque of the "picadores," a grotesque but a sadder sight. Five poor men in rags, who, for the sake of two or three reals, allowed themselves to be mounted on donkeys and receive the charge of the bull. Here they come in close phalanx, cheered by at least 5000 people; the five donkeys with their ears well forward, and their tails set closely between their legs; the ragged "picadores," without saddle or bridle, riding with a jaunty air, and a grim smile on their dirty faces, that was comical in the extreme. The gates are opened again, and the bull goes to work. He charges them at once, but they are so closely packed that they resist the shock and the bull retires. He has broken a leg of one of the poor animals, but the riders tie it up with a handkerchief, and continue marching slowly round, keeping well together as their only chance. A few more charges and down they all go. The men run for their lives and leap the barriers, and the donkeys are thrown up in the air!'

Do not think, however, that the masses of the Spanish people waste their time or substance on these barbaric displays. I know better. The working people of Spain, especially the agriculturists, are kind, clement, industrious, just, sober, and courteous. I do not accept one half of the stories told of their superstition and cruelty. I have read in the 'Times' an

account given by an English barrister, who, near
Murcia, was attending to some very litigious business,
and who was seized and nearly beaten to a jelly by
the peasantry and people. He was, according to his
account, believed to be a child-thief, and he asserts
that the reason for his seizure was that the common
people believe that children are stolen for their entrails,
which are used to grease the telegraph wires! After
reading his account, and from some communication
with the people in and around Murcia, I believe that
the lawyer was seized by private parties interested in
closing his legal career in Spain.

The working people of Spain I have commended.
They must not be confounded with the riff-raff of the
cities, or with the effete, corrupt, gangrened 'snobs
and nobs,' by which I mean the hidalgos and nobility
who have lived on the industry and production of the
honest and forbearing 'common people' so long, and
who are now either absent from Spain when her trial
comes, or else hatching plots in the interest of Isabella
or Don Carlos, or some one else, to frustrate the
majestic will of the people. Of that, when I come to
politics. I am tempted even now to speak of these
things, for I have been reminded at every turn in
Spain that her grandees—those who fawned on the
Queen in power, and who ran away from Spain when
she was 'turned out'—do not represent the staunch
and generous elements of the Spanish character. If I
may be pardoned for again referring to Washington
Irving, I find, in looking at the Spanish people, what he
described—that the severity of the scenery, which I have
endeavoured to depicture, is in unison with the attributes of the people. He said that he better understood
the proud, hardy, frugal, and abstemious Spaniard, his
manly defiance of hardships and contempt of effeminate indulgences, since he had seen the country the

Spaniard inhabited! There is a stern and simple sublimity about these dry, calcined, white and yellow, sun-dried and heated mountains, which makes one feel that the people who have subdued them are worthy of a better fate than to be ridden by a *blasé* aristocracy, booted and spurred for their subjection. Five per cent. of the Spanish aristocracy and plutocracy may be trusted; ninety-five per cent. are a mass of putrescence. The revolution ought to bury it in lime.

At Murcia we spent an afternoon agreeably in visiting a model country villa, the residence of a former English Minister to Spain, Lord Howden. His Lordship does not live there now, but at Bayonne. He is an eccentric man in some respects, and not unpleasantly peculiar in his love for an out-of-door life and displays of fruit and flower. Besides, he demonstrated his independence when advanced in years by marrying a beautiful Spanish actress, with whom he fell in love. It was January loving May; and, as a consequence, January made of his home here a perpetual May. We visited the ex-minister's place in a sort of Noah's ark of a carriage, very like an obsolete omnibus. We crossed the River Segura, and over a rough road, through fields luxuriant with the coming harvest. We passed up to the gardens, through an avenue of palms, and then under the interlacing boughs of the plane-trees, making a close arbour, until we find ourselves intoxicated with the fragrance of red and pink roses which, in profusion, border the garden-paths in every direction. The red pomegranate was in full blush, and overhung the rose-trees. A great green arbour of wire is near, under which is an immense reservoir of water. The donkey which pumped it up for the little canals which everywhere intersect the grounds, had the satisfaction himself of circulating within a pleasant and entrancing spot. No wonder the

palms flourished here; their feet are so damp and
their heads so hot, according to the theory of their
cultivation.

Within the house we found every evidence of wealth
and of that refinement which taste alone can give.
Marble pavements, vari-coloured, on the first floor,
and on the upper floor black and white veined marble,
or rare porcelain, gave to this abode an air of delicacy,
light, and coolness just suited to a summer abode,
amidst fragrant flowers and delicious fruit. Here, too,
we find English engravings of fox-hunts and races,
and of scenes in the desert; handsome marble baths;
a library rich in choice volumes, and from its windows a view over the demesne; and, to crown the
whole, rare vases filled with fresh flowers, like those in
Vallambrosa's house at Cannes; and, outside, a piazza
curtained with an immense blooming passion vine,
with fringes of flowering oleanders outside the verandahs.

About the grounds peasants were stripping the
mulberry-trees for the silk-worms. We went to see
these short-lived workers weave their beautiful tombs.
They were in the second story of an outhouse, ranged
on planks a foot apart with little stacks of hay, up to
the ceiling. The cocoons were already forming for
these; others were living heartily on the leaves, which
were upon another scaffolding. They were all very
busy. The business pays well in this part of Spain.

We take one more look at the beautiful villa; present
our thanks to the actress's cousin, who was our conductor, and is bailiff of his lordship; and bow to the
marble medallions of Francis I. of France and Anna
of Poictiers, which adorn the face of the mansion. We
wonder, too, why his lordship, in completing this
Claude Melnotte picture of a villa, has not created a
' clear lake,' in which to reflect and double these floral

beauties for his theatrical wife, and for his romantic
life. We pass by fields where peasants are ploughing
with the Egyptian plough of three thousand years
ago; we see them scattering seed as they follow the
plough; and after them a bevy of birds, chattering,
fluttering, and getting a nice supper from the seed
sown. We admire the picturesque costume of the
peasants—white shirt and white pants to the knee, and
very loose, and a brilliant girdle or sash. We thus
close up the week and have the promise of a Sabbath
rest. It will be a luxury; for Sabbaths are rare in
journeys through European cities.

CHAPTER XV.

ELCHE, ALICANTE, AND VALENCIA.

'The palmtree-cinctured city stands,
    Bright white beneath, as heaven bright blue
    Above it.'

IN leaving Murcia for Alicante, our way at first lay through narrow streets; only one vehicle at a time can run in them. It required the 'mozo,' or coach boy, to be on foot half the time, to guide the leader through the narrow defiles. Indeed, this mozo performed the duty of whipper-up at every hill, with wonderful agility and to the horror of all the donkey-drivers in the streets. Having horses, tandem, they could not be driven from the box; and the mozo would leap to the ground and make the horses dash and the fire fly from the pavement. Indeed, the French definition of speed in driving: '*brûler le pavé,*' to burn the pavement, is as applicable to the diligence as to a voiture, to a macadamized road as to a boulder-paved street. Thus, up and down, and out of town, we dashed along; now towns appeared hugging rocky mountains, on which old castles frowned loftily, like the baron above his vassals; now mountains as bleak as Vesuvius; now in the blue sky, the pale, whitish moon, much like a speck of cloud, and hardly seen—in crescent form—symbolizing the decay of Mohammedan power under Christian effulgence, appeared; now, on the road, the carts, much like a butcher boy's in New York, being two-wheeled vehicles, with a large Jersey round top to

each, and sides made with canes, appeared; then, and
often, the white mules of Murcia, tasselled with red,
and half shaved of their hair, bearing their burdens;
and the splendid oxen, in their scarlet head-gear; and
during these fresh experiences our driver or his mozo
sang their prolonged, improvised song about palm-
trees and their fruit, or about the mountains so rich
in streams, or about almonds, wheat, and what not.
While the song was loud and jocund, new phases of
vegetable beauty gratified our eyes. It was positive
relief. The eye was tired of the everlasting volcanic appearance. We had reached the very climax of aridity
and desolation; we had counted the black basalt
mountains on the horizon, and the ferruginous calcareous hills. We had turned over and over again
the volume of Nature, looking at its illustrations of
geology, every leaf a twisted rock of melted granite, or
of mottled limestone. Lo! at once, as if by magic,
the Moorish villas with the twisted columns, and the
old Moorish water-wheels pumping for water by
donkey power, appear again; then the prickly pear in
bloom about them; then the gardens of red pepper;
then the fig, very forward; then the peasants in their
tambourine hats and short, loose pants, split up to the
knees and opulent with silver buttons; then a church
with figures of the Saviour and Virgin in front; then
some peasants on the road flourishing their gay mantles; and then, hot and blistering in the sun, the blue
and copper domes of a splendid city shine in the
air! Palms surround it—palms plumed and beautiful.
Surrounding these palms, and hemming them in with
their arid and thirsty forms, are the mountains. We
may find gardens of perennial verdure, but we never
lose the sight of the bleak Sierras. They are relieved
only by the black, basaltic-fused granite, which Pluto,
from the inner rind of the earth, has pushed up into

the light, as ambassadors from the ever-during shades to the blue heavens.

This city, into which we are ushered, under the crack of whip, and with wonder and delight, is none other than Elche—celebrated in Spain for its forests of palms. Indeed, the river here, whose empty bed we crossed, is actually drunk dry by the palms. The Moors made these plans of irrigation, and faithfully the Spanish practise them. The best way to see those contrivances, and the palm woods they foster, is not to stray along the river, where there are few; but go into the forests, as the Doctor did, and see how they are planted in rows and watered in groups. A large river is thus used. The palms are not here for beauty, but for fruit. It is a business. We went up into the cathedral tower of Santa Maria—that of the blue dome! From it, we see first—Alicante on the sea, with its towers of blue and brass; then the sea and ships ten miles off; then the castles on the intervening mountains, and the chateaux lining the roads; then, to the west, the calcined mountains a-glow in the furnace at white heat; then, on the north and east, and bending round to the south, beyond the flat Moorish roofs of the city, are the straight, tall spires of the million palms, each from thirty to fifty feet high, whose stems are feathered with curving leaves, and golden with flower and fruit!

We have a weakness for this royal tree. We have admired it at St. Remo and Bordighera, whence the Pope obtains the palms which are used in sacred service, and which he blesses for his flock. We have seen them growing in the dust of Nice, and very pretty and tropical they seem there! We have seen great colonnades and arcades of them, in the Jardin d'Essai, near Algiers, where every variety is to be found. We have seen at that African garden, what we never shall see in their native soil—the cocoa and betel-nut palms; both tall, slim, graceful, beautiful; the cocoa palm making for the native a thatched abode, leaning from the shore towards the sea, and dropping its fruit even in the waves, to be distributed among the isles; while the betel-palm, with its green sheaf of leaves, topping a slim stem of six inches diameter, rises into the sunny air, forty feet! We have seen other palms, not so graceful, growing all through northern Africa, their native home; but nowhere have we seen anything comparable to these palm forests of Elche! It is the sun which here gives them their commanding altitude and saccharine fruit, but it is the sun aided by his handicraftsman, water!—water utilized by skilful irrigation.

We are in the driest climate of Europe. All writers agree in this. Medical men say, that here is the spot for bronchial ailments. We go to Alicante, where there are 25,000 people, living in the same clear, dry air; and living, not because they have rain from heaven, but water from the earth. A splendid spring has supplied irrigation for Alicante for a thousand years; but how utterly dusty, hot, and white the intervening country seems. Only here and there is there a green spot, and even then the green has a thirsty aspect. The trees are all set in holes to catch and hold what water there may be given. It has rained but three times this winter and spring at Alicante, consequently there is no crop of oats and corn, although the fields are all ploughed and the seed grain is all in. There is a splendid old chateau on the mountain, seven hundred feet high, which overlooks Alicante. It is thoroughly fortified. It seems clean cut out of the sky. No mist or moisture obscures anything. The blue of the heavens is intense and bright. The shore is shallow, and bathing is common and con-

venient; even in winter the water being about 60°.
Alicante is a great resort for invalids, but medical men
recommend them to go out of the town under the
shadow of the mountains, for sometimes the sea makes
the air moist and relaxing.

Our landlord welcomed us warmly at Alicante. He
told us that he knew we were Americans; for, as he
said, he had a brother in Brazil! When we told him
that we were from the United States of *North* America,
he mentioned in a lively way, ' *Yorick*,' (alas!) '*Bosty*,'
and ' *Feedelph*,' as cities, to show us his familiarity with
our beloved land. He was a good landlord, however,
and did his duty to us. We asked him, 'why Alicante
was so dry?' He said that it was not so *very* dry, for
he had found water by digging.' 'Where?' 'At his
garden in another part of the city—would we go and
see it? It was green and beautiful.' We said: 'Why
do you not bore for artesian wells?' Not understanding this, he said that he had gone down one hundred
feet, and found a fine well; but he afterwards explained,
that the idea of an artesian well was novel; it had
never been broached here. The rocks about Alicante,
properly terraced, would smile like those at Nice or
Mentone, if only watered. Not a blade of grass grows
here of itself. A few fatty plants, more like dusty
chips, were found among the rocks. What we remarked going out of Carthagena is to be seen here
also. The peasant has ploughed up 'lots' of lime-
stones, ashes, and rubbish, and having planted his oats
or barley, waits during the long summer glare of sun
for the rain, expecting some scattered blades to appear;
or, while waiting for the crop which may come once
in three or four years, he relieves the dusty fields by
watching and watering an olive orchard of dwarfed
growth, or some fig trees of doubtful life. These he
waters from a well, whose waters are lifted by the wheel

or Moorish Noria with donkey power; or sometimes, as we have seen it, pumped by a whole family of mother and daughters; and they call this—the garden of Europe! Sun and earth do not make gardens. Eden had four rivers in it, and if Adam did not irrigate, then the perfection of Paradise was found in that refinement of Art which miraculously distributed its 'honey dew' on flowers and fruit, without the sweat of unfallen man, or the worry of untempted woman. If this land about Alicante is a garden, for ever fragrant, flowering and fruitful, as pictured, what is Corsica? Here the bee would starve. He could not improve each shining hour, for the lack of flowers. Not even the thistle grows upon the clayish, herbageless fields.

We leave Alicante in the morning, and still ascend, according to our barometer, on the way to Valencia. When one thousand three hundred feet above the sea, we begin to see pomegranates and figs in plenty, and vines, too, which require less water. But there is this paradox with this strange land, that as you rise you find more water. We actually see a stream or so, running from the mountains. True, the land is bleak still, but we find tamarisk and poplars; and, circled with vegetation, a unique town called Sax (rock I suppose, from the Latin, saxum), springs out of its green setting, and rising 1000 feet or more into the air, looks down baronially on a company of flat roofs cringing at its base for protection. They look like little frightened brothers getting on the safe side of a big brother. This castle of Sax seemed peaked in the air; in fact, it is just hung on the sharp edge of a pinnacle! As we rise still, the clouds seem on a level with us; the mountains seem less, and some pines appear. We are on the plateau, I suppose. It grows cooler. We can perceive Valencia afar off—fifteen miles. We know that we are nearing Valencia, for the rice fields, either green or covered with water, are becoming common. We are in the irrigated vale which surrounds and supports the great city of Valencia. The rice fields contribute more to the life than to the health of the city. Every mountain top is castled, and running streams appear all about us. The apricot adorns the earth, and the soil is now changing from white to red. The orange grows in shrubs and not luxuriant. The carouba is an immense tree. It grows anywhere almost, seeming to like dust. It is cultivated for its bean, which is given as food to cattle. The crops are splendid, especially of grain. The peasants are plucking out, with their hands, every weed from the grain, or catching worms in a sort of black bag at the end of a pole, which they pass along or thrust through the lines of wheat or barley, with a sort of sleight-of-hand. Now, English cottages, neatly thatched with straw, and whitewashed, attract the eye; especially, as they are festooned in flowers. Every field is edged with a furrow, or ditch, for irrigation. One may begin to see that the famous plain of Valencia is indeed an Eden. It uses up a river; quietly absorbing it in every way; but so noiselessly that you hardly wonder at its diversion from the bed. Now the villas of the rich appear; and their gardens, glorified by the sunbeams as well as fructified by water, make of this plain the paragon which is far-famed. Hedges—made of the white and red wild rose—here and there appear ; and within their enclosure, beautiful houses, tinted with yellow and azure. So that you see now what is really necessary to constitute beauty. Allow me to illustrate : To create the eclectic beauty of the ancients it is not only necessary that the eye of the Ideal should be lucid and lustrous, the features proportionate and rounded, the hair like that of one of Murillo's Madonnas—the very dream of loveliness—the form graceful and lithe, and the step stately and free; but it requires an internal organism of health and ribs of symmetry and strength. You cannot have a Venus without bones. To this form she must come at last; as from this she began to grow. So this Eden of Valencia so vaunted and so beautiful, is made by the very qualities of the ungainly rocks and rugged mountains—those ribs of nature— which protrude so unpleasantly in the distant sky. The soil and the water come from thence; the skilful hand of the Moor gave them culture and direction, and the Spaniard does not neglect the lessons of his old enemy. It may not be easy to analyse loveliness in the abstract, but it may be satisfactory to know how and under what conditions loveliness may be reproduced. Language has been exhausted by visitors to describe the refinement everywhere around Valencia. One of the Spanish archbishops is quoted as saying that not only was each church a museum, each season another spring, each field a beautiful garden, but their united attractions suggest to us some happy spot in the lovely vale of Tempe. Others have likened it to the gardens of the Hesperides, where the fragrant flowers and golden fruit are companions on the same trees, guarded by no dragons, unless those be such who are conciliated by a few reales. Indeed, the ancients placed their elysium somewhere hereabouts.

I could take my readers up into a high place—not as a certain wily potentate did, into a high mountain, but the tower of the cathedral, and there, one by one, point at these vegetable splendours of Valentia vale. Beyond the mosque-shaped domes, which only need the crescent to carry them back four hundred years; outside the tall houses, whose windows are shaded with the matting curtains so peculiar to this part of Spain; over the fifty church roofs of blue tiles and glistening copper; outside the great Moorish gates or towers,

once used for a parliament, when Valencia had her
kingdom apart from Castile, and now used as a prison ;
far beyond the old walls which shut in over a hundred
thousand people, lies, in the enduring emerald of perennial spring, this lovely Elysium, this paragon of
gardens, this terrestrial Paradise! Spenser's Knight
of Courtesy, Sir Callidore, had many hardships to under-
go before he found his bower of bliss, where maidens
pressed from the growing grapes the wine he quaffed,
and with their roseate 'wine press' (fingers, to wit)
presented to him goblets purple with the light of love.
In some sort, we are doing Sir Callidore in faerie
land. We have gone through the volcanic debris and
calcined desolation of this south-eastern coast of Spain,
to find at the end—our Valencian Bower of Bliss. We
drink to the beauties of its balconies in the sun-warmed
wine of its Vega. We have approached it with gradual
step. Not all at once, but from absolute sterility to
sickly clover and stunted vines ; from dusty fig trees
and scrubby oranges ; from rocks full of geothermal
heat, radiating in vain upon land where no water is,
and where no green life springs, we come at last to a
vale, through which a whole river, as large as the
Thames or Wabash, percolates, every drop utilized
and every gush making its oil for the olive, its gold
for the orange, its vermilion for the pomegranate, and
its petal for the magnolia! Water! Water! Water!
We are, as bodies, eighty per cent. water. Plants have
more. What water is, in its analysis, we know. Therefore, Americans! rejoice, I say rejoice in your Mis-
sissippis, Missouris, Connecticuts, Sacramentos, and
Hudsons. Rejoice in your mountains, and clouds,
and rains. You will never know, till the Great
Drought which will follow the Great Evaporation
under the influence of the Great Conflagration, what
water is in the great economies. Do not, my Ameri-

can brother, waste your water by too free an admixture of it, with other elements. Use it for horticulture, agriculture, and navigation.

If my writing from Valencia seems too jocund, remember that it is the City of Mirth! The peasants are honest, buoyant and pleasure-seeking. Their music is not too sedate; their dresses are gay as the most theatric could wish; the domes of their churches are dressed in cerulean hues; their streets are twisted, as if they had drunk whole rivulets of the dark-red wines of the plains around; their mountains, with true Spanish pride, trick themselves, morning and evening, in 'trailing clouds of glory,' and their river—the honest, laborious Turia—does it all! Without it, there would be no Eden here, and no corn and oil, and no people.

The Turia is spanned by five bridges. The people of Valencia ought thus to honour it with these arches of triumph, although there is no water to run under them! To be honest, if not poetical; to be homely, if not elegant—the plain of Valencia 'sops' up the river, and the Cardinal de Retz wrote well when he saw the results of this effort: '*Toute la champagne, qui est émaillée d'un million de différentes fleurs qui flattent la vue, y exhale un million d'odeurs différentes qui charment l'odorat.*' And all these wonders of flowers and fragrance are made by a discreet use of water, under a latitude of 38°, where it never snows, and hardly ever rains! No wonder the medical world send invalids hither for a dry, tonic climate. 'Winter sunbeams' I am looking for. I have found them here in white rays, undecomposed by the prism, for there is hardly enough moisture for a rainbow. The winds which blow here, even from the north or west, lose their moisture or their cold before they salute Valencia. Here, if anywhere out of the Riviera, we find the conditions which the Father of Medicine, Hippocrates himself, prescribed as essential to good climate and good health. Here we find what another doctor, —one William Shakespeare—enunciated when he made the mad Dane say, 'I am but mad north-north-west; when the wind is *southerly* I know a hawk from a handsaw.' When Wellington marched his armies through Spain, and when the French did the same, the endurance they displayed was utterly wonderful, for they had this dry climate; while in Algiers, under the moist climate, or with the sirocco blowing on them, they found fatigue, sickness, death, and not unfrequently death by suicide. On one march, under the sirocco, eleven of General Bugeaud's soldiers committed suicide. That southern wind was not of the kind Shakespeare referred to as sanitary. But winds have much to do with health, and the shelter given from harsh northern blasts to Alicante and Valencia by the mountains north of them, draws out the life of man to its largest limit. It conquers even winter— that enemy of old age, *inimicior senibus hyems*—by its genial sunbeams, and its dry, bright atmosphere.

I do not know how better to end my sojourn here than by recalling the contrasts which have led me step by step to this spot of luxuriance. They are to be found in the mixture of rough and smooth, fertility and sterility, fire-burned mountains and water-fed valleys—the granite shooting up in great black jets through the fantastic limestone mountains, and making with castles and towns a picturesque confusion; as it were the moon, first as science knows it, utterly crisp and dead—ashes, ashes, ashes—and then, the same orb all at once enchanted and enchanting, as if under a lover's fancy, and making an Eden of Earth! This is the last analysis of our trip through south-eastern Spain. We have lingered amidst this garden, naming its flowers, like our first mother. We have wondered at its wealth of luxuriance. Marvelling more, we have thought it strange that a people, so ingenious and industrious as the Spanish labourer and peasant, should so long have submitted to the rule of aristocratic vampires. These, sucking their blood, eating up their substance, building palaces out of their industry, have, like the nobles and aristocrats of the French Revolution, deserted their benefactors and left their land a prey to whatever of political riot and disease may come upon it, under the conflicts of faction and party. But I should reserve all political thoughts till I reach Madrid, and there learn more precisely the present situation and prospects of the Spanish nation.

The Alhambra.

CHAPTER XVI.

GRENADA—ANDALUSIA.

Te Deum Laudamus! was up the Alcala sung:
Down from the Alhambra's minarets were all the crescents flung;
The arms thereon of Arragon they with Castile display;
One king comes in, in triumph—one weeping goes away.'

Spanish Ballad.

IF a tourist be at Madrid, tnere is but one good way to reach Grenada, namely, by rail to Menzibar, nearly due south, and arriving after twelve hours, breakfast there, and thence by diligence to Grenada, arriving about ten at night. I am particular to state this, for the guide books are blind or faulty. The latter part of the trip by diligence is very inspiring. If you are particular, get a seat as we did, in the 'berlina,' which is Spanish for *coupé*. The way the eight or more mules are made to travel, under the direction of the conductor or mayoral, lash of postilion, and hurrah of driver—and especially up hill, where the tug comes— would astonish an English or American Jehu.

The diligence is very heavy, but its rattle is lively. The mules are shaven half way down. Their tails, too, are half shaved, with a tuft of hair left at their roots, spreading out into a sort of moustache at the *os coccygis*. This shaving of the mule is said to be a sanitary process. It prevents cutaneous disorders and keeps him cool. The way that 'meek child of misery' is lampooned by driver, postilion, and director, and sometimes by a passenger who hires an outside seat in lieu of the conductor, who retires within, demonstrates that the barbering has speed for its object. To secure speed the pachydermatous outside of the mule is rendered, by shearing, somewhat sensible to the lash and whip stock. The mule is said to be the favourite animal in Spain, but he is treated very harshly. He is suited to the acclivities and declivities of the country, and his stubbornness and resignation, his endurance and imperturbability seem to be suited to the Spanish race. I know that even outside of Spain he has been abused. He is sometimes called an ass. But he boasts that if he has an ass for his father, he has a horse for his mother! How he was abused in America during our civil war! Yet the war paths were macadamized with his bones. The phosphorescence from their decay led many a brigadier to glory. I used to think it hard that legislators offered resolutions of thanks to so many brigadiers and other generals, while never a one was tendered to the mule.

In Spain a good mule is worth more than a horse.
The best mule will bring three or four hundred dollars,
while the best horse generally brings two hundred.
The minimum price is nearly in the same proportion.
Most of the mules come from the Pyrenees, or France.
Prim drives a team of six, in gorgeous ruddy accoutre-
ments. They are the coach horses, as an Irishman
would say, of the best families. Sometimes their
coats are clipped or shaved fantastically, zebra style, or
in spots; Gipsies do this well. As we travel we may
see them armed, like Atropos, with shears, emblematic
of their profession as mule barbers.

On our route to Grenada, justice compels us to say,
that the whipping of the mules has not so much to do
with our speed as the hallooing. Every mule has a
name. The name generally is resonant. It ends in
an *a* or an *o*. The opportunity of exercising the
os rotundum is never neglected. Our driver had a
knack of running his fist into his ear, so that he was
not stunned at his own horrid howling. Our leading
mule was named 'Romero.' My old friend, the
Mexican Minister, would be shocked to hear the
variety of intonations and expletives wherewith his
musical name was sounded and accompanied. The
general tone rises at the ending of the word, thus :—
Ro-me-r-o-o-o-*o-o*-o-O!! The particular mule addressed by name, generally signifies his possession of
ears, for he 'gets up.' When we came to pass
through the narrow streets of the towns—Jaen, for
example — and with our team of eight, when the
immense diligence undertook to turn at right angles,
and that, too, in streets so narrow that the wheels
grazed the houses on either side, ah! there and then
was skill worthy of a charioteer in the Olympian
games. To crowd the eight mules into one, to make
that one gallop and fly round in a hurry, required

a *finesse* and *élan* accomplished only by the postilion afoot with lash, by the conductor with a magazine of stones, and by the united and turbulent hallooing of all the three persons employed—at all the eight mules by name, at the same time! The dark-eyed señoritas stopped watering their flowers in the balconies to gaze after us; beggars forgot to show their sores and whine their plaints; the cobbler in the basement waxed curious and gazed after us as if it were his last, last look. I could never become accustomed to the incessant hubbub and beating of the animals. I believe that the Spaniard thinks that his voice is ever sweet to the mule, even when raised to a screech, and that his whacks with the butt end of his whip are oats and refreshment. Was it not Irving who remarked that dry blows serve in lieu of provender in Spain, for all beasts of burden? We stopped at the famous city of Jaen for dinner, sauntered up the high hill to the cathedral, and took a survey of the scene. It was not the historic associations, or the old towers of the church, that seemed most peculiar to Jaen; but her caballeros. What groups of lazy dignity; how they did seem to stand, as if idleness alone were honourable, the only effort being to light a fresh cigarette, or gaze after the stranger. I introduce to the reader one of these groups.

In going from Madrid to Grenada you pass through La Mancha. It is apparently very bleak and uninteresting. It is not so dry and calcareous, or calcined, as the country between Murcia and Carthagena; but is not comparable to the magnificent country between Grenada and Cordova, or Cordova and Seville. These latter are the best parts of Spain, and do justice to the reputation given to Andalusia by Moor and Christian, as a terrestrial paradise. We do not see much else, however, between Madrid and Grenada, than valleys of grain and mountains of sterility. Few trees are visible, but the people are everywhere interesting; not alone because they are the result of a mixture of Celt, Greek, Phœnician, Carthaginian and Roman of the elder day, with the native Iberian; not alone because the Goths and Vandals here mingled their tough spirits and heroic blood with these races, but because the Moor—the child of the Orient—has himself freely mixed with each and all, and transfused into the dark, lustrous eye and lithe figure of the Spaniard, the imaginative, poetical, and luxuriant qualities of the East.

The oriental character of Spain is everywhere observable. Spain has always been apart from Europe. The Moors overran the peninsula from Gibraltar to the Pyrenees, and were only stopped from overrunning France by the battle of Tours. They left their impress not alone in their alcazars, their houses, their palaces, their systems of horticulture, agriculture, pisciculture and irrigation, but in the very blood and bones of the Spaniard. The traveller is surprised now and then at finding so many fair-faced and even light and red-haired people in Spain, especially in Castile. He notices it, because it is exceptional, as he thinks. These are signs of Gothic invasion and conquest. But the Moors have been here within three hundred and seventy years. In the year of the discovery of America, Isabella and Ferdinand received from Boabdil the key of the Alhambra. This strange race, full of science, learning, grace, chivalry and dash—who at one stroke struck off the crowns from the kings of all the nations, from India to Morocco—subsided like an ebbing wave.

The bull-fight, the most obnoxious institution yet to be seen in Spain, was one of the legacies of the Moor, and seems to me very incongruous with the refinement of that civilization which Europe in the dark ages came to learn in the schools and universities at Cordova, Toledo, Seville, and Grenada. But the bull-fight was the brutal, sanguinary side of the Moorish character. Wherever you go in Andalusia, you will find the radiance of this brilliant though dusky people.

The Spanish alamedas, or public promenades, show a wonderful variety of people and costume. The white-kilted Valencian is a picture; and the velvet clad Andalusian is another. The lady decked with her unvarying dark mantilla, and the grave gentleman without avocation, in his Spanish cloak hiding his tatters (so common in America—I mean the cloak— a quarter of a century ago) thrown over the left shoulder in a grand way; not to speak of the Majo, or Spanish exquisite, who is called here *lechugino*, signifying a sucking pig and a small lettuce; and the Spanish students whom we meet every time we walk in and around the Alhambra precincts, and some-times, as on Sunday last, in groups, with their caps worn in a jaunty way and cloaks lined with the most inflammatory red, and always showing the lining— these give evidence that for display the Spaniard will do anything. It is either Irving or Ford who says that he will rob his larder, and eke out a scanty meal with a few vegetables, in order to furnish his wife with a graceful mantilla and himself with a dark capa!

A new era of progressive ideas is dawning for Spain. This is observable not alone in the free discussions of the Cortes, but from the better order prevalent throughout the country. We have seen no drunken-ness, no rows, no fights anywhere. We have had occasion to make this observation, especially between Grenada and Madrid. There is little or no evidence of repressive measures. True, we see soldiers sauntering, gun in hand, along the highway between Jaen and Grenada, but they are continued here now from habit. There have been no robberies for some time. The bandits are as scarce as the contrabandistas, of whom Irving tells so many stories of forty years ago. Even the gipsies, about whom Matteo, Irving's valet, used to tell horrors, are as well behaved as the Arabs and Kabyles of Algiers. We have been among them, and can testify to their good conduct, nice homes and fascinating dances.

In the winter of 1852, I saw Washington Irving for the first and only time—*Virgilium tantum vidi*—and I well remember that he said to me, 'If you would taste the Orient with a dash of Arabian spice, you may do it in Spain. Go to Andalusia. Go, as I was accustomed to go, on horseback, through its mountains and valleys; and, above all, see the monuments of Moorish elegance and grandeur in the Mosque of Cordova, the Alcazar of Seville, and the Alhambra at Grenada.' I then made a pilgrim's vow, at some day to see the mountains, fortresses, castles, gardens, palaces and homes, to which the genius of Irving has added an enchantment, which the Moorish architect, the Arab story-teller, the Spanish poet and the monastic historian did not and could not bestow.

The builder of the Alhambra was an illustrious captain, a great prince, a good king, and as married as Solomon. His monuments remain about as Irving saw them forty years ago! The Moorish character still remains, although there is so much that is changed. Even since Irving was here, what changes! What changes in and around the Alhambra! A fortnight goes by here so delightfully, that it is more like a dream than reality, and leaves little time for writing commentaries. One of these changes, however, is so striking, that I may be permitted to record it.

Last Sunday, three thousand republican volunteers
were under arms, training with the manual—marching,
countermarching, double-quick, mark time—and all
reviewed by a republican alcalde; and this, too, within
the royal walls of the Alhambra! Here, where in old
Yusef's day forty thousand soldiers could be sustained
within these rough, red walls, and where four hundred
thousand people slept in the city under their protection; here, where once issued to and from the gate
of justice, to the plaza of cisterns, Moorish squadrons
of horse and soldiers afoot, with banners flying and
scimetars drawn, white guards and black guards; here,
where thronged to high mass the conquering host of
Isabella and Ferdinand, in 1492;—here where Columbus waited on their Majesties so wearily—when
from these halls, his requests refused, 'indignantly did
he toward the ocean bend his way, and shaking from
his feet the dust of Spain;' here, whither he was recalled
to fresh energy in his enterprise;—where the cuirassed
knight and silken courtier, the grand cardinal and
great captain, mitred prelate and shaven monk, bowed
at one altar with King and Queen, to thank God for
their victory over Boabdil and for the taking of the
fortress—here, on last Sunday, on this very plaza, I
mingled with the republican throng of many thousands, who were practising their tactics for the struggle
of the future!

It is something that Yusef or Irving could never
have anticipated. A republican muster of volunteers
in front of the palace of Charles V., and within the
shadow of the ruddy towers of the Moorish fortress!
The spell is broken indeed! We are living in an era
of transitions. Creeds die, and prejudices give way.
Old monuments like these around Grenada may remain, but the foundations of the social fabric are
stirred by popular tremblings. Like a wild courser,

leaping along the highway, and down the bye way, or like one of those splendid equestrian effigies which Art has copied from Nature in this clime of the Arab steed—fit to symbolize the war horse of Job—Progress shakes its mane, and thunders over the pavement, it may be in unbridled freedom; but the chariot follows in a smooth and even course, and the goal is approached in safety! May the Progress of Spain find a similar realization at the end of the course! Certainly the crowds upon the plaza preserve the best order. All the city have wandered up to those heights this beautiful Sunday afternoon. The companies are manœuvring and a band is playing. Water-carriers are singing their *agua fresca*—'fresh water!' *Quien quiere agua?*—'Who wants water?'—and vending it to drinkers at a quarto a glass. Everybody here is thirsty, but no one drinks anything but the crystal, cool water which comes down from the Sierras. The English drink beer; the Germans swallow their lager; the French drink their absinthe; the Americans their whisky and bitters; but the Spaniards, as a people, drink water! Their air is so dry and exhilarating, and their wines are so rich, that water is to them indispensable. Everybody, men, women, and children, are drinking it on this Sabbath day. The wheel at the cistern, immortalized by Irving, is going briskly; the carriers fill their little wooden casks, fix them on their backs, and sing away, *Agua! Agua!* A few pea-nut peddlers also appear.

Where I sit, upon the stone bench near the wall, are some half-dozen señoras and señoritas, dressed fashionably. They are of the better class, and have come up, as I infer from their manner, to laugh at the republicans. The awkwardness of the volunteers seems to them, just now, very funny. Beggars ply their vocation, and exhibit their *argumentum ad*

misericordiam, with woful plaints and saddest 07ulula-
tion.   Beautiful women—without bonnets, all in black
mantillas, only a veil of that hue upon their glossy
hair, and having unmistakably the dark Moorish eye,
saunter about with *nonchalant* air.   These are 'Spain's
dark glancing daughters angelically kind,' whom Byron
found at Cadiz, and whom we shall find all through
Andalusia.

All are strangers to me except one or two of
the volunteers, keepers of the Alhambra, who have recognised me before as a republican 'and a brother,'
with whom I gossip much amid the palace walls,
and by whom I am now introduced to many others.
Presently we see the captain of a company whose face
is familiar.   It is Captain Mariano, of the Hotel
'Washington Irving.'   He marches by with his company of 'boastful but brave Andalusians,' and salutes
us with '*Viva Republica !*' '*Viva America !*' and
we respond.   Ghost of Boabdil, the Yellow-bearded!
The tears you shed on leaving the Alhambra, and for
which your cruel mother reproached you, may well
flow afresh!

They certainly would, if Boabdil could have seen a
company of a hundred and more of these republicans
drinking, on the invitation of the writer, from the
spiggot fixed adroitly to the leg of a dead porker,
whose skin was plump with the Valdepeñas vintage,
and drinking to the Spanish Republic, on an American
model!   And when cheers went up from the gardens
of the Fonda, for America! her Minister! and a
federal republic for Spain! were they not followed by
the 'goblet's tributary round,' from the hogskins above
mentioned?   And when the writer,—I should say
speaker,—responded, in a modest way of course, to
the salutations, on behalf of some forty millions of
American republicans,—the shade of Isabella of Cas-

tile must have sighed for the ill-spent *bijouterie* wherewith she encouraged Columbus to discover so republican a world as America. The scene is worth
remembering. It indicates the changes here since
1829, when Irving lived in the Alhambra.

In consideration of the peculiarity of this phase
of my experience at the Alhambra, I propose to do
two things in this volume, for which I deliberately
turn my back to the critical lash. One is the publish-
ing my bill of fare, and the expense thereof, of this
republican festivity, that the future American when
he does likewise, may count the cost. Here it is :—

	Reales.
Convite republicanos federales,—4 arobas (hogs' skins) de vino	168
Nan (9) asistunas y 9 incurtidos (slices)	48
Segars	36
9 libras de solchilson	144
Comidà de los officiales	96
	492

Secondly : I produce the remarks made by the
author on that occasion. Who reported them,
modesty declines to mention. They are reported as
they were spoken, on the Spanish model. Here is the
report :—

With many cheers for the republic, their captain,
and the American Minister, Mr. Hale—who arrived
while they were assembling—the wine was passed and
the hilarity began. After the company had enjoyed the
hospitalities, Captain Mariano introduced to them
the gentleman who had invited them to the entertainment. He was received by the company with many
vivas, and spoke as follows :—

'Señores! Republicanos! I speak from the hospitable gardens of the 'Washington Irving' Hotel. It
bears the name of the most honoured of American

republicans. (Vivas.) His name is not less known in the republic of letters than in the American republic. I regret that I cannot acknowledge your courtesy and sentiments in behalf of my country, its Minister, and myself, the humblest of its representatives, in your own magnificent language. I shall ask my Irish friend, Señor Maurice Mullone, to translate my words. They will not lose, but gain much, by his translation, and into your own tongue. Your language is called the eldest son of the Latin, and from distant days through many vicissitudes—from the great Republic of Rome to the latest free utterance of your republican members of the constituent Cortes, this language of Cicero has syllabled the aspirations and preserved the laws of republican liberty. (Vivas.) It is a gorgeous vehicle for the conveyance of truth. It may be perverted. It has been even here. But here, where I have seen, within those walls of the Alhambra, so many elegant effigies of dead dynasties — of Moorish absolutism in its barbaric pomp and delicate refinement, and Spanish royalty in its most arrogant pretensions and aggravated exactions; here in Grenada, where repose the bones of Ferdinand and Isabella, who aided Columbus in the discovery of a new hemisphere, as the home of republics, and who are, therefore and thereby, made lustrous in history ; here, where, if anywhere, the signs of royal power have the fascinating glamour of the past ; here, I have seen to-day, under arms, in front of the unfinished palace of Charles V., and under the shadow of the old dismantled tower of the Alhambra, three thousand volunteer soldiers of a federal Spanish Government. (Great vivas! bravos!) While by the policy of the American republic, the American people do not intervene with arms in the affairs of foreign nations; while the American Minister cannot, with propriety, answer the partisan salutation you have tendered, yet I say to you, speaking consciously the unanimous voice of my country, that there are forty millions of republicans there, of all sects and parties, extending their hands, as they have extended their example, to welcome the birth of a Spanish republic! (Vivas.) More than that, there are twelve other republics of the New World which would lift up their voice in your own grand language for a new order in their mother country. (Vivas.) You have cried "Long live the Republic!" (Vivas.) Do not despair of the republic! You have no king. (Cries of "No.") You have no queen. (Loud cries of "No, no!" and vivas.) You are now a republic! You may have heard of the man who was astonished when told that he spoke prose. You may be astonished when I tell you that you are now under a republic (vivas), and yet you live! You earn your wages. Your young señoritas are still winsome, winning, and being won. (Laughter.) Your señoras will still embrace you and present you children. (Laughter.) And yet all this under a republic! This can be continued. Organize your system; and then select your chief; not alone because he is a general, but because he is a citizen— honest, patriotic, and intelligent. Call him what you please; but make him not supreme; only the executive of your supreme will, expressed through your provincial organisms, public opinion, and a constituted federal order. Thus you will make the republic, now provisional and national at Madrid, in your Cortes, federal throughout each province of your historic land! (Vivas.) You saluted me with the cry: "Long live the *Federal* Republic." A federal republic is rational, for every land and for each hemisphere. A republic not federal would lead, as the French Republic led, to the lantern and the guillotine! Liberty herself might be the first victim! A federal republic implies personal liberty, consisting with social order and public spirit. In a federal republic there is a foedus, a league, a bund of States; each State sovereign over its home-concerns; having its provincial legislature, its ancient customs and franchises, unimpaired by central power, whether that central power be consolidated in an executive tyrant of one head, or a legislative tyrant of many heads. To attain such a republic requires moderation with freedom. You have already made progress in commercial and industrial freedom. You have already freedom of discussion and of opinion, in speech and press, and freedom of soul and body. You can perpetuate these only by self-imposed restraints. Your vegas, which lie below us, are warmed by the sun, but they are tempered by the snows of the sierras above them. Your harvests come as well from the warm solar breath as from the melted snows. It is so with your Liberties. Heaven gives you enthusiasm. It is in your warm hearts. Reason gives you the coolness of moderation, by which to temper enthusiasm. Joining enthusiasm with reason, Liberty results. Under her reign, your plains will be green and golden with fruitful industry, your homes happy, and your republic a realization of your most splendid hopes! To restrain freedom by moderation, avoid the excesses incident to revolutions, frown upon infidel and rash counsels among yourselves; reserve the ballot and keep it pure; reserve the freedom of the press, and keep it rational and fair: the right to worship God without secular hindrance; the right of life, liberty, and the pursuit of happiness; and to conserve these, constitute your republic, not as a tyrannical, consolidated unity, but as a democratic, decentralized diversity in unity, *E pluribus Unum*—in fine, a federal republic! (Vivas.)

 'Your mountains are rich in every kind of precious

material, especially marbles. Build your temple out
of the various marbles of different colours, hewn, they
may be, by different hands, and of different sizes ; but
let them all be fitly joined together, and the founda
tions so firm, and the arches so keyed, that no con-
vulsions of the passionate populace, and no reaction
of king-craft, shall shake them from their proper
places. You have it in your power thus to create, out
of dissimilar materials and interests, a federate unity.
If, however, your elected rulers prefer a monarchy
—(murmurs)—bide your time and struggle rather
with rational, than with violent methods. Civil war,
Spaniards, is the grave of liberty !

'If they give you the federal republic—which God
help !—guard it with vestal vigilance, for it is a more
precious legacy than all yonder monuments of Moorish
luxury or Spanish regality—(vivas)—which fill this
atmosphere with enchantment. Such a republic is
the United States, under its written Constitution.
May your Cortes make for you such an organic law.
(Vivas.) They cannot do better than the Eternal
God ! He has made the planets above us—each having
a sphere, and all circling round a central orb. Valencia,
Valladolid, Aragon, Toledo, Estremadura, Murcia,
Catalonia, Grenada — each having an ancient order,
established customs, provincial rights, and peculiar
methods for their observance, may revolve like the
planets, in their own circles, and in safety around a
federal central luminary ; which, without consuming,
will illumine each and all. Thus the central power
may be preserved, without aggrandizing itself at the
cost or loss of provincial independence and personal
liberty. (Great vivas.) Then you will not be harmed
in person, nor be taxed in property unless you consent
to it (vivas), by your own representative, voted for by
yourself. Thus, each Grenadian will become not less

14

a Spaniard, though always a Grenadian, and ever a man!
He will become, as the American republican is—or
ought to be under our Constitution—a sovereign,
divinely anointed and crowned, and bearing his own
sceptre, by the grace of God! (Vivas.)

'In the meridian of Roman power, your own Andalusia gave to the world Adrian, Theodosius, and Trajan. They were born at Seville. Your great Gothic
kings were elected by your people; but they wore the
purple often careless of the people they ruled. Your
Spanish kings, Charles, Philip, Ferdinand—in whose
jewels Spain shines in history—all these belong to the
dead past. They are dust. Their swords and sceptres
at best were emblematic of rude and absolute sway.
Their thrones were erected in the petrification of the
human heart. When the people have empire here—
every man bearing the fasces of the republic in the
procession of power—then a new epoch will dawn for
Spain! Before its splendours will pale all the glories of
royalty! I salute you, Señores, as from the great
American Federation; and drink to the permanent
establishment of the Empire of the People, whose reign
may it be more beautiful than the Alhambra, and more
enduring than the marbles of your kings!' (Prolonged
vivas.)

It is needless to say, that the evening thus celebrated
in the gardens of the Fonda, was joyous beyond expression. It is treated by the author as a happy phase of
Andalusian hilarity; but the earnestness of the audience,
and their rapt attention to the lessons and metaphors
of Federation, show that there is an anxiety to learn,
as well as to practise, the lessons of moderation and
liberty. From that evening forward, he never ceased
to watch the varying phases of Spanish politics. The
sequel will show how faithfully he has recorded facts,
and with what foresight he has reasoned upon them.

For Spain, all liberty-loving people—whether under a polity republican in form or not—lift up their voice in the prayer implied in the Byronic line :—

'When shall her olive branch be free from blight ! '

It is not for me to write more than general impressions of the Alhambra. Where Irving has done so much, it would be a rude hand that would touch the canvas. Still there are changes—other than political —since he lived here, which I may notice. Only *one* person, among all those whom Irving describes, remains. You remember Zia Antonia (Aunt Antonia), who had charge of the Alhambra, and received its visitors and dispensed its favours and flowers, who lived in a corner of the palace, with her nephew and niece, and with whom Irving lived, as it were, in her household. She is not the survivor. She is dead. Her niece, too, the damsel Dolores, who had as many arch-ways as the Alhambra itself, whose ' bright looks merited a merrier name,' the heiress of her aunt and the *fiancée* of her cousin, she and her husband are gone to the grave. She left two deaf and dumb daughters, and one survives, the heiress of the aunt's possessions, ' consisting of certain ruinous tenements in the fortress,' as Irving then described them, but now, I am glad to say, in good repair. But she resides in Malaga, and is not here. This solitary granddaughter of his talkative hostess is absent and ' mute.'

The one survivor here is none other than that ' tall, meagre varlet, whose rusty brown cloak was intended to conceal the ragged state of his nether garments '— whom Irving found lounging in the sunshine at the gateway of the domains, gossiping with a soldier— none other than that '*hijo de la Alhambra*,' Mateo Ximenes. Irving adopted him into his service as valet, cicerone, guide, guard, and historiographic squire ;

heard his marvellous stories; pictured his quaint, superserviceable zeal with most magical touch ; made even his grandfather, the legendary tailor, an historical study — in fine, he was the gossiping Figaro of the Alhambra, and to Irving a very Sancho Panza in his search after its romances, as well as his Asmodeus, who uncovered the roofs of Grenada for his study of its domestic life.

Mateo is yet living. I had the honour of congratulating him on his seventy-seventh birthday. He is a little the worse for wear and time. His head is well sprinkled with gray. He wears a jacket yet, after the Spanish manner, and a nice Andalusian hat of velvet, but he is no longer able to do the duties of cicerone. His son José inherits that office. Mateo makes (by proxy) ribbons of various colours upon a loom in his house. He was a ribbon weaver, as his father was, and carries on the work still by deputy. He loves to talk about Irving, and is very proud of the immortality secured him by the American author's pen.

Much else has changed in the Alhambra. That angling in the sky for swallows and martlets which Irving so graphically pictured, as the employment of the school urchins, is now obsolete. The ' sons of the Alhambra' allow the birds to live and guard their fruit! The first court, called *de la Alberca*, or Fish Pond, in which Irving used to bathe, is still clear, cool, and inviting. It is surrounded with a lovely hedge of myrtles, clipped square, and is full of gold fish. But the frogs which croaked there are gone. The saloons upon the right of the corridors of this court were once occupied by the Sultanas. The rooms have been much changed since the Moorish days, and since Irving was here. The archives here collected, and which Irving read, disappeared in 1860. There is a recess in the wall, where I saw two splendid Moorish vases, enamelled in blue, white, and gold. They were found full of Moorish gold, since Irving's time, and there is that much fact for the foundation of his beautiful tales of the 'Moor's Legacy' and the 'Two Discreet Statues.' I did not learn that any Arabic writing in a sandal-wood box, or any wreath of golden myrtle, were concomitants with, or necessary to the discovery.

My courier, a month ago, found in the Court of Lions, seated by the fountain, an old turbaned man. He was none of the visionary enchanted Arabs of whom Mateo used to inform Irving, but a Turkish general, a tourist. He was found by the courier, beating his breast and shedding tears over those relics and ruins of his race. From my observation, in Africa, I can testify, that what Irving says in reference to the sighing of the Moors for this, their terrestrial paradise, is true. They believe, even yet, that Spain will fall, and that they will be restored to these their old homes. They preserve not only the old keys of their houses, and the titles of their property, but their lineage, so as to claim their own in their good time coming. One of our friends in Algiers, Mustapha, indulges this hope. But the sight of three thousand republicans marching within the walls of the Alhambra, and the vision of Spain with free speech, free press, and free soul, would dissipate their dreams.

One word more about the Alhambra. Ravishing as an architectural study, the purpose for which it was built makes it morally suggestive only of degradation to human nature. The inner life of the harem must have been made up of bickerings, jealousies, devilries, spites, miseries, tragedies, bloodthirstiness, and blood-guiltiness; the fear of the mother for her child; the rivalry of a new wife made intense by the birth of a new prince; the vigilance of mutes, eunuchs, and spies; and the insatiate sensuality of the Moorish tigers. No large range of fancy is needed to picture these, as the real 'tales of the Alhambra.'

From one of the towers—the bell tower—in looking down upon the court below, I saw many galley-slaves enjoying themselves in their prison. I am told that they are here for ten years each, and that the life they live is rather pleasant; so much so, that some try to return as soon as they get out! But their life is inexpressible contentment compared to the 'nuptial joy' of the beauties of the harem! It is especially happy in the comparison, if those beauties happened to be what they so often were, Christian maidens of rare culture and loveliness, who were subjected to a life worse than slavery or death. When the tourist wanders through the Alhambra courts, admires its Arabesques and fountains, pillars and turrets, halls and towers—all the fairy architecture of this refined place and race—he must not forget to be disenchanted in one respect. If I had the humorous skill of Artemus Ward among the Mormons, I might draw aside the veil from this grossly sensuous Moorish race, and its elegant and effeminate temples of lust; and, if in no other way, illustrate its Diabolism. As it is, the only illustration I can furnish is that in the engraving of the 'Moorish door of the Alhambra.' It will enable the reader to judge of the exquisite elegance of the architecture, within whose walls and halls was concealed the highest refinement of sensuality.

CHAPTER XVII.

*FROM GRENADA TO MALAGA — THROUGH THE
SIERRAS — ARCADIA.*

> ' Jewel of the mountain ring,
> City of perpetual spring,
> City that the sea still kisses,
> Where the wind is dowered with kisses;
> From the starry jasmine flowers,
> And the thousand orange bowers.'

ON the route from Valencia to Madrid, I lost my companion, Dr. Bennet. He left us to go to the south. We saw him for only an hour afterwards, amidst the Alhambra courts. Having made the circuit of Andalusia, we will soon return to Madrid. Then and there, under the tuition of fresh events, and with more experience of Spanish life and character, I will try and make out a diagnosis of the condition of the Spanish body politic.

Meanwhile will you allow me to show how easy it is to go from Grenada to Malaga, and how easy it is to go out of the latter place, under the impulsion of military displays?

We leave Grenada in the cars, and without much deflection from a right line west, burst through the mountains, after crossing the splendid Vega. We go in a direction, as if we were going direct to Cadiz or Gibraltar. We leave the superb towers where Irving lived and dreamed; where Boabdil held high carnival with wives and eunuchs after the Moslem method; where Isabella the Catholic held high mass; we leave behind us the Pinos bridge, from which Columbus returned at the request of Isabella, to hear that Spain would help him in his designs on a new world; we leave the mountains of Elvira, around whose feet the Moors and Spaniards fought for seven hundred years; we leave the superb sepulchres of Spanish and Moorish royalty and priesthood, and in two hours we are at Loja, and in ten minutes after, in the coupé of a diligence. We try to recall the scattered and splendid memories which make this land of Grenada so romantic. But who can indulge in such luxuries of association when our driver gallops his mules, with their jingle of bells and with terrific outcries through town and country, through narrow streets and up perilous mountain roads? Who can follow the labyrinths of historical lore, when he is winding with easy grace and safety up and down splendid mountain roads? But all this has been told by a hundred tourists in their experiences of the Continental diligence.

I have never read, however, a description of what I saw before entering Malaga. Indeed, I despair of conveying the impression I received. It was not the wild mountain barrier which shelters Malaga to the northeast. Those mountains so grand and lofty, beyond and above which tower the Sierra Nevadas, ever in view to the south of Grenada, and even at Seville ever in view, would astonish of themselves; but a description of mountain scenery loses its charm by oft-repetition. The route lifts us nearly as high as these snow mountains. The air is cool, and we 'wrap the drapery' of our mantles about us, and gaze out upon the mountains below us in wonder, love, and praise. The blue sea appears between the mountains far below. Here is the charm! These mountains for twenty miles around Malaga, are covered with—snow? No. With olives? No. With dust, ashes, rocks, shrubs? No, only with grape-vines. The sides—all sides—from the bottom of the valleys to the summits, are grape-covered. These vines look fresh and advanced. The sides of the mountains are ridged by the rains, and as far as you can see,—mountain below and below mountain—mountains in groups and ranges—the hills on mountains turning to and from the sun—are all covered with grapes! grapes! grapes! Only one little patch of olives is seen far off; all the rest of the vegetable life is—the grape—grape —grape. Not a blade of grass, or a fibre of moss or lichen; the grapes monopolize the reddish soil, and capture every fugitive rain drop. The soil, such as it is, is worked to an extreme nicety of cultivation. The vines are stubby and set in regular holes, some five feet apart. As far as the eye can pierce, from this lofty perch, the red sides of the mountains are specked with the emerald vines. If you would estimate the vines, or the acres of them even go to the commercial statistics of Malaga, and see what a yield is here.

Not that all Malaga wine is from grapes here grown and pressed. No, nearly all the wines of commerce from here and elsewhere are fabricated. The greater part of the wines of Spain and France are indebted to the potato for their fine spirit and fruity life.

I am not skilled in grape culture; and cannot technically testify to the modes employed around Malaga. Nor can I affirm that any of the rules applicable to the mountains here, would suit various American localities. Whether drains should be made; how far apart the plants should be; whether sub-soiling is required or cross ploughing; whether the soil should be comminuted and mixed by the spade; whether it is best to start a vineyard with cuttings or with plants; how to support the vines,—these and other matters depend so much on the locality; but one thing even a tyro may detect, that in the fabrication, adulteration, and commerce of wines, the frauds are enormous.

The potato, that simple esculent, becomes an important agent. The potato it is that, if it does not give colour in the cup to the wine, at least does give the carbuncle to the nose. That homely Celtic diet, how it gushes and bubbles to the beaker's brim, beaded with rubies! But how the wines, which owe their sparkle and spirit to it, are made before the homely admixture, is told so graphically by an old Ohio friend, William J. Flagg (who has by reason of his 'Longworth' connection, a right to speak), in a recent volume, that I am tempted to make one 'cutting' from it. He says that in the wine press factory of La Tour, whence issues the good Médoc, men tramp out the grapes with their unwashed feet. Again, in the process of stirring up the mass during fermentation, naked men go into the wine vat, chin deep. Drink deep, brothers, of the flowing Bordolais and Burgundy!

But I do not like to disenchant people. Let us admire everything. It is worth while to admire in Andalusia, without being critical. Even the peasant we meet on our downward way into Malaga—or the muleteer in his leather leggings, or all be-buttoned with his cotton pantaloons, very short and loose, and velvet hat, and gay foulards under it—is an object of admiration. Does he not wear them all with such a grace? On our downward path we are obstructed by a crowd of leather-legged muleteers and leather-headed teamsters. They stop the diligence. Is it revolution or brigandage? Neither. We are told that the big cart of the teamsters, loaded from the bottom thereof—which touches the ground almost—to the round top thereof, has been upset and tumbled down the mountain, with its three mules tandem! It is hard to believe it, as the team is up on the road again, and the cart put together, but immovable. The country people have helped to put the cart together and to reload. We are requested, with our eight mules, to help in pulling it up the mountain. Our *mayor* assents. He unhitches six mules, and with halloo and screech, whip and push, the heights are gained.

As we approach Malaga, the mountains are dotted white with villas. How clean, sweet, and un-Espagnol they shine under the glowing sun! Nowhere else in Spain have we seen so many beautified spectacles of country life. We still approach, on our spiral downward path, toward Malaga. The country beneath us looks like a raised map of a Swiss canton! Now! See! all at a glance! Malaga itself! Around it are greenish and golden square plots of land—evidence of cultivation and of crops; the green of fields unreaped; the gold of fields harvested; and both floating in a flood of sunlight! Malaga is still ten miles off; more than an hour. I count for a little time the peaks of mountains in view—before we go down into the valley. These peaked mountains, which look so far below, look very high to us, when *we* are below. If a grand glacier could be changed into rocks, and the rocks abraded into smoothness, then seamed by rains, then decked with vines, and then dotted with blanched villas—it would seem to be the country immediately around and above Malaga. But it is not round *these* acclivities we move. A valley is between them and us. We approach Malaga from the north-east. These physical phenomena are on the north-west. Between them and us is a river; a river without a name, save 'River of the City,' and without water; for it is a bed of gravel, used in dry times as a highway, and walled to protect the city through which it runs from sudden, devastating winter floods. The river is used up, to grow the grape and other fruits, before it reaches Malaga. It appears not in the stream, but in the greenery of the plain above the city and for many miles out. As Malaga has over

109,000 people, they must be fed—aye, and watered. Wine will not do. Raisins are not made of calcined rock or ashy dust. Given the water, and then the white grape, which the youngster of New York makes into wine beneath his molars, grows into translucent beauty! Given the water, and hence the miracle! It becomes the sweet muscatel which makes Malaga magnificent under her canopies of grey mountain and cerulean heaven! It is this toothsome grape, made without any other sugar than the sunbeam melts into the juicy branches, and again, without any aid but its own chemical qualities, made into the raisins which the children of many lands roll as sweet morsels under their tongues—it is this that gives to Malaga its commerce and its importance. If Malaga had no history, her million boxes of raisins per annum would sweeten her memory in the hearts of childhood! If neither Phœnician nor Goth; if neither Scipio nor Tarik, Roman nor Moor; if neither the first Ferdinand of Spain nor my Bonapartist General of Ajaccio—Sebastiani; if neither the French Loveredo nor the Spanish Espartero, had taken, besieged, conquered, ravaged, or held Malaga, its name would be glorious to the connoisseur of the 'mountain' wines; and its Lagrimas, or ruby tears, which drop from the unpressed grape, would fill the goblet of its fame!

Aside from the vines and wines, Malaga is of interest as a sanitary resort. Bronchitis and laryngitis, accompanied by loss of voice, are here cured completely, according to Dr. Francis's book. The atmosphere is dry and warm. Fires are seldom used even in winter. It is, therefore, the Paradise of clergymen, lawyers, and orators, whose throats have been damaged by overmuch speaking. The thermometer is rarely below 60 degrees. The sun is very warm during the daytime in summer, but the mountains of snow are near. The winds lose their harshness before they reach Malaga. The latitude is a little above 36 degrees. It is warmer than Algiers, because protected by mountains from the north, which Algiers is not. People live to a good old age here and hereabouts. There is a proverb that in this part of Andalusia 'old men of eighty are chickens!' The climate is drier than Rome or Naples, but not so dry as Nice and Mentone. Still it is free from the harsh mistral of the Riviera. It is warmer in the winter than Madeira by 2 degrees, than Rome or Pisa by 6 degrees, than Nice by 7 degrees, than Cadiz or Valencia by 3 degrees, than Pau by 10 degrees, than Lisbon by 5 degrees; and is colder in winter than Malta by 1 degree, than Cairo by 6 degrees; and, like Corsica, it enjoys much equability of temperature. Its nights are most deliciously cool. Indigo, cotton, and sugar grow here, and nowhere else in Europe. We see that the wheat fields have already been cut, and May is not over! Hence I may affirm that 'Winter Sunbeams,' for which I have made such diligent search, may here be found in golden profusion. There is only one drawback—the river-bed is a daily and nightly nuisance. I cannot explain further. The starry jasmine and orange flowers, with which I have decorated the head of this chapter, are not intended to apply to this part of the description. The people of Malaga ought to be ashamed of themselves. They make the empty bed of the river a filthy sty—I beg pardon—pigs would hardly do worse. The people wait for a winter freshet to wash out the river-bed into the tideless sea; and your readers can infer the results! If I were to winter here, I should seek a home upon the hills or under the mountains, in sight of the sweet blue sea, and afar from the ill-flavoured port!

It was not without interest, however, apart from the

health, wine, and grape questions, that I regarded
Malaga. Its fertile vega, nine miles wide and twice as
long, was very beautiful to the eye, as its dry, bright
air was very grateful to the lungs. Its cypresses and
lemons, its few palms and much dust, its empty river-
bed, by no means as fragrant as its alameda; its cathedral, once a Moorish mosque, then a Gothic church,
and now of mixed Corinthian, lifting its earthquake-
shaken towers into the clear air 350 feet—all these are
for the tourist to note; but it was not these, or any
of these, that stopped our diligence on the outskirts of
the city.

We had rumours of insurrection all the day of
our sojourn at Malaga. The Cortes had just voted
down, by over 100 majority, the proposition for a
republic! Castelar, the eminent Republican orator,
whose words of fire had burned into the hearts of the
Malagueros, had just made his last splendid protest
against kings and their craft. Malaga rang with
.praises of a republic. Malaga remembered her martyrs of last fall. Far out upon the vine-hills, we heard
that Malaga would rise and pronounce—nay, had risen
and pronounced—for the republic. Hence the stop
at the gate of the city; hence the eager query of the
diligence passengers: 'Is there trouble in the city?'
'*No.*' 'Is any expected?' '*No se sabe*,' with a
shrug. 'What do you think, Señor?' '*Sabe Dios,
quien sabe?*' God knows—who can tell?

We all take our places in the diligence and ramble
into the city. Soldiers are plentiful as blackberries.
Being a stranger, and having two ladies to convoy,
and remembering how last fall a party of Americans
were fired on by the government dastards here, I
conclude, after taking advice and after sending out
some scouts, that it was best to leave on the next
day. My scouts reported that the city was as omi-

nously still as death; that it was generally lively, but not now; that the music, so often heard in Southern Spain, and which bears the name of Malagneno,—that gipsy, Moorish, oriental air, that most lamentable of laments, — was heard no more and nowhere; that the Alameda, generally crowded during the moonlit May evenings with bewitching Malaganenas, was now deserted, save here and there a suspicious group of citizens, watched covertly by detectives and dogged by Dogberry,—by watchmen with lantern hid and pike concealed! Another report came that, by the morning train, flocks of people had left the city, citizens and strangers; in fact, that the hotel, as we had occasion to notice, was empty, and that the cry was still 'they go;' further, that the republican committee had abdicated all responsibility if troubles came, and that proclamations inflammatory and pacificatory were on all the walls, and read by quiet groups; that General Thomas, an Englishman, or of English descent, was in command, and was a determined man, and had said that he would shoot and kill, on the first outbreak, in the most miscellaneous way; and that there was an unusually deep feeling as a consequence of these provoking threats and preparations. The soldiers were kept ready, not alone in the city, but in the castle, for the first popular demonstration. My courier had been finding out where the American flag floated, and where our eagle perched, so that we might, in case of danger, retreat under their ample fold and wing.

We prepared to go to bed under the excitement. As our rooms looked out on the Alameda, there was little sleep done by us or any one. Far down the avenues could be seen stealthy groups. Now and then is discerned the flash of a match, to light a cigarette; but no more. The moon came out, and so did the cavalry. Two battalions clattered and thundered over the city. Their iron hoofs made the fire fly from the pavements, and their swords gleamed and glistened in the moonlight. Then came the steady tramp, tramp of the soldiers, and then at every moment, from every part, the whistle of the watchman. Not a sound was made, save from the horses' hoofs, muffled tramps, and whistling signals. This went on all night; but 'not a gun was heard, not a funeral note.'

Next morning everybody seemed careworn and sleepy. A mist obscured the mountains above. That old Moorish castle, near the hill of the Pharos, is called the Alcazaba. Its Puerta de la Cava is renowned, if not in history, in legend, as the scene of the suicide of Count Julian's daughter, whose woes brought on the Moorish invasion, and whose Iliad has been sung in prose by Irving. This castle is hid under a veil, even as Irving dropped over its rigid outlines the drapery of his genius. As we drive to the depot, we perceive hundreds of soldiers about the train. There comes at last a sense of relief, when our party is fairly in the cars and we rattle away from the insurrectionary town. The mist lifts a little. We see a streak of sunlight on a bleak, bright mountain ahead of us. We pass by gardens of immense fig-trees. The mountains begin to shine white. We are in the vine-hills again. The vines are very, very plentiful. The grain-crops are in. Cactus, oleander, orange, and pomegranate—all these appear; but the grape is still king in this republican province. Bacchus is here a democrat! He dominates by numbers. We pass establishments where the raisins are prepared for the market. As three-fourths of this trade is with the United States, it is of interest to know that the muscatel and *uva larga* are most used. The grape-stalk is cut partly through, and then the grape dries under the sweetest of sun-glows. These make the best raisins. The common sort are called *Lexias*, and are perfected by being dipped in a lye made of burnt vine-tendrils. The green grapes, whose seeds shine through the clear skin, as if in emerald amber, are sent to England and America in jars. I do not mean that other fruits are not of Malaga. The finest oranges and almonds here abound, and as we dash off from the vicinity of this fruit-abounding place, we have a chance to observe why they so abound. It is the same story I told you of Alicante, Valencia, and Murcia. It is the same story which made Milianah another Damascus, and Grenada peerless,—with the rarest elegance of cultivation—it is irrigation. 'Water is the WAHAN of Creation,' saith the Buddhist.

I said much about this element of Spanish wealth, but did not give you the facts and foundation of the system. Having now seen under my own eyes the principal irrigation works of Spain, I feel more competent to write about it. In Valencia, from the rivers Turia and Jucar, there are 56,810 acres under irrigation; in Murcia, from the Segura, 25,915; at Orihuela, from the Segura, 50,318; at Elche (place of plumy palms), 40,010; Alicante, 9139; at Granada, from the Darro and Genil, 46,930 acres. These works, whose results at least I have witnessed, comprise nearly one-half of the irrigation area of Spain. The total quantity of irrigated land is 374,269 acres. The best works are those of Valencia, Murcia, and Granada. They are the oldest, being made by the Moors about the year 800. On the Spanish conquest and division of land, the rules of the Saracens about water were re-established. Some of these works are made of masonry, in which rain is collected, as at Alicante. The principal rivers of Spain, like the Tagus and Guadalquivir, run to the sea with but little utility as motive or creative power. But almost anything, and in any quantity, of vegetable beauty and utility can be raised in Spain with this water power. Peppers and peaches, apricots and apples, olives and oranges, sugar and citron, cotton and corn, potatoes and pears; and never less than two crops a year, and sometimes four! Of course, irrigation enhances the value of the land. In Castellon good irrigated ground is $700 per acre, dry ground only $50. In Murcia $3000 per acre is given for the good ground there irrigated. In Valencia it averages from $1000 to $2000 per acre. Irrigation adds 1200 per cent. to the valuation of the land. The water is an article of lease, trade, sale, auction, for it is as indispensable as the land. It is the same as on the Riviera. Around Cannes, Nice, and Mentone the water is sold by the cubic foot, or per second! The owner of a rivulet is rich. There is no poetry in calling the mountain streams *silvery*. It is fact. On some of the government works, they let out the water per cubic foot per annum, as boards of public works in America do for milling purposes. The same is done at the foot of the Alps in Italy.

Of course, there must be many difficulties in the division of the water among claimants. Water is mobile. It is a leveller, and gets through all the smallest crevices to its position. It is a very litigious element—jealous of its ancient and natural rights. The Moors used to have courts for water cases. These met and yet meet in some portions of Spain —to hear and decide complaints in the olden way and in the open air. Pedro has done, for example, as many an English or American miller, according to the law-books, has done—practised hydraulics on the sly—*i.e.* dammed up a few inches, or undammed a few, for a little decrease of his neighbour's water

power and a little increase of his own. ' Good! But
Pedro's neighbour grows wroth. He goes to the court,
and there the dusty titles from the dusky Moorish days
are examined, and the case decided, when the damming
and undamming are supposed to be suspended; supposed I say.

The country we pass through from Malaga to Cordova, first in a westerly and then in a northerly direction,
is well worth crossing the Atlantic to see. It is interesting, not alone for its illustration of the power of water
and vegetation, but for the railroad views in the very
midst, or heart, or innermost core of the immense
Sierras which the road penetrates. The mountains
ever hover on the horizon's edge as we travel west,
much as we saw them at Alicante, parched, white, and
rugged; but the valleys smile with fields and fruitage,
and the streams which made them smile, are fringed
with the same vegetation that we saw in Algiers. Indeed, the flora here is all African. The people have the
Moorish tastes, but they have besides a dash of chivalric romance. The red kirtles of the women are picturesque. Here and there we see quite a tableau.
Yonder, under a tamarisk, upon the bank of a stream,
dignified with aloes, and red with oleanders (for they
are in full blaze here and now), is a group of peasants.
They are tending their cattle and sheep. A small
boy is tinkling his guitar, while the bells of the animals
in response make the scene alluring enough to be
Arcadian. We observe some Indian corn, which
makes an American feel at home; then, at Alora and
other places,—country seats for Malaga merchants,—
we perceive palms, and oranges, and pomegranates,
and apricots, and cypresses, and Swiss cottages. These
are all made by irrigation; cottages and all! How
deftly the water is turned in and on. The peasants
have here short, white trowsers (legs half bare), which

are very loose and split up. These pantaloons have furnished the Spanish student with a symbol of his— purse. Here we·see a shepherd with a white crook. He looks patriarchal. He is watching his sheep, and his donkey is watching him! Now *that* is Arcadian! As we think of it, he swings his pastoral emblem in a jolly way at the engineer of the locomotive. A donkey in Arcadia is tolerable; but a locomotive is not to be endured. It disenchants us of Sylvanus and Faunus. It gives a new 'colour' to Pan's pipe.

As we approach the mountains, we wonder how we are to get through, the red rocks look so formidable! Soon we are in a gorge, then a tunnel, and then dash we go into the cool air and thick black- ness, under a mountain several thousand feet high! Then out of the tunnel, and lo! we stretch our necks out and look up! There is seen only a little patch of blue, which breaks down from the heavens, struggling through craggy peaks and ragged rocks! The moun- tains here are honey-combed and water-worn, as in Corsica. They are iron-painted. Some have the form of palisades; some have every phantasy of shape. It seems as if the engineer who laid out this route had begun at the nadir and had worked up toward the zenith, and that we were dropped somehow into the midst of wonderful defiles and necromantic halls, surrounded by spires and walls of illimitable height and grotesque for- mation.

It was worth risking an insurrection, to see these marvellous mountains, and how the art and science of man have overcome, or rather *under*-come them! But for this tunnelling, no being—not even the goat or chamois—could ever have got into, or got out of these profound depths in the innermost penetralia of these natural temples! Science has lifted the rocky veil; the adytum has been pierced!

As we proceed, we also ascend. The vegetation changes. From gardens to rocks; from the palm, whose fruit are luxuries, to the scrubby palm of Africa, and the prickly broom, whose yellow flower so often appeared in Algiers, until at last we reach Teba. There the French Empress has a splendid estate. It is not in view; no matter. The road rises still. We are on terraces, and above us are mountains, themselves terraces. Then long reaches of level plain appear between the mountains. And as we go up and on, all at once we feel that we have passed the grand sierra which hugs the Mediterranean coast from Gibraltar to Almeria, that we are out on the open plain of Andalusia, whose plateau is swept by the breezes from the Atlantic without hindrance of mountain; that we are on a wide, high prairie, whose northern bound is the Sierra Morena, defending it from the blast; near whose feet is the Athens of the dark ages, Cordova; down whose gentle western slope the Guadalquiver (pronounced *Waddle-kiv-eer*), red with the fresh mountain mould, runs to commercial, historic Cadiz, and past thy stately towers, O glorious, courtly Seville! Here, then, we have Andalusia—the Tarshish of the Jew and the Paradise of the Moor! Here, again, came the Augustan era of literature to the turbaned people, and a golden age of chivalry, inspired by a perpetual religious warfare! Here is the land of the bandit, contrabandist, the dancer, the bull-fighter, and the muleteer. Here is the land of song and story about love and war. Here were the great captains; nay, as I live, here and now, from this depot of Montillo, I see the town where the 'Great Captain' Cordova himself was born. Mantillo, whose wines are like liquid amber, with no admixture of Celtic esculent—sipped by connoisseurs the world around—here, indeed, is its romantic tower and castle, protecting its town of 15,000, whose fame is preserved more by the fragrance of its liquid amber than by the glory of its great soldier.

I know that half of Andalusia is in a state of nature; its soil so fertile and its sun so warm; its waters so skilfully trained to work, and its coasts so grand in harbour and beauty. Yet half the land is given over to the palmetto, the oleander, the lentisk, and licorice. Its aromatic shrubbery may well grow where the rude old Egyptian plough does its imperfect labour. The Moorish pump, with its wheel of jars, turned by donkey or mule, does the work in a day which an American pump would do in an hour! But no! Twang your guitar, happy Andalusian! On with the festive dance, Majo of Cordova! Snap your castanet, dark-eyed daughter of Seville! Let Love's rebeck resound! Let Fandango hold high festival! What have hydraulic olive-presses, centrifugal pumps, and threshing-machines to do with thee? You repeat to-day, in habit and custom, what your Moorish predecessors repeated in words more than deeds: 'Ojala!' 'GOD WORKS FOR US!' Let us be resigned!

How we reached Cordova; what we saw at Seville; and what between these ancient homes of Oriental power and Spanish grandeur, and these present abodes of luxury and beauty, may be reserved for another chapter. We certainly have ceased to wonder at Irving's remark that Andalusia is a garden, while Mancha is a desert. Its undulating hills, grain-covered, remind us of Ohio; its treeless plains remind us of Illinois; its flora of Algiers; its fountains of the Riviera; but its distant castellated mountains, its romantic, towered towns, its alhambras and alcazars, its vine-covered acclivities and palm-decked villas, its donkeys, its fruits, its marbles, its costumes, its songs, its dances, and its Orientalism, are all its own!

Having found a place of repose, with all the 'Sun-beams' required for tonic and illumination, I should like to be free to take a glance all around! I have traversed the Provinces of Madrid, Mancha, Jaen, Cordova, Grenada, Malaga, and Seville; besides sojourning in other cities on the way. I have already written about Murcia, Alicante, and Valencia. While allowing much for the historical interest and present attractions of Northern Spain which I have not seen, especially at Burgos, the capital of Castile; Barcelona, the capital of Catalonia; and Saragossa, the capital of Arragon—I think that I have already seen the best of Spain. I can form a fair idea of its soil, climate, productions, people, and geography. We have been in no hurry through Spain. Notwithstanding that the political entanglement still continues, and that the chair of state is hovering, like the coffin of Mahomet, between heaven and earth—between republicanism and royalty —that is, it is about to take the form of a Serrano-Prim-Regency—we have experienced no difficulties.

One reason why I have lingered in Spain, into the summer, is that it is the pleasantest country to be in; I mean for climate and temperature. We have not as yet had a hot, or even a warm day. Even at Malaga, which is on the level of the sea, there was a gratefully cool breeze at evening, which made the climate more agreeable than some other experiences. Spain, along its Mediterranean coast-line, from Barcelona to Cadiz, is tonical and delightful in its climate. Murcia and Malaga are unparalleled for dry, exhilarating air. Spain, in the interior, is a lofty plateau. It averages over two thousand feet above the sea. It is treeless and in summer burnt. In winter it is cold and dreary, but as we have found, in spring, and so far, one day into summer, it is cool and bracing. Spain is the land in which to be re-oxygenated. Everywhere the water is

good; better for the tourist—as a matter of health—
than the wines. The peptic effects of the climate I
have already referred to. The natives dip their bread
in oil; or, making holes in the loaf, fill and soak it
with oil. Then they swallow with a relish, and I hope,
digest it. Spain is free from malaria, except in the
rice lands about Valencia. In 'Winter Sunbeams,' of
which I have been in search, and whose glow stimu-
lates, southern Spain is rich. I do not mean to dispraise Italy or disparage France. The sierras of Spain
do the work of protection from the Northern 'eager,
nipping airs,' which the Alps and Apennines do for
the Riviera. The peninsular configuration of Spain,
with its mountains running east and west, shelters the
shores.

There are some most select and delightful influences
in Spain—I call them intellectual and moral tonics.
I wish I could enumerate them all, so as to impress
the reader as I have been impressed, and so as to dis-
possess him of many prejudices which I once had.
There is the skyey influence, which has balm in its air
and its light. There are the associations of the sea-
coast and the country. There is the grand architecture
and historic renown of the Spanish cities—the fossils
and remains of antique civilisations—Phœnician,
Roman, Gothic, and Moorish. There are the masterpieces of art—superior art—art of the time of
Charles V. and of Leo X.; the genius of Murillo
and Velasquez—illustrated in the Museum and the
Cathedral. There are to be found here the grandest
Gothic minsters, enshrining the most exquisite relics
of religion as well as those of the painter and sculptor
—minsters where God is worshipped with a state so
solemn and a fear so awful that His presence cannot
but hallow the shrines. Then, again, here are the
influences of the authors—Cervantes at the head of

the roll—who have peopled Spain with their *personæ* and poetry. All these influences make Spain a museum, a library, an asylum, a retreat, delightful for invalid or misanthrope, for traveller or student, for artist or author, for old men or maidens!

But why are there so few who travel hither and here? The Englishman and American rush over Europe and omit Spain. The reason for this omission some years ago is explicable and apparent. Then Spain had no railroads, and her other roads were unsafe. To cross the Pyrenees then, was to go out of Europe. It was to eat nothing but garlic. It was to leave the land of cookery and France. 'God sends Spain meat, but the devil cooks it,' said the French. The snobbish Englishman says: 'Cawn't get the "Times" there, ye know, nor Ba-asses pale ale, ye know! Nothing but a bloody bull fight in the cities, and howwid, wascally wobbers in the country, ye know.' An American abroad is either a noodle, a doodle, a scholar, or a business man. The two former scarcely ever go to Spain. The American scholar has made Spain glorious in English literature; and the business man goes thither either for wines and raisins, or to consult about—Cuba!

The mountains of ashy hue and the valleys of perennial green, the wide prairie and the rugged rocks, the dash out of the vineyards of Malaga into the lovely vega of Grenada, or from the glorious Alcazar of Seville to the many-pillared mosque of Cordova, from the Elysium of Valencia to the Sahara of Mancha; all these, a quarter of a century ago, had no attraction, and, as yet, have but little charm for the dilettante tourist. Not a few, however, are prevented from travelling in Spain by groundless fears of political disturbances. The land is volcanic in political as in other respects. Our minister testifies to hundreds of letters, as to whether it is safe to venture this spring and summer into Spain. He answers, as I do, 'Yes!' No safer place. If there be danger, it is sure to be sounded beforehand. The Spanish are very pronounced. They are incarnate pronunciamentos dressed in soldier clothes, and a trumpet is always noising their political vicissitudes. I have been most agreeably disappointed at the perfect order, safety, and pleasure of Spanish travelling. Whether by diligence, rail, or carriage, it is the same. The worst that can be said is that the butter is bad. My worst profanity on its appearance has been: 'Oh! Lard! Oh! Lard!' But you soon become used to it, or to its absence. The trouble of travelling in Spain has been much lessened, and the pleasure much enhanced, since the revolution of last autumn. There are now no octroi dues. Figuerola and Serrano have effected this, along with other free-trade reforms. Your trunks are not opened at the gates of cities. The railroad people are civil. The railroads are generally as fast, and as comfortable and safe, as those of America. The trains wait longer at the stations, but the cars are excellent and comfortable. The inns—whether you call them fondas and posadas, or ventas—might be bettered; but they form no objection to the travelling here. You have all the comforts you require, and more than you expect. You have always good fruit, such as raisins, oranges, and apricots. The wines are of all qualities, and you may select. The Swiss are opening hotels in the principal places. One Swiss company has several hotels. This arrangement I found convenient; for when I grew short of funds, or negligent in my exchequer, I paid my bills, for example, for Seville and Cordova, at—Madrid! The money is easily understood, as the old Spanish milled dollar is the real standard. Generally, the people count in reals; one real being over four cents, or about four reals to a franc. The peseta—a common silver coin—is four reals. It resembles the old twenty-cent piece which used to be seen in America, and which used to slide in for a smooth 'quarter of a dollar.' It is a little smaller than the quarter, and without the pillars of Hercules—those grand symbols of *ancient* Iberian power! The cent is called a quarto, and like its double—'dos quartos'—it is copper. The gold coinage is like the American, or as ours used to be in the good old days. There is a one-hundred-real piece, equal to five dollars; about the English sovereign. It would not be hard to change the Spanish coinage into the decimal system. Already the French decimal has been introduced for measures. The old yard, league, and quart are not altogether out of vogue with the people or peasantry, but, under official rule and proclamation, they are fast becoming obsolete.

It is the custom of Englishmen, especially, and Americans too, to underrate Spain, and depreciate the Spanish people. Englishmen say that they—with the Duke—won all the battles here against the French, and that the Spaniards dodged the dangers and ran. I am not going to say that the Spaniards by their own unaided force, at that epoch, did full justice to their former prowess and fame. But I do say, from observation of French, German, Italian, and other troops, and by comparison with the Spanish, that I do not see any difference, in either appearance or discipline, to the disadvantage of the Spanish. We are apt to take as a standard the officers, grandees, or *élite* of a people. As to Spain, I admit that there does seem to be a degeneracy among the better (?) classes. Compared to the portraits and pictures I see—and not relying on the magniloquent histories I read—there is an enormous falling off. The Cids and 'Great Captains' are now *raræ aves*. But in the living towns of Spain—on the sea coast, as at Cadiz, Malaga, Carthagena, Valencia, and Barcelona; or, in such places as Seville and Madrid—the people look and act equal to any work in the field, whether with a hoe or with a bayonet! If there transpires here what may happen on the election of the king—*i.e.* a civil war—you will see that Andalusia, from Grenada to Malaga, will be foremost in the fight for self-government without a king. All the peasants and tradespeople whom I have met are republicans. They understand what that means too. A rough peasant whom I met in the little town of Santi Ponce, near Seville, told me that all his townspeople were republicans; that they contemptuously sent about his business the day before a fellow who came there to peddle monarchical newspapers; and that they read the republican papers and understood their own affairs. The free speech, free education, and free press of Spain have made a wonderful revolution. Everywhere the boys in the cities, or rather the blind men and poor women, especially at Malaga, Madrid, and Seville, are crying their papers, just as if they were in London or New York. But I do not wish to vaticinate about politics, at least not till I get to Madrid, where the cauldron is boiling and bubbling, and where Serrano as Regent is soon to be installed.

But it is a little interesting, if not significant, that in the village of Santi Ponce—occupying the site of the magnificent city of Italica, whose ruins I will hereafter describe—the city where Trajan, Adrian, and Theodosius were born—founded by Scipio Africanus for his veterans; a city once adorned with sumptuous edifices, and which was so fond of imperialism and Rome that it sought to lose its character of a free municipium by becoming a Roman colonia,—it is not a little significant that the people here are all republicans! A bright-eyed girl, of one of the huts where we bought some old Roman coins and mosaics, sang a song in a jolly way to us, to show us the feeling of her vicinage, the burden of which was :—'The republic we seek we will have. If they don't like it, they may swallow it!' That is a free but correct translation, though I am not sure of my Spanish.

CHAPTER XVIII.

SUMMER IN SPAIN. SEVILLE, SCENES, SOUNDS, AND SENTIMENTS.

'Fair is proud Seville!   Let her country boast
Her strength, her wealth, her site of ancient days.
Beneath soft eve's consenting star
Fandango twirls his jocund castanet.'—BYRON.

THIS is the first day of summer!   This spring, however, has had no chill of winter. It lingers into summer, and preserves its vernal charms.   Like a coy maiden—'only not divine'—spring, by her backward ways, not only attracts, but in such a way as to repel from any exertion, especially that of writing.   We rather live and breathe than work and move.   It is the very season of Flora yet, and in her own primal beauty and favoured clime.

I knew, by reading and repute, of the heat of the summer sun of Spain.   The dog days are as fierce as the fabled hounds which ate their keeper.   No drinking of water, no screening of head and eye with umbrella or green glasses, no awnings in city or shades in country, no stone wall or enormous head gear, can temper the heats of Spanish midsummer.   The sun roasts, fries, and bakes you, as well as the already calcined soil.   Was it not Sancho who put the curds into the Don's helmet?   It must have been in summer, for they melted so fluently that the Don began to think that his brains were running.   The soil cracks and gapes all athirst. The rivers have died out, for want of water.   The grass, where irrigation is not, is as shrivelled as an old woman's hand after a hard day's washing, and as brown as sienna. The olive turns pale with heat and dust. The donkeys, almost alone of the animals, imperturbably plod their meek and weary way. The heated traveller rushes into every venta, and the porous earthen refrigerator is emptied to cool his parched tongue. The proverb has it that the sun is the cloak of the poor: '*Es el sol la capa de los pobres.*' I think that proverb was made for winter, and is no good proverb. Proverbs should be applicable to all time and seasons. I, therefore, contest it, on the same principle that I would have contested Sir Roger de Coverley's will. He made the will in winter, and left all his servants (bless him!) greatcoats; but he died (alas!) in summer, when greatcoats were useless! The Spaniard boasts, not only that the first language was his, but that the sun first shone over this old capital of Toledo! I will not contest that, but simply say that not as yet has the sun, with Spanish courtesy, taken off his sombrero. We, therefore, linger. When Apollo begins to shoot his arrows, we shall retreat.

The reader may find us following the meanderings of the Guadalquivir, and about, like Kalif Walid, or San Ferdinand, to capture Seville. In order to reach Seville from Malaga by cars, we had to go above one hundred miles north to Cordova, and thence as far south-west to Seville. We have now accomplished the sight of both those ancient capitals. With Jaen and Grenada they make up the four kingdoms known as Andalusia, so sedulously hidden by the crafty Phœnician merchants under the (still) inexplicable *nominis umbra* of Tarshish, so tenaciously held by the Moors, and so splendidly glorified by Spanish poetry and romance with all the hues of the Orient! This is Andalusia! I was disappointed in Jaen and Cordova,

but not in Grenada or Seville.   Of Grenada I had
read much, in Irving and elsewhere.   My impressions
had become indelibly photographed.   Of Seville I had
formed, if I may thus express it, a sensuous ideal!   It
was to me the gorgeous East and the fruitful South.
It was the epitome of Asia and Africa.   Here, too, was
to be found the severe taste of the Goth, with the
elegant refinement of the Moor and the haughty
grandeur of the Spaniard.   Here were all the affluence
of nature and the skill of art.   I was not disappointed.
But how can I reproduce on paper my impressions—
and under circumstances that almost forbid the group-
ing of incidents and thoughts; how cast a reminiscence
even of a week on these absorbing scenes.

Perhaps the best order is chronological, and the best
narrative is the simplest.   On the railway we had the
courtesy extended to us of seats in the saloon of a
reserved car.   This was done by the kindness of the
engineer and inspector of the roads in Andalusia.   He
was a cultivated and communicative man.   He had
relatives in Malaga whom we knew.   To them we
were primarily indebted for this valued courtesy.   We
learned everything from him about railroads and
politics. The railroads of Spain do not pay well. The
charges are high, the running is slow; but the travel
is safe.   You never have an apprehension about danger
of life or limb.   There has been, even as recently as
last autumn, apprehension from another source.   You
will always see most respectably dressed police soldiers
about the stations.   They are uniformed in cocked
hats and Quaker-cut coats, white pants and high
boots, looking like General Washington as a gendarme.   They are observable at all the depots and
along all the routes in Spain, because, some time
ago, on some even of these main thoroughfares,
there were brigands.   Yes, indeed.   They attacked

the cars here even this winter. They do not throw the cars off the track. They are not diabolical enough for that. Even they are too chivalric and Andalusian for that. But they move down on a depot, salute and present arms politely to the conductors, firemen, and engineers, and, courteously leaving the passengers intact, proceed to examine the freight, bills, and cars, and appropriate what may strike their fancy as useful. Like the robbers described by M. Huc, in his book on Thibet, they are the very pink of verbal urbanity and predatory politeness. You remember how the Thibetan brigand procured—oh, call it not 'stole'—a horse and cloak. 'Venerable elder brother,' said he to the elderly chevalier, 'I am fatigued and footsore with my journey; wilt thou have the happiness to dismount?' or, 'The sun is hot, very hot; your cloak, my honoured lord. Is it not too warm for thee and the season?' Thus the Andalusian robber.

Our friend, the inspector of railroads, said that the brigandage here is nearly obsolete; but if there be a civil war, it will arise again. 'Will there be a civil war?' we inquire. 'No, not yet. If no outbreak yesterday in Cadiz or Malaga, when the vote for monarchy was carried in the Cortes, then it will not come, at least not yet.' 'When then?' 'When the time comes for choosing the king. Just now, the proposition for a regency, and the probability that one of the popular leaders of the revolution will be regent, the full powers of a republican president being given to him as regent by the Constitution, take away the immediate motive for a republican uprising.' 'Who is likely to be king?' I ask. 'We have one candidate in Seville, where lives the Duke of Montpensier. He is now in Lisbon. He has money. His money made the revolution. He has suffered for the cause, and was exiled by the Queen.' Our friend, I thought, expected the Duke to obtain the crown. 'Then,' he continued, 'there is the ex-Prince of the Asturias—Isabella's son and heir; but he stands little chance just now. Some think General Prim desires it, and full a half-dozen others.' 'What of Serrano?' I inquired. 'He is very persuasive and popular.' 'What of Montpensier?' 'He has shown himself too much of a poltroon. The people like pluck.' 'What sort of a statesman is Figuerola, the Finance Minister?' 'Ah! he has a splendid intellect, is a thorough theorist, and yet is a practical reformer. He has taken office to carry ou his ideas. He believes, as "New Spain" does, in commercial freedom and unrestricted intercourse. He has a difficult problem, as he has cut off many sources of revenue, and, with all his economies, he is in a deficiency. He is lampooned and abused, and has a hard time, especially with Catalonia, Prim's province, where there is so much done in manufactures. There the people are protectionists. General Prim has encouraged them to cry down free trade; but being republicans, although they want to have all people tributary to them by buying their stuffs in Barcelona, they will be likely to rebel.' I learned from my companion that he favoured free intercourse. Said he: 'What is the use of my business—railroading—if we do not trade with other nations? These railroads are suffering because of bad laws; yet they were made mostly by foreign capital, and to reach out of Spain and Madrid, on a system, to every nation. The English own several roads. The French have built many. If we do not allow French steamers to come into our port without burdensome tonnage and port dues, and French traders to cross with their goods over and into our borders and interior, what is the use of railways? Home travel will not keep up the roads. We shall relapse into the old mule and donkey system.'

Spain is making progress in many ways not noted by other nations. She will yet refute Buckle's theory of her improgressiveness, absurdly based on her religion, earthquakes, and climate.

As we talked, the cars rolled us within view of Seville. We perceive it a long way off. Its many churches shine in tower, dome, and spire from afar. Its Giraldi, or cathedral tower—once that of a mosque—lifts the city from the plain, as St. Peter's lifts Rome from the Campagna. For seven hundred years this Giraldi, with its whirling vane—made up of a female figure and a flag—has played its demagogic part in the face of high Heaven; for is it not changing with every wind, as the *aura popularis?* Now it holds with Moor, now with Christian, and, regardless of all other influences, it has been true only in one thing, fidelity to its own whims. We have little time to see this rare tower from the cars. We find that we are in a lively city. Donkeys and venders of vegetables (see sketch) abound. The crowds at the depot literally *beggar* all description. Beggars abound; they indicate a prosperous city. I do not regard beggars as a sign of adversity. A goodly lot of them may be found in desolate places, but they generally congregate where there is something to beg for. The Spanish language is such a fit vehicle for a moaning tone that the beggars use that intonation even when there is no hope of obtaining alms. They whine for the very 'luxury of grief.' At railroad stations, I have seen the beggars thirty yards off, peeping through the palings at you in the cars, and then and there, utterly hopeless of response, making their piteous appeals to the Señorito for 'carita.' So sad are their melancholy tones that you feel reproached because you do not leap from the cars, break through or over the fence and fill their outstretched palms! When you do render them a service, what an outpouring it is? At Jaen, when our diligence stopped, we had about forty round us. I adopted a new plan, I picked out the most conscientious-looking person, a fine-eyed old señora of about eighty, and giving her some silver, begged her, as she loved her kind, to distribute it according to the needs of the crowd. I picked out, luckily, like a good drover, the leader of the herd. She started down the street to the fountain, the motley miserables following with murmurs of admiration. There, deputing one of the number to go and get her silver changed into coppers, she distributed them fairly. She came back to thank us for the trust we had reposed in her. At the Alhambra in Grenada there are some half-dozen gentlemen beggars of the tender years of seventy and upwards. These you invariably meet. They represent the Genius of the place. One evening while sitting all alone on the stone seat, near the wall, in the Plaza de los Algibes (or place of cisterns), one of these venerable local genii approached. He made his plaint. The nightingales were singing in the elm groves near; the fountains were plashing musically around; the dim twilight, creeping up the mountain, barely revealed features which Murillo would have loved to paint (for who can paint a beggar like Murillo?) The time and place were favourable to his prayer. I ransacked my pockets for coppers, but being out of coppers I gave him a peseta, a silver coin. You should have seen him; he kissed the coin; the water wells up into his eyes. Remember, this was the plaza of wells. Perhaps he had been in direst distress, who knows? He calls over the list of saints and invokes them on my head. His fervency makes me almost join in the water business. He invoked the sweet Saviour to bless me, and finally hobbles away with streaming eyes, covering the coin with his labial delights! Next morning, when I was looking out of the balcony of my Fonda, in the first dawn of the day, I saw below my venerable friend in a fight with two other elders. Their united ages were about two hundred and fifty years. One had knocked off my beggar's hat. Behind its turned-up, well worn velvet brim he, like others, carried his money and papers. The coppers rolled round in profusion. There was a noisy row then; not all noise either. It was not appeased but increased by the appearance of a couple of beggar women—female Methuselahs—on the scene. I came to the conclusion that I had expended my sympathy the evening before a little prodigally; but I will say this, my man fought nobly. This is the last battle of the Alhambra!

But there is no need that the traveller in Spain should be over-troubled with beggars. If he is recognised as a foreigner, he is sure to be confiscated to some extent. Why not? If, however, he learns to say: ' *Perdone Usted por Dios, Hermano !* ' ' For God's sake, my brother, will you excuse me?' The beggar will cease to whine his petition. All your negatives, even if polyglot and multiplied, from a crisp English 'No'— to a fine Castilian ' Na-d-a '—will not avail like this gentlemanly appeal to the chivalric mendicant. The philosophy of it is this: Every Spaniard is Moorish, Oriental, grandiose. The feudal system was never in Spain to degrade. Every one, the poorest, is as good as another. He feels it. If you respect his feelings, he respects you. Even the beggars consider themselves ' *Caballeros de Dios* '—the gentlemen of God!

How we escaped from the gangs at the Seville depot I hardly know. No one can be rough with these 'gentlemen of God' without exposing himself to the charge of being vulgar. Soon we are in our hotel in the Plaza de Magdalena. It is situated in a square, surrounded by palaces and decked with orange trees.

Fountains are in the centre, and the ladies of Seville
are already out and about for the twilight stroll. How
beautiful and sweet all seems! Our balcony—in fact,
our rooms are a part of the balcony—overlook the
plaza. So far as one can be in a house and out of
doors at the same time, we are. The streets of Seville
are narrow; though the Alamedas are wide, shaded,
and fine. There is one peculiarity here: canvas is
spread from roof to roof, shading the streets. I never
saw a gayer place than Seville. The fountains seem
to be more sparkling than anywhere else. In winter,
it is said to be wet, though it has no snow or ice. The
climate is dry. The houses, erected a thousand years
ago by the Moors, have never been harmed by frost
or much wasted by time. The city has a look of
Bagdad or some other Eastern city. I said in a preceding chapter that it recalled Damascus. The fountains made me think of that. The houses are made to
suit the climate; the narrow, winding streets, canvass-
covered and cool; the wide spacious houses, with their
Moorish courts, filled with gardens of flowers and
fountains; the iron-grated shutterless windows, protected by an *estera* or awning; the open-worked iron-
grate, partly gilt; the Moorish azuelos, or clean blue
tilings; the stem-like pillars of the court; the court
itself covered in summer by an awning; these not only
give the idea of Oriental luxury, tell tales of the
thousand and one nights, and lull the senses in delicious dreams, but convey the impression also of comfort, strength, and seclusion.

Byron said that Seville was famous for oranges and
women. I might add for its river and its fountains,
its *fêtes* and bull-fights; its cathedral, Alcazar, and
Alameda; its Roman ruins, its museum of Murillos,
its palatial tobacco manufactory, its Moorish memories
and municipal Nodus, and its former fame as a mart

of commerce and colonies. Its river, the Guadalquivir, is nothing of itself. I do not mean that either.
To an American, used to grand rivers, it is not an
imposing stream. Its waters were painted red with the
soil, for it seemed full with a freshet. Its banks are
low, but well walled in the city. It often overflows,
not only over the meadows, but into the city, as high-
water marks testify. It seems to wander about where
it pleases. It furrows out its way through the Andalusian plains to the sea. It may leave Seville, on one
side, some day, as it did the ancient Italica. Unlike
most of the rivers of Spain, it debouches in Spain and
not in Portugal. One great reason why Spain has
sought to annex Portugal, and why its king may
possibly yet unite Portugal with Spain, is that the
Tagus, Duero, and Minho empty themselves into
the Atlantic in *other* territory. It is the Mississippi
outlet question on a small scale; for those streams
might under sufficient protection and commercial
interests be made navigable. The Guadalquivir was
navigable as far as Cordova in Roman days. We
perceive now lying at the wharfs of Seville, opposite
the Montpensier Palace and the Alameda—which extends along its banks—many steamers. You may go
to Cadiz in these steamers, and thence by the same
line to London. The barges look very poor. They
take you back to the early days of the canal; for they
are like the old canal scows. The English brought
steamers here and they superseded the barges; though
it is claimed by Spanish pride that as early as 1543 a
steamer was launched at Barcelona—the first steamer
of the world! The Spanish officials opposed the
matter, and it died. It was left to John Fitch and
Robert Fulton to accomplish what Spanish ingenuity
endeavoured, and what Spanish stupidity foiled. The
Guadalquivir is not a poetical river to look at.

Spenser never would have called it into his fluvial symposium in Faerie land had he known it in reality. It sounds mellifluously. How its name glides glibly from the liquid larynx and trilling tongue! I believe it means, literally, 'pellucid stream!' So might the riled Missouri—red and yellow with two thousand miles of rushing—claim the same clear, silvery significance. The Guadalquivir seems to echo the sense of silver music. But it is not only turbid but dull in its flow. Its way is made through those level plains which mark one of the seven zones from East to West which divide Spain. It has all the size but not the beauty of the Tagus, which flows in my view and with a lively tune too, round this ancient city of Toledo. It has been credited with all the poetry of the Tagus, which was, according to Spanish grandiloquence, sanded with gold and embedded in flowers, while along its enamelled banks the nightingale sang his madrigals to the blushing rose. But it has what the Tagus has not—a mirage! By atmospheric refraction, glare of sun, and clouds of vapour, it seemed to the Moors that demons were playing tricks along the Guadalquivir. Armies, cities, and combats appeared and then evanished. They called it the Devil's Water. Ducks, cattle, donkeys, and sheep, here and there are found along its marshy banks, and some sickly inhabitants, but few villages. The Guadalquivir has hardly the merit of the Tagus, which turns many a mill. Yet must I not forget that once at Seville, inland though it be, and by means of the Guadalquivir, a powerful guild of merchants lived.

It was from hence, rather than from Cadiz, that the great discoveries of Columbus and his collaborateurs in navigation experienced that attention and spirit of adventure which made Spain, in the sixteenth century, so rich in silver and gold! The loss of the Spanish

colonies has decreased its importance. Before the time
of the fifth Charles it was the capital of Spain; and even
yet, with its 125,000 people, and its civil, provincial,
and military importance, it is not unworthy of its olden
fame. Its ecclesiastical rule reaches across the straits
to Ceuta, whither it has followed the Moors; to Cadiz
and Malaga, to Teneriffe and the Canaries. In earliest
days Seville and Cordova were rivals. The former stood
by Cæsar, and the latter by Pompey. Consequently,
when Cæsar triumphed, he stood by Seville, although
its people were more Punic than Roman. Seville was
even then rich and grand; but it had a rival near, not
Cordova, but Italica, where Roman emperors were
born. We visited its ruins, and will presently write of
it. Seville met the fate of other Spanish cities. The
Goths, who became as luxurious as those they over-
came, made of it a capital in the sixth century. I saw
in the Armoury at Madrid the gold crown of a
Gothic king, found amidst the ruins of Seville. The
Moors conquered the Goths. The same wild fanaticism
which deluged the then known world from Scinde to
Tetuan swept over beautiful Seville. It came—

> ' Like a cloud of locusts, whom the South
> Wafts from the plains of wasted Africa,
> A countless multitude they came:
> Syrian, Moor, Saracen, Greek renegade,
> Persian, Copt, and Tartar, in one band
> Of erring faith conjoined.'

But the wave left no unpleasing *débris*. On the contrary, the waters receded to show the rarest city of the
Occident. At one time tributary to Damascus, at
another to Cordova, then under independent sheiks,
and once a republic of Moors, it finally became the
scene of the most romantic and fierce of the wars
between Moor and Spaniard. It fell before St. Ferdinand six hundred years ago. And to-day this marvel

of history—this gem worn in Phœnician, Roman, Gothic, Moorish, and Spanish diadems—is poetically heralded to the present and the future as distinguished for its —oranges and women !

If you would see both those celebrities to advantage drive down the banks of the river, under the shade of the great plane trees; for Seville, like every Spanish city, has its alamedas. You may see the oranges peeping through the iron gratings of the Montpensier Palace; or, if you please, you may see their golden orbs glorify the old walls of the gardens of the Alcazar. They cannot rival those of Blidah, in Africa, or surpass those of Nice, though I would not dispraise their quality or beauty. The trees begin to bear in six years. The fruit grows richer for twenty years; then it fails. In March the blooms come out. In October the oranges begin to be gilded. They are then picked for commerce. They never grow larger after they colour. In spring the aroma from the orange-trees makes Seville sweet-smelling to satiety. The Seville people will not eat oranges till March, nor it is said, after sunset. The vendors in the streets cry them almost as volubly and musically as the time-honoured watchmen cry the hours of the night, and the condition of the weather. The cry is, 'Oranges—sweet as honey.'

As to the other celebrity of Seville—I mean *the women*—has not the cry gone up for many a year, '*mas dulces que almibar*,'—sweeter than honey, or the honeycomb. But, as there is no chronological or other order for the treatment of this most exquisite of Seville delicacies, I will reserve it till I see them in the national dances—under the brilliant light, moving to the telegraphic click of the castanet, the twanging tinkle of the guitar, and the mournfully sweet roundelay of the gauna. Anchorite you may be; but I defy you

to go beneath the flower-decked balconies by day, and
look up; or by moonlight pass the iron bars through
which the lover whispers his passion, and look in;
or pass down the Alameda, where the Orient-eyed
daughters of the Seville aristocracy are rolling in their
escutcheoned carriages, or, mounted on their magnificent Barbaries, witch the world with their graceful
horsemanship—I defy you to see those specimens of
the Andalusian fair without thinking of a thousand
romances of the days of chivalry, when Christian
knights fought for the Moslem Zaydis and Fatimas of
the Moorish harem; or of the times when henna-
tinctured fingers, partly opening the lattice, peeped
through the jalousy down upon furtive lover, or the
gay world from which they were excluded.

I said that the Seville women should be seen in the
Andalusian dances. You may not see the Spanish
dances at the theatre. The dance of the Spanish
theatre you can as well see at Paris, London, or New
York. Spain is still the land of the bolero and the
fandango, and these used to be a part of every play;
but playing after the Spanish method is at an end.
'Lope de Vega,' and 'Calderon' have given way to
Italian opera or French pieces. I would have gone
often to the theatre if I could have seen the genuine
tragico-comical hidalgo, in boots and bluster, spread
his large quantity of rhetorical butter over his thin
piece of artistic bread. Twice only to the theatre did
I go; once to hear 'La Belle Helène,' in Spanish, and
the Greek heroes never had so Spanish a chance to
swagger. Offenbach would have been delighted, for
they did it well. I also heard Tamberlik in Italian
opera; he is a favourite in Madrid. The audience presented him with a silver crown, and I fancy the audience did not pay for the crown. The bull-fight attracts
the Spaniard almost exclusively; yet, in Andalusia,

and in Seville especially, the national, inimitable spirit-
inspiring dance, called 'baile,' still survives without
theatrical help.   The castanet will stir a Spaniard even
more quickly than a handsome toss of a horse and
picador by a splendid bull.   We longed to see this
dance, not in theatric display, but danced by Majo
and Maja—the exquisites of either sex, dressed in their
native costumes.   We had already seen the gipsy dances
at Grenada.   The dances of the gipsy are not unlike
those we saw in Africa by Arab and Kabyle, and are
not very unlike the Spanish dances we saw at Seville.
These dances and these dancers have not changed
since the Roman days.   Tambourine, guitar, and
castanet, were described in the classics long before
Cervantes described their effect as like the quicksilver
of the five senses.   Hence, I conclude, from what I
have read and seen, that all these dances issued from
the Orient at a remote period of antiquity, and they
are not unlike each other in kind any more than in
origin.

We found that an arrangement could be made for a
funcion by our paying for the refreshments. (*Funcion*
is the word.   A funcion is the assemblage for a dance
in Spain.)   A funcion was, therefore, prepared at a
hall in one of the narrow streets of Seville, some miles
from our hotel.   We went about ten o'clock.   The
room is full of both sexes.   The men are smoking
their cigarettes.   That they do in every place.   We
are used to it.   A funcion is no exceptional place, any
more than the cars or the dining-room.   The women
are lively, and not all of them young.   Quadrilles are
under way as we enter.   Between the quadrilles four
señoritas dance the national dances.   They are dressed
in short Andalusian kirtles, pretty well flounced, very
gay, either crimson or yellow; bodice over the hip, and
a head-dress or cap coquettishly covering the chignon

behind, with pendants of ribbon rings. A huge gilt comb, stuck in jauntily on one side, ornaments the back hair. In one dance where there was 'a proposal' of marriage, the little, short, narrow, black silk mantilla is added for coquettish display. These dances begin by a loud screaming wail of a song, of which I have often spoken, the verses ending rather musically, in a tremulous prolonged quaver of—ahs. Then the guitar follows; then the dance is constant. The step is light, the motions are very quick, the whirl of body and poise of foot, the sway, the mien, the grace—these are indescribable. Did you ever see the little foot of an Andalusian dancing girl? In Mexico? No, sir. That will not do. In Lima, you say? Well, Lima has its satin slipper neatly filled. I will not quarrel as to Lima. The indigenous article in its neatest, smallest, plumpest finesse of a foot is to be seen only in Andalusia or in Seville; and that too by microscopic observation. How it twinkles! how it hides! What a new meaning to this little dancing verse :—

> 'Her feet beneath her petticoat,
> Like little mice, stole in and out
>   As if they feared the light;
> And oh! to see them dance you'd say
> No sun upon an Easter day
>   Was half so fine a sight!'

But *time* is called! Time in dancing 'is of the essence,' as lawyers say; and these petite feet keep it exactly. The 'limbs' have not so much to do with these dances as the rest of the body; but all is decorous. There is no ill-meaning. These dances have a history and historians. I will not dwell on their peculiarities. The most graceful girl, Fatima—a Moorish name—was one whom I christened 'Little Fatty.' She could walk on her toes, as if she had no fleshy avoirdupois to upbear. Although she has evidently made her ivory teeth do much execution, yet her ivory castanets do more, notwithstanding her plumpness. At this *funcion* we have a band, but the performers also sing as they play. They make the building ring with their wailing songs. The dances conclude with the famous *Ole*, a dance celebrated by Martial, who was a Spaniard, and by Horace also. The master of ceremonies has this dance performed immediately before us. I have a good chance, as an interested antiquary—antiquaries are not always averse to Terpsichore and her devotees—to study the spirit of the scene. As the señorita concludes her last step, hiding one foot somewhere, and with the other poising herself on one toe, her oleaginous rotundity in the air, with head back and arms waving, she astounds me by dashing her spotless handkerchief into my lap! I had read of Seville that—

> ‘The men are fire—the women tow,
>   Puff! comes the devil—away they go.’

And spelling tow, *toe*, I realized one half the couplet; but of the handkerchief business,—I had not read of that in Petronius or Scaliger. This is a new stanza in the poetry of motion! This is an æsthetical climax which requires explanation. With much embarrassment—not unpleasant—I ask my companion, ‘What must I do?’ ‘Do?’ ‘Yes; must I throw it back?’ Here was innocence—paradisaical, before-the-fall innocence! ‘No, no!’ ‘Will she come for it?’ ‘Never!’ ‘Goodness! Well?’ ‘Well?’ ‘What then?’ ‘Put something in it; silver will do; not gold. Then you must go up and present it to her, in your best style!’ I looked for little ‘Fatty.’ She had curled up on a footstool to save her clothes—at the foot of either her mother or a duenna; she looked like one of Velasquez’s dwarfs. Was I afraid? No—never, &c. I

boldly mustered—my *muchas gracias*—for the honour,
&c.; and with half-a-dozen chinking pesetas within the
cambric, I laid my tribute in her lap! As I bowed a
lovely crimson was remarked overspreading my in-
genuous face! 'Fatty' wreathed her adipose and
pretty features into dimples and smiles; and—I—retired. A wreath of dimples is so mixed a metaphor
that I use it to show that my embarrassment remains.
Fair, fat, fatty Fatima, farewell for ever.

I do not say that all the women of Seville are either
fair or fat, or deserve to be associated with honied
oranges. I saw a company of three thousand coming
out of the tobacco manufactory, and I did not see
anything very sweet or remarkable in their features or
conduct. They belonged to the lower classes, and
live from hand to mouth. The Government uses one
of the most splendid buildings, an old palace, for this
monopoly. In it they employ the number of females
I have named. These women are renowned less for
the liveliness of their lives and expression of features
than for the pliancy and piquancy of their tongues.
Let the forward soldier, who hangs about the portal
to see them come forth at evening, as they do in
droves, salute one, beware! It is understood that the
new Government is going to abolish this monopoly of
the tobacco business. They would do well to abolish it.
Thereby they will set us and others a good example.
In America the Government undertakes printing
presses, speculates in cotton fields, and runs railroads.
Where they will run to before they get through we
shall see some day. They are all running sores on the
body politic.

The Cathedral of Seville is hardly surpassed in the
Catholic world. It is next to St. Peter's. The riches
of a great mercantile community, at the time when
the galleons of Spain were freighted with the silver and

gold of the new hemisphere, were lavished upon this
splendid temple.   How to picture its Gothic gloom,
its numerous naves, its grand organs, its double rows
of immense pillars, its gorgeous chapels ; how to
picture one chapel only, lighted with the sacred tapers,
and glittering with stars on a blue firmament, coun-
terparts of the floral decorations upon the altar ; how
to limn to the eye the vision of St. Anthony, the ap-
parition of the Infant Jesus to the monk, or the
Guardian Angel, each by Murillo—would it not require something of the graphic grace of Murillo's own
pencil?   The latter picture is, to me, next to another
of Murillo's—the 'Washing of the Diseased by the
Virgin,' which was stolen hence by Soult, and afterwards returned by France to Madrid—the most significant of all the pictures which I have ever seen.
I would hardly except the 'Transfiguration,' by
Raphael.   I have seen all the genuine Murillos at Madrid, Seville, Granada, and at the Louvre, and I confess
to a new delight at every new study of his works.

Murillo was born at Seville about the beginning of
the seventeenth century.   He made his native city
famous.   It is only, however, within a few years that
his bronze monument has been erected before the
Museum, where are gathered so many of his genuine
works.   He was the painter of feminine and infantile
beauty.   Ford says that his first pictures were cold,
his second warm, his third dim, misty, and spiritual.
His drawing was most conspicuous in the first, his
colour in the second, and his ethereal grace in the last.
The vapoury, exquisite, ill-defined glory of the hair
of his 'conceptions' is rivalled by no touches of art
comparable to them.   It is objected that he lacks the
sublime and unearthly ; that his children are to be
seen in Seville, and are not types of the infant Saviour
before whom the Magi bowed ; that his saints are

Andalusians, and his Madonnas señoras of Seville. But no one denies the magic grace and blending colour which gives to his lines and forms a naturalness which captivates the soul of the simple as well as the connoisseur. His 'Artist's Dream' and its sequel, which we saw in the Museum of San Fernando at Madrid, are more famous than others of his works, perhaps because they have ever had the light to display them. The ether in which they are painted has been permitted to come down in a golden shower for their exhibition. But this dim and grand Cathedral is hardly the place to show them to advantage. Besides, when we saw them there, the light was much curtained by the heavy gold-trimmed velvet hangings.

These hangings were just put up, for the next day was the celebration of Corpus Christi. The guide assured us that the hangings were a present made by the merchants of Cadiz and Seville, and cost 32,000 dollars. This cicerone had an eye to pecuniary values. He gave us an appraisal of each Murillo in Spanish dollars. One he valued at 500,000 dollars. While following him we saw the elevation of the two patron saints of Seville, San Laureano, and San Isidore. They were brothers, and in the religious wars led the Christians. They were successively made archbishops. They are represented, even by those who do not agree with their creed, as men of great intellectual force and acumen. Their figures of silver were lifted by means of ropes and pulleys to their places, for the ceremonies of to-morrow. Rightly to have described this supreme wonder of cathedral architecture, with its many-coloured marbles and richly-hued windows, its illuminated volumes and finely carved choir, one should be entirely alone in the great hush of its stony heart. The seventeen splendid entrances should be closed, the

16

world shut out, and the twenty-three chapels should
be unbarred, that the eye might be nearer to the rich
adornings, sculptures and paintings within the sacred
precincts. The ninety-three painted windows should
shed their choicest dim, religious light. Here in these
aisles, where the uncontaminated effluence of God is
not tainted with the impurities of mortality; here,
under the forms of the sacrificed Saviour and beati-
fied Virgin, with the cross garlanded in enduring
marble, or chased in silver made of the first offerings
of Columbus from our New World; here, at the twilight hour, rendered even more dusky by the dim light
of the Cathedral—here Murillo used to wander, ponder,
and dream. What unpainted imaginings were his!
Before one picture here—a descent from the Cross,
by Campana, a pupil of Angelo—he used to stand in
reverential reverie until his eyes swam in tears; until,
in rapt vision, he almost waited for the holy men to
complete the work of taking our blessed Saviour from
the tree. It was before this picture that he desired to
be buried; as before it, there came to him in rapt
vision—

> ' The progeny immortal
> Of Painting, Sculpture, and rapt Poesy,
> And Arts, though unimagined, yet to be ! '

But how can I picture the infinite variety of art,
taste, and wealth here gathered. Whither shall I turn?
To the colossal St. Christopher, bearing the infant
Saviour over the stream? To the historic silver keys
of Seville in the sacristy? To the tomb of Mendoza?
To the palisades of pipes in the great organs? To the
marble medallions? To lofty-vaulted roofs? Or shall
I not rather await the great ceremony of to-morrow,
when the living masses throng along these aisles, listen
to the symphonies of the organs, and the chaunts of
human praise? I cannot elect what to do where the

confusion is so interesting and the interest is so charming.   As I walk along, thoughtful and silent, in the
grand temple, toward the front portal, listening to the
music which now begins as the prelude of to-morrow,
my eye catches on the pavement a view of two caravels at my feet! This is singular! I look closely.
Colon,—Columbus! As an American—a Columbian,
and as a navigator, having been in every continent,
and having once lived in a city of that name—I step
back! Columbus! No; it cannot be.   I saw his tomb
in Cuba. I know that his body is buried in that old grey
ivied cathedral at the Havana.   I look closely ; for the
letters are worn.   On with my glasses! Down on my
knees! Off with the dust! Yes; it is Columbus; but
it is only the tomb of his son—Fernando Colon.   I
read: 'To Castile and Leon Columbus gave a new
world!' Here, too, is the epitaph of the son for the
father; very touching. But that the father was so pre-
eminent, the son would not be in the shade ; for he
was a rare man and scholar.   Within this cathedral
is preserved his library of 18,000 volumes, and with
them the log-book written by the hand of the father,
and his volume on the 'Imaging of a New World;'
together with his *a priori* proof from Scripture of its
existence.   Where his son lived there is now a village,
whither Sevillians go on Sundays for cheap wine.
The village is, named Gelduba, and gives to the
descendants of Columbus the title of Count.   The
family sepulchre is there.

Meanwhile, we have forgotten that this cathedral is
founded on a mosque.   The tower — Giralda — is
Mohammedan.   It has for its vane a woman holding a
metallic flag, called the Labaro, or banner of Constantine.   This blows about according to the wind.   Of
course, many jokes are made about its feminine fickleness.   Although the woman is named 'Faith,' weighs

twenty-five hundred weight, and is fourteen feet high, yet she is moved as easily as a child's bladder-balloon by every zephyr! The illustration displays what I would say. Up this tower we walked—up—up—350 feet. The ascent is easy. The plane is gently inclined. It is a good deal easier to go up than to learn the architect's algebra, for will you believe it? we have found in this architect the originator of that branch of mathematics. His name is Jaber. He was a Moor. He made this tower out of an 'unknown quantity' of Christian and Roman statuary hereabouts. His invariable formula was X plus Y equivalent to nothing. He came to that in the end. Here—to this lofty pinnacle which old Jaber, by 'contracting' his mathematics, built so extensively — we climb. Here we are amid the monstrous bells. From this point the Mohammedan used to cry the muezzin. The same tower now summons the Christian to prayers, and never with more energy than with those bells at this festival season. What a splendid view we have from the tower. Skim the horizon around. You see the mountains of Morena afar; then Almaden, or its mountain vicinity, famed for quicksilver; then, between the mountains and your position, a plain fruitful with harvest ready for the sickle; then the Guadalquivir, across which is the suburb of Triaria; and beyond that, and beyond the green olive hills, is the village where the unhonoured Cortez lived and died, where the honoured son of Columbus lived and died, and where many an old king and emperor was born and died. Sweep round with the walls of the city. Run your vision from the Palace of Montpensier, or from the Tower of Gold, or from the Alcazar and its gardens of orange, citron, pomegranate, and roses—if your eye will run from these three attractions which lie together near the river, thence from this point of triple interest, you may range

round the walls. You may wonder at the tenacity of
the Roman cement and the Moorish brick, one upon
the other, which constitute even yet an impregnable
fortification. Observe the old gates of the city. You
will in your range pass through many suburbs; but
the old walls mark the old city yet. The view within
the walls is that of an Oriental city. The house-roofs
are tiled with the grey tile, a little mossed, while the
roofs and domes of the churches are, some of them,
blue-tiled. The city is compact, and interspersed with
greenery, but it does not attract the eye so much from
this lofty point as the objects immediately beneath.
If you can quit following the pigeons and hawks which
have made these towers so populous, and who flit in
and out among angles and corbeilles, pinnacles and
eaves, look calmly, or, if you cannot, let your head
swim down dizzily, as mine did, upon the orange court
below — down upon the fountains where the pious
Moslems used to wash before prayer—down upon the
Moorish walls, square buttresses, truncated pillars, and
globe-shaped decorations — down upon the walls of
the great Moorish palace, the Alcazar—down upon the
Exchange, where three hundred years ago merchants
met to discuss the health of the Inca, the shares of
Potosi, the news from the Havana (as yet they do now,
and quite briskly), where they counted their gains from
Chile and Peru, Mexico and Costa Rica—down upon
the bull-ring which has so often resounded with
plaudits to the real Spanish hero, the matador, and
which next Sunday is to be crowded in honour of the
festal season—and down upon the Plaza Santo Tomas—
where is seen the shop of Figaro—made immortal by
the lyre (spell it right, lyre) as the barber of Seville!
I think, after such a flight, we may rest at Figaro's.
Let him gossip of the señoritas, of his vicinage, and of
Don Juan, whose house, by the bye, we saw too, near

Figaro's, not a fictitious Don or house either, but a much more authentic person and mansion than that of the barber!

The next day found us at the Cathedral to see the FÊTE and the PROCESSION moving thence. It was a rainy day; but the canopies were across the narrow streets, and the procession, with its emblems and fraternities, is gathering at the Cathedral. We are there before 10 o'clock. The Cathedral is crowded. The tapestries adorn every pillar and nook. The organs are responding to the music of the choir. The music of the boys is seraphic—female voices they seem in sweetness. The throng presses to the east end. Here at this end is a richly endowed octave, and pictures treating of the conception, one by Murillo! Here the dignitaries, choristers, &c., assemble. Here the ceremonies are proceeding; and, as one exhibition of joy for the risen Saviour, are the 'Seises,' or Sixes. These are twelve little children, apparelled like pages in the time of Philip III., in blue, white, and red, innocent as the children Christ blessed, and they danced a beautiful yet solemn dance, with castanets accompanying sacred music! The effect of this symbol of jubilee did not strike us as in any degree either ill-timed or inappropriate. It is one of the expressions of the happy Andalusian heart in its own favourite way, and is no more to be carped at than the music of the violin, organ or flute, being but another mode or accompaniment of lyrical or hymnic expression. The procession of fraternities is made up of the most substantial men. Each bears a lighted wax candle. The military and civic processions join, then the Archbishop, followed by the priestly order; and thus, with music of bands and amid crowds in every balcony and along the streets, all in their best attire, and hardly exempt from the rain which the awnings do not altogether avert, the procession moves through the principal avenues. These avenues are decked in canopies and with hangings from every house. In the evening Seville is illuminated. All the gaiety of this lively people, notwithstanding the rain, shows itself on this grand occasion of Corpus Christi. It is said that the day is not kept so gorgeously as it used to be here. But I do not see how it could be more august or imposing. It is said, again, that the ceremonies of Corpus Christi are conducted here with a solemnity and grandeur second only to their celebration at Rome! I can well believe it.

I have had much to do to keep my pen from a premature description of the Alcazar, or house of Cæsar. It is on the site of the Roman prætorium. It was built for a Moorish king, but has been so altered—so gothified, or modernised—the ceilings have been so renewed, and so much has been added by the Spanish kings and queens, that it is hard to tell which is the Moorish work and which is its reproduction. Here Charles V. was married. Here the Philips introduced the royal portraits into the building, fishes into the ponds, new tropical trees into the gardens, and fresh fountains through all the walks. The palace has been whitewashed, and the aqueducts injured; but much has been recently restored. The grandest hall is that of the ambassadors. It is on the grand scale what the Alhambra is in miniature. We are shown where Pedro the Cruel killed his brother; also a painting of four skulls where he hanged four venal judges. Pedro deserved his name. He was in the habit of murdering almost anybody; when he could find no one else handy, he used to select a few rich Jews and burn them to keep his hand in. When he took a fancy to a young lady, and she jilted him, he burned her to a cinder. Only one lady, whose portrait is preserved in

the Alcazar—the beauteous Padilla—ruled this monster. The story of Pedro is her story.

But as some one laid flowers on Nero's grave, so
Pedro has found his defenders. Voltaire is among
them. Pedro may not have been so bad as he is
painted. Lockhart has several ballads about him.
It may be said in extenuation that he came to the
throne in bad times. His domestic relations are
illustrated in the lives of his father, King Alonzo, and
his mother, and the lady Guzman, whom his father
afterwards made queen, and whose sons he made
princes. On his father's death, Pedro caused the
Lady Guzman to be beheaded in the castle of Tala-
veyra. Her powerful sons then began the struggle
for revenge. One of them, however, Don Fadrique,
made peace with Pedro, and came on invitation to
a tournament in Seville,

'For plenar court and knightly sport within the listed ring.'

It is the old story in which the imperious Padilla
played her part.

'My lady craves a New Year's gift;
Thy head, methinks, may serve the shift.'

The head was given to Padilla, and she gave it to
a mastiff to devour; meanwhile leaning out of her
painted bower to see the mastiff play. We know the
history of Padilla. The opera tells it. It is sung in
the sweetest of music. For Padilla, Pedro deserted or
rather murdered his wife—the unhappy Blanche of
Bourbon whose plaint Lockhart so sweetly sings:—

'The crown they put upon my head was a crown of blood and sighs,
God grant me soon another crown more precious in the skies!'

It was this Pedro the Cruel whom the Black Prince
came with the English and Gascons to fight. Old

Froissart tells that story. Its sequel was the violent death of Pedro by the hand of his illegitimate brother Henry. Walter Scott has a ballad about this. So that literature has much to do with the immortality of infamy belonging to that age and dynasty.

Never was there a more beautiful domain than this Alcazar. It is simply horrible to associate with it those scenes and days of perfidy, cruelty, and murder. It is, however, like the Alhambra, blood-stained and lust-tainted. It likewise resembles the Alhambra in its palatial decoration ; only the halls are more extensive. The gardens are the most beautiful in Europe. Orange, box, and myrtle form the walls. Labyrinths, coats of arms, and other quaint shapes and devices, appear in the clipt vegetation. All through the fairy realm the odour of the asahar fills the air. In the lower garden is an azulejo—a domed Moorish kiosk. Under the palace are gloomy apartments, once used as bathing-rooms and prisons. Here we found the relics of the Roman days lying loosely around. In the terrace of the palace, above the gardens—you may wander in and out—into the Alhambra-tinted rooms, and out on the balconies into the fragrant air. When it is all over, you may wonder at the jumble of architecture and civilizations—wonder what all this was meant for, who paid for it, and why Pedro the Cruel ever lived; and then you may pay four reals to the porter as consideration for the suggestion of your thoughts.

The reader will do me the credit to admit that I have not generally affected the archæologist. Under great temptations — I have hitherto refrained. A scholarly friend writes to me—from amidst the comforts of a New York mansion in mid winter—that he could not scold me for being as poetic and pulmonary in the South of Europe, as a fashionable clergyman.

He advised me to resort to antiquarian research,
and let the large family of American Leatherlungs
blow out and away; observing that men now shine
like the bust of Brutus by absence; he recommended me to improve that absence by exhuming
at Ancona a MS. proving that the mole of Hadrian
was only a wart on the imperial physiognomy, or by
digging up at Ostia a silver buckle with the inscription Sus. per Coll.—proving that 'gallowses' were
worn, as well as erected, by the Roman patricians,
tempore Gallieni.

Although thus advised to rush after ruins, and
forget, in historic doubt, the dirt of the 'living
present,' I have abstained. Although Roman ruins
are so numerous in Britain, France, Algiers, and
especially Spain ; and, although I have been near them,
and tempted, yet I thought it preferable, if I desired
to exercise my faculties as excavator or annotator, to
search for the relics at Rome itself, or read the
epitaphs of departed greatness, without the delusive
gloss of distance and doubt. I would, myself, prefer
to rummage among the catacombs of the Eternal
City, but not being able to get to Rome, the next
best place for illustrations of Roman Imperialism
is at Seville, or rather Italica, within sight of Seville.
Were not three Emperors born there? Were not
the Cæsars especially fond of the place? Was it
not the pet—and why should it not be?—the pet of
the wits, who indulged in too free a use of the
pasquinade (excuse the anachronism)? or I should
rather say, the Martial-ade, and, as a consequence,
were exiled to this other Rome? And is there anything more illustrative of Roman greatness and power
than the fact that here, in (then) far-off Iberia, the
mistress of the world ruled while she refined the people,
who here illustrated by letters and art the lessons

which the mistress taught with so much potential persuasion?

So, leaving Seville and its museums, its fête-days and its bovine fights, its Oriental languor and Gothic grandeur, its dances and demoiselles, its alcazars and alamedas, we take a long, dusty ride to the Roman ruins of Italica. Crossing the river, glancing at the crowd on the bridge who are watching the seining below for a drowned man, stopping at the Venta des Estrelles—'Star Tavern'—where we are served with the rarest muscatel, native to the country, tasting, however, of the pig-skin, and the tar on its inside, in which it is bottled—passing across the plain where once the Guadalquivir ran before it took a fancy for the Seville vicinity, we arrive at a little village, called Sante Ponce. This village is literally on the top of Italica. We stop at once at the amphitheatre. Before we venture within let us ponder. It is hard to believe that so much dust is collected above these forums, theatres, palaces, houses, and temples. Where, O conscript fathers! are the sumptuous edifices now? Not as a school declamation do I ask it; but there, here, under and around, is all that is left of this once proud municipium. One mosaic pavement, fast disappearing by being torn up by tourists, and one little patch of fresco, are all that remain of your sumptuousness. The proud city where Trajan, Adrian, and Theodosius were born—where Scipio Africanus built homes for his veterans; where their eagles were borne in many a triumphal procession; the Goth and the Moor have abandoned, and even the Guadalquivir deserted. Your palaces have been quarries for yonder rival city. A few great stones lie around, relics of your greatness, but the lizard and the gipsy lurk amidst the broken fragments!

Some marble statues have been found by excavation

under the olive orchards. We saw in the Museum at
Seville and at the Alcazar some broken monuments
of Italican art. Several of these represent Augustus
Cæsar, several Hercules, some Trajan and Theodosius.
Every specimen seemed to be either a foot or leg without
a body, a breast without a trunk, a head without a body,
or a bust without a nose. By putting this and that
together, one might form a complete human body, the
nose excepted. I cannot distinctly affirm that I ever met
with one single piece of authentic ancient sculpture
with a nose. It is sad, but so it is. To those who
are seeking through the chisel the perpetuation of
their image and features; to the ambitious statesmen
and soldiers of America and other lands, I would say:
Ponder the lesson which is taught by Italica and the
ages! If you insist on being handed down in brass,
very well. Brass may do for you! But if in marble,
you will go down the corridors of time noiselessly, in
noseless if not nameless marble. For when your
noses are lost, what is there to mark your heroic
quality? When that characteristic protuberance has
become pulverized, and your artist has neglected to
sculp on the marble the name of 'John Smythe, of
Smytheville, brigadier and congressman!' what is
there of consideration for the outlay of your green-
backs or the satisfaction of Smytheville? Again,
what is there for the renown of Smytheville? And
to you, my coloured American brother, a word: You
are seeking fame, honour, office. You, too, like
Scipio Africanus, Cæsar, and Pompey, desire to be
handed down in enduring marble. When thou
knowest that even Roman noses are abraded by time,
let not the brutal Conservative, sneering at your
features through the medium of science, taunt your
effigy in the great future you are seeking. Therefore, my brother, be not ambitious of such immor-

talization. Your full-length figure will turn out a bust. Your bust will be noseless; your name will be in dust; your fame in ashes. Few marbles survive in perfection.

> ' Can Volume, Pillar, Pile, preserve thee Great ?
> If not, do not trust the noseless statue.'

Italica teaches this lesson ; we are all mortal ; mortality is dust; dust is unpleasant. We found it so ; for the very dust, golden with historic memories— perhaps the dust of martyred Christians who died in the amphitheatre—ah! that brings me to the spot where my moralizing began. Into the amphitheatre we go. It is the Coliseum over again. But to reach the arena we creep under vaults. We go in where the gladiators and wild beasts went in. There is a musty, damp smell, and plenty of moss about these vaults. There are streams running yet, after two thousand years. They bubble up here and there. They show good masonry. Adrian made the reservoirs and aqueducts. The fountains where the gladiators bathed and the room for the prisoners are here yet. Grass grows in the circle and upon the seats. The latter are broken, but they are as marked as if filled yesterday by ten thousand spectators. We see a few stones with inscriptions. We have come thus far without a guide. Directly we see other visitors convoyed by an old man wearing a sash and carrying a gun. He is the warder. He sends after us his young nephew, Pedro. We had already read a writing, posted upon a fragment of a ruin, advising us that the Commission of Monuments in Seville had selected Gregory Ximenes, as the guide and philosopher of Italica. (The Ximenes—from the great Cardinal to Irving's 'Son of Alhambra'—are famous in Spain.) The nephew, however, did well for a while. I questioned him : 'Pedro, what did they do here in the old times?' He has his lesson well. 'Six things, señor.' 'What?' 'Racing.' 'What racing —horse-racing or human racing?' 'Chariot racing.' 'Good.' 'Next, they pitched quoits, fought with fists, filled the arena with water and had naval battles,' or, as he called it, simulacia of fights; 'then they fought wild beasts.' He had enumerated five—very well. He said that the gladiators used to wash in the rooms below, and that the well was full yet, twelve feet deep. He pointed out the cages for the lions. Sitting amidst the arena, on a stone, with the innocent, yellow, purple, and red flowers growing about us, the nephew told us the most sanguinary stories, and showed us the actual teeth of the wild beasts. He told us how some of the gladiators preferred to suffer death in their cells rather than to endure the ignominy of fighting with beasts for their lives! The old warder then, having dismissed the tourists, came to us. He had his gun, a cartouche box, and a bran new leather belt, with a large medal, and on the latter these signs: 'No. 8 Do.' Now, this mystery I had seen all through Seville. It was on flags, in processions, in churches, over shops, in ventas, over the City Hall—everywhere, No. 8 Do! It was as mysterious as S. T. XX 1860. Was it an advertise-ment—a provocation to curiosity? Did it mean bitters, ale, or was number eight cabalistic and signi-ficant of Carlyle's religion? Do! This ancient warder solved my doubt. Seville was ever faithful to Alonzo, son of its saviour from the Moors. Alonzo was a learned fool. He gave Seville this badge. It is called El Nodo. It means '*No-m'ha dexa-do,*' and that means 'It has not deserted me.' *Madexa* is an old Spanish word for *knot.* Nodus is Latin for knot. Thus Seville happened to hit on the Phœnician merchant-mark, the Nodus Herculis. The figure 8 represents the knot. This mark in Phœnicia meant commerce.

Seville, without intending it, reproduced this emblem of commercial adventure. How she illustrated it history tells; for at one time she had 400,000 population.

Our warder, Gregory, was a man of much interest, in his own eyes. He was a little of a wag, too. He had been under fire, and under Prim. He had been to Rome, and had the cross of San Fernando. He conducted us through the vaults, then unlocked a cupboard and showed us, to his own surprise, first, a beautiful big lizard called Lagardos, which crept within, and then, to ours, a perfect piece of Roman mosaic—the only remnant, as he said, of the decoration of the building. It was then boxed up, to preserve it. He next led us out of the theatre into his hut among the tumbled rocks. The ivy grew round the ruins. Hollyhocks, onions, verbenas, snails, pinks, and grapevines—these were scattered amidst the fragments. His little hut was without ventilation. We asked him if he lived there alone. 'Yes, he had no wife now; his gun was his wife.' He was not so well off as Robinson Crusoe, he said, who had a man Friday. His hut was pretty well charred, as he made a fire inside, and the smoke had only the door for a chimney. I asked him what he knew of Crusoe. He had read it in Spanish. It was his favourite. He believed it. I asked him if he had read of Don Quixote. 'Oh! yes, señor!' But he added—I translate him literally—'I do not hold to the truth of all that book.' He said it was in some respects like Sinbad. He had read Sinbad, and did not accept it as verity. It is a fact, in Spain, or at least in La Mancha, that the common people believe implicitly in the Don. They have not a doubt about Sancho. In verification that he was a real man, a peasant told me that he had seen in La Mancha one of Sancho's descendants. He looked like the pictures of Sancho, and he was a good, devoted friend, fat, jolly, and selfish withal.

Gregory showed us with pride his verbenas and pinks, his nectarines and grapes, growing amidst the blocks of Roman ruin. He was accustomed to kill rabbits and partridges amidst the ruins and fields, for a supper now and then, but kept his gun near him, in these times, as he hinted bravely, for other purposes. He said all the town was Republican. He was for Prim. I asked him if he had ever been married. Yes; but he had lost his Eve some time ago. Wrapping his handkerchief around his head, he started out, over the wheat fields and through the olive orchards, to show us the Roman baths, the forum, and the ancient city. The latter is only partially exhumed. Its stones are now mansions of Seville. Ximenes said he had not travelled much himself, but his picture had gone round the world. Some photographist had taken him amidst the ruins of the Roman Empire! We clambered after him over and under ground, through brambles and weeds, and saw what was to be seen of ancient glory. That was but little. Only two relics did we bring away—a piece of mosaic and a Roman coin. But we have materials for future meditation; for we know that here, where the grain and flax, the wine and olive, now grow, the thoughtless peasants make ruin and decay tributary to human wants. They draw on what seems a bankrupt treasury of a proud and defunct empire—which the Cæsars themselves had helped to build—for the oil which they mix with their salads and the bread which they saturate with the oil. Imperial Cæsar's dust—you know what Hamlet says.

On our way to Seville we kept to the slope of the hills. We stopped at Montpensier's country palace and secured entrance. It is on the Calle Real, or Castileja de la Cuesta. The view is very fine; but we

did not go there for the view. Here it was that Cortez
lived and died. In 1547, aged sixty-three, the con-
queror of Mexico—a broken man—yielded up his
spirit to his God. His bones were removed to Mexico,
the scene of his glory and crimes. Yet he is a hero in
Spain. Montpensier has one room of his palace altogether dedicated to Mexico. There is here a picture
of his brother, the Prince de Joinville, bombarding
Vera Cruz; a *fac simile* of the Act of Independence of
Mexico in 1821; the sword of Iturbide; a fine view
of Queretaro and Jalapa; a portrait of the ill-starred
Count of Bourbon, massacred in Sonora some years
ago: several perfect likenesses and portraits of Cortez,
and one of Columbus—a wonderful picture, taken in
1506, fourteen years after he discovered the New
World. He is represented as bald-headed, with great
perceptive faculties, a wide space between the eyes,
large nose, and keenest eyes; but he looks like a sad
and wearied man. Was the portrait taken after he experienced the ingratitude of Spain?

CHAPTER XIX.

A PRINCELY AND ECCLESIASTICAL CAPITAL—TOLEDO.

> 'Whilom upon his banks did legions throng,
>   Of Moor and knight in mailed splendour drest.
>   The Paynim turban and the Christian crest,
>   Mixed on the bleeding stream, by floating hosts oppressed.'
>   SOUTHEY.

HAVE a ring of cities, full gemmed, in my memory, which deserve to be displayed. Carthagena, Alicante, Valencia, Murcia, Grenada, Malaga, Seville, and others, I have endeavoured to illustrate. I have not yet spoken of Toledo, Cordova, Aranjuez, Madrid, and Saragossa. For the sake of completeness, I owe something of description and much of praise to these olden seats of princely and ecclesiastical power. I have omitted to follow my own rule, to photograph impressions in the momentary sunbeam. Omitting to do so, and a few days intervening, my recollection becomes entangled; but it is the tangle of flowers in a luxuriant but neglected garden. Where is there to be found a more fruitful source of blooming memories than at Cordova and Toledo? Where has there been so much of regality as upon the banks of the Tagus, amidst the pastures, forests, and palaces, of the modern seat of royalty, Aranjuez? And then there is so much to be gleaned at Madrid. The prevailing political excitement prevented my observing much that was interesting

The Alcazar is high above the rocky banks of the
river. You may drive up nearly to its portals. A
balustrade, decorated with large stone 'cannon balls,'
lines the upward roads. We pass stone statues of
Gothic kings, very gray and, as usual with all old
statues, without noses. How quaintly these effigies
stand, frosted over with the rime of years. Not less
motionless are the lazy, live Toledans at midday,
snugly snoring in the shadow and angles of the great
square of the palace. Nothing seems to work hereabouts but the river. The Tagus generally has not
been considered worth a—dam—but here there is one.
The stream roars and plashes, and turns the slow,
creaking old water-wheels, as it did when the Moors
lived here. It is refreshing to see the river run. We
look through the rocky defiles whence it issues. Then
it seems to please itself by spreading out and winding
over the Huertas del Rey—the meadows to the West,
where it works less noisily but as valuably—after the
manner of other waters in Spanish soil. But then, as
if tired of the effort, it wanders lazily off toward the
West. Its waters are fretted by no commerce. They
are content to empty themselves into the ocean, near
the capital of Portugal.

Many a romantic incident, however, is associated
with its flow. Have you not read Southey's poem—
'The Last of the Goths'—and Irving's description of
Florinda, who was so fair and yet so frail (the frailty,
however, Southey, as well as Irving, denies), and who
was seen by Roderick, the last King of the Goths,
in her bath by its banks—the same Roderick who
brought so much trouble on Spain, nay, its downfall,
by his rude wooing. That will do for one story of the
Tagus. Besides, the Cid Campeador was once Governor in Toledo. That fact alone is a romance!
Was he not the hammer of Thor for strength and the

scimitar of Saladin for chivalry? Is not his name a
poem? Around it cohere sword, and cuirass, and
buckler, guitar, castanet, and beauty! Nay, his name
is a volume of gold-clasped lyrics of war and love!
Far down upon the green Huertas, on which I look
from the Alcazar towers, the Cid once convened a
Cortes. It met upon the banks of the river, near
which are yet to be found some genuine ruins, amidst
which some poor peasants live. These are the ruins
of the proverbial 'Castle in Spain.' This castle, it is
said by Ford, duly confirmed by Murray, was built in
the air, by Galafre, a king, as a home for the beautiful
Galiana, his daughter, whom he loved passing well.
Cynical criticism has it that no such king was ever
crowned; that no such king ever had a daughter;
and that no such king ever constructed a castle for
any supposed, or non-existent, or other daughter!
Hence I argue that there is good foundation for this
ethereal architecture! It is also added that Charles
Martel courted this apochryphal daughter of a king
who never was, who lived in a castle constructed in
phantasy; and that Charles, in a fight which never
took place, slew a rival, who it is generally believed
was not killed, inasmuch as Charles never came to
Spain and the rivalry was a myth! Is not this the
very fifth essence of Spanish romance? But the
Tagus has had something besides ideal castles upon
its banks. The direst conflicts of that long religious
war between Christian and Moor here took place.
They furnish the argument of many a roundelay.

But let us clamber up *into* the Alcazar. It is
refurnishing, if not rebuilding. The workmen tell us
that before the recent revolution it was intended to be
fitted up for a college. Once a palace and a fortress,
it is now—a nondescript. It has felt the improving
hands of Alonzo, Charles, and Philip. Façades, rooms,

and staircases were built by them, and afterwards used by paupers; for it was changed into a weaving factory for the poor to work in. The French, in the Napoleonic wars, made it a barrack. They mutilated and nearly burned it down. Kings have been born here. From its towers, overlooking the regions round about, through which the Tagus flows, queens have gazed down upon the tide of battle, upon which the fate of kingdoms depended. From its lofty point is seen the circle of olive-clad hills beyond the dull, dead town, the ruins of the Roman circus in the plain, and the yellowish, whitish walls of the city, here and there surmounted by the still numerous spires of the churches. Through its court the birds—the only live and jocund inhabitants—chirp the slow hours away. Let us hence, Come. There is something so offensively defunct here that it would be better to go at once to the Cathedral. There we are sure of seeing, at least, mimic life, illustrated by the old paintings and tombs, which prove that men have once stirred the dust of this heavily bound, seven sleepers of a city.

Rumbling down and up through the narrow winding of the city streets, followed by beggars, and still gazed at by the roused population, we reach the principal square of Toledo. It is called Zocodover. I have not studied the philology of the term; but I think it means the plaza of laziness. It used to be the resort of all the devil-may-care, clever, proud, swindling, and gambling Don Whiskerandos of the ancient regime. Still, on a warm day, it is revisited by the ghosts of those departed worthies, which lingering haunt the shady ends of cool stone seats.

Come along further! We may see something which speaks of the olden time, and of the Jews who then suffered, if indeed there ever was a time in

Spain when they did not suffer. The Jews of Spain
were of the highest quality. Toledo was their
Western Jerusalem. I have seen an elaborate paper
in Latin, preserved by the Toledan ecclesiastics, which
purports to be a protest against holding the Toledan
Jews responsible for the Crucifixion of Christ. They
prove that they were of the ten tribes transported
by Nebuchadnezzar to Spain; and not only assert
an *alibi*, but insist that when they were consulted as
to the Crucifixion, they searched the Scriptures and
found that Jesus was the Christ; that he fulfilled the
prophecies of their sacred books; and that hence they
advised against the 'deep damnation of his taking off.'
Whether this is a fictitious document, or a ruse of the
Jews when hard pressed by the Christians—or what,
I am not to judge.

Two synagogues remain to attest what this class
once were. The older one, built in the ninth century, was almost destroyed by a mob, and then
changed into a church. The French made it a store-
house. The soil below the pavement is from Mount
Zion, and the beams of the ceiling are cedars from
Lebanon. Its aisles and arches, pillars and patterns,
mark an era of great Hebrew taste and wealth. The
other synagogue was built in the fourteenth century,
by one Samuel Levi—the rich treasurer of Pedro the
Cruel. It had Moorish arches, an artesonada roof,
honeycomb cornices — all Oriental, bespeaking not
only a peculiar people, but a people peculiarly Oriental
in their architecture. These are but the spoiled and
degraded memorials of a race which clung with such
tenacity to its faith that no fires of persecution could
sever the bond. Ever despoiled and ever filling their
coffers; children of peace, and making wealth by its
pursuit; coming to this, their favourite Tarshish—the
golden America of their hopes—from the persecutions

of the far East; driven by Roman Emperors from Italy in Pagan days; cut off by 'scythes of revenge' through Moorish hate; in goods mulcted and in person decapitated; driven from place to place by Catholic and Protestant, they ever clung to the horns of their ancient altar, and bore the oracles of God, as in the ark of the covenant, down through the dark and bloody ages, until in this bright noon of a new era they are at length enfranchised even in England and in Spain. In England they have at length the right of suffrage and that of sitting in Parliament. In Spain they are accountable to none but Jehovah for their religious convictions. In America they have ever been free; and, being free, how they have thriven! A wonderful race; Semitic it is called, from its ancestor Shem; Oriental it is, ever persistent, energetic, and brilliant. What music has it not composed and sung since its harp was taken down from the willows of Babylon? What eloquence, what art, what letters, what genius in journalism, in finance, and in statesmanship, has it not illustrated? Let the Hebrews of France, Germany, and England attest. If I mistake not, the late Premier of England came from the loins of one of the tribes of this Spanish 'Tarshish'! Now that Spain is free, and the Jews so opulent, would it not be worthy of their high descent and rich estate if they rescued 'Juderia' in Toledo, and its two despoiled synagogues from degradation and decay?

Now for the cathedral, which towers above all, proudly eminent! We alight at its gates. Before we enter the temple itself, observe upon the walls of its cloistered avenues pictures of grinning Moors. They are cutting up Christians with scimitars, or dragging them to disgrace and death. This was a lively preface to the stone-bound volume we are soon to open. But even these vivacious frescoes are already

beginning to lose their colours, and the turbaned and
breeched forms their lineaments.   But here, along this
court—upon the left, or outside—is an orangery, and
the orange trees are intermixed with laurel and cypress.
That surely shows life sweet and fragrant.   The birds
make their nests here.   They sing us a hymn from the
cypress as we enter the heavy, bronzed portals.   We
tread over marble pavements, on which are written the
virtues and station of those whose ashes repose beneath.
We observe sculptures and paintings, 'an innumerable host,' altars and tombs, jasper steps and alabaster
forms, bronze doors and carved wood-work, bas-reliefs
and cloisters.   There are, amidst this opulent labyrinth
of art, two conspicuous objects.   One is the superior
painted windows.   At sunset they glow like jewels!
They are justly celebrated as marvels.   The other is
the Gothic Respaldos of the fourteenth century.   This
is the boast of Toledo.   It consists of sculptures in
white   marble—almost   yellow   now.   Columns   and
cherubs surround St. Raphael in full figure, with his
head downwards and wings stretched, flying out from
the marble clouds!   It cost—excuse me—but it is
reckoned at 200,000 ducats!   I omit the dollar, as
that detracts from its poetry and subtracts from its
age.   This is the most extensive piece of sculptured
marble that I have ever seen.   It extends to the vault
of the Cathedral.

But why should I disturb the dust of old Toledo?
Why linger amidst these petrified abodes of the great
Mendoza and Ximenes—those redoubtable old ecclesiastical knights, whose swords were as potential as
their croziers?   'Their swords are rust; their bodies
dust; their souls are with the saints—I trust.'   No
renaissance in style or sculpture can arouse them to
glory again.   Spain has left them in the rear.   The
orators of the Cortes—the Figueras and Castellars—

take the lessons of these elder primates only to frame
a new code, instinct with the present and fit for the
political future.   Nay, to the credit of hierarchical
Spain be it said that her bishops and archbishops have
sought places in the Cortes, and have shivered many a
lance with the ardent orators of that forum. His Grace
of Santiago, foremost in the arena, brings his learned
eloquence to bear in support of the unity of his religion, and its paramount authority in connection with
the State.   So, in England, their Graces of Canter-
bury and Dublin, and the bishops of Derry and St.
David's, Tuam and Peterborough, and more especially
the last named, have within the week held the Lords
entranced by their learning and eloquence in defence,
or in derogation, of the Irish Episcopal establishment.   Omitting all discussion as to the merits of
such contests—to state the question is all the logic
required for an American—it is a bright augury of a
better era when schoolman and churchman step forth
as it were, from the unseen world into the arena of
reason, to combat error, defend truth, or even to
uphold the wrongs which time and power have crys-
tallized into institutions and vested rights!

But the mind is so framed as to be reverent of the
past and of the dead.   It requires an effort, sometimes
a convulsion, to free it from the slavery of mere symbols.   There is a just medium between this reverence
and its iconoclastic enemy.   Spain is struggling for it.
The best Catholics have helped the struggle.   Indeed,
there is no other religion in Spain except the Catholic,
and there is no probability of any other just now.   If
the Protestant would stop progress and promote re-
'action, let him do as some of the 'weaker brethren'
are doing, taking advantage of the new code of toleration, and send into Spain inflammatory tracts and
fanatical colporteurs; both abusive of the existing

faith, and so Protestant in excessive zealotry as scarcely
to be Christian in charity or practice.

Turn we again to the world of symbol and faith in
the great Toledan temple. No bickerings of politicians,
no polemics of theologians, disturb the repose of its
solemn aisles. All about us are effigies of the departed.
Yonder chapel celebrates the victory of Ximenes, when
he took Oran from the Moors. We have visited in
Africa the scene of his exploits. Here is the sword of
Alonzo VI., who conquered Toledo; there is the same
cross which the great Cardinal Mendoza elevated in
presence of Ferdinand and Isabella on the captured
Alhambra! Come with me into the library of this
cathedral! How cool is 'the still air of delightful
studies'! How dim the light! Here are volumes
gigantic in size, and ponderous with Greek, Latin,
and Arabic lore! Here are Talmuds and Korans;
illuminated Bibles, the gifts of kings; missals, whose
pages were once turned by the hands of an emperor
of two hemispheres, and printed Italian books as well
(six thousand), and all reposing unread under dust.
Has not a new volume been opened by Spain? Hope
illuminates and progress peruses it. Paintings there are
of Virgin and Child, Saviour and saint, holy families,
in this great cathedral, but I miss the ever graceful
Murillos, which make the Cathedral of Seville as
entrancing as a thought of Heaven.

Yet, with all these relics of the past, what is it the
stranger recalls when he thinks, if he ever does think,
of Toledo? Nay, I make the question more specific.
I ask of the Buckeyes—who have a city christened
after this proud capital of Spain—what is it you think
of when you recall old Toledo? Is it her grand
primacy, her cathedral, her theocracy? Is it her piety?
Is it her Hebrew name—derived from Toledoth—signifying 'city of generations;' the refuge of the children

of Israel when Nebuchadnezzar took Jerusalem? Is
it her strange Hebraic history — her having been
turned over by the Jews to the Moors, because of
Gothic persecution, and afterwards, when they were
persecuted by the Moors, turned back again by the
Jews to the Christians? Do you think of Leo-
vigildo, the king, who consolidated the Gothic power
and made Toledo its capital? Is it that Wamba
and the Cid once ruled here? Is it that you recall
the magnificent charities and hospitals here once existing? Is it her rugged cliffs of gneiss or her magnificent veins of granite? Is it her splendid plaza
for loungers, her Alameda, her Alcazar, her romances,
her Châteaux en Espagne? Is it her primitive donkey
power? For know, that Toledo, being on a high
rock and being without water, and the old Moorish
aqueducts, or 'roads of silver,' being destroyed, has
depended this many a year on donkey power for
water. The donkeys carry the jars in their panniers,
and thirsty Toledo is thus satisfied. Is it this system
of *elevators* which recalls your Spanish namesake?
Is it the 'great wars which make ambition virtue'?
Both Toledos have had them. Old Toledo was the
subject of many a fray, bloody and bitter, as your
Maumee Valley was when mad Anthony Wayne
waged his Indian warfare, and as new Toledo was when,
as disputed ground in the 'Wolverine war' between
Ohio and Michigan, she witnessed the destruction of
water-melons and corn whisky! The sweat which
then flowed, and the feathers which were then ruined,
are known to the old inhabitants of Ohio. Then I
was a youth, but I have the recollection of hearing
valiant colonels, in my own native Muskingum hills,
addressing the militia drawn up round them in hollow
squares, inspiring them to rescue the realm of quinine
and hoop-poles from the grasp of the insatiate Michi-

ganders! The recollection makes my heart tremble. Ah! that was a war, whose Cid no poet has dared yet to celebrate! The passions then engendered even yet vibrate in the 'cornstalks' of the Maumee Valley! A remarkable war! when soldiers retreated before a foe which was not pursuing, and ran through almost impassable swamps, guided by the battle-fires of their own flaming eyes! The dead and wounded of that war were never counted. Both sides fought for a boundary-line, and both ran that line with the same exactitude and compass. Their lines were both straight. I said I was but a boy then; but the tympanum of my ear even now, at this distance and age, echoes to the rataplan of that sanguinary war. And yet, I venture to say, that it is not this history, that is the vinculum of memory, which connects these historic spots of similar name!

Why then was Toledo, in America, thus called? Who christened it? What for? Where and how do Americans get their magnificent names for village, town, and city? What is now the association between the old and the new? To answer these queries I have searched history, science, geology, river, people—associations of all kinds—fruitlessly. Stay! From the war to the warrior, from the warrior to his weapon, there are but one, two, three steps! Eureka!

Toledo, Spain, and Toledo, Ohio, each has a *Blade!* True, one is a newspaper and the other a sword. But when the Toledan of either city would recall its name-sake, is it not because he is situated like Aladdin, who, when he courted the beautiful princess, found a flash-ing blade between them! *Eureka!* I should be untrue to my vocation of a tourist if I did not describe the visit I made to the manufactory of these world-famed blades. Down the steeps of the rock-built city, out upon the road and over the plain to the south-west, and on the bank of the Tagus, scarcely a mile, we drive to the celebrated *Fabrica de Armas*. That huge rectangular pile you see before you was erected by Charles III. in 1788. It has a chapel dedicated to Santa Barbara, who patronizes arms. The sign over the door of the factory indicates that the arms are called *blancas*. Why white I do not know, except from the sheen of the metal when made into the blade. We enter within the gate. Our first salutation is an unexpected challenge from a pet brindle goat—who gave us a 'Mee-hah-hah!' We are next saluted at the door by an *employé*, who was more courteous. He takes care to let us know that *he* made the Toledan blades which were sent to the Paris Exhibition. He himself had the honour of taking them there. The pride of the blade has entered his Castilian soul! He first invites us to a room where there is a collection of finished blades. He takes one up. It is of fine temper and mirrored-polished. He tries it. It curls up like a watch spring or a glistening snake. It flies straight again! He hacks a hard log of iron or steel with it. No dent is left on the keen edge. The sword fills Falstaff's definition of a good bilboa, 'in the circumference of a peck hilt to point, heel to head.'

What is the secret of this refinement of steel? Our conductor says the peculiar water. But there is some secret behind. It is not the metal; that is English. It is not the fire; that is the same old Promethean spark. Is there any secret in the handling? They tell us not; but I noticed at the forges the most painstaking labour and diligence. The secret of the *Damascus* blade, or of its manufacture, ought to be, if anywhere, known here. Indeed, the Moors first taught the Toledans to make this blade. Knowing that fact, and putting the unknown down as the work of Oriental enchantment, we are content, without inquiry

further, to follow our guide ; first, into the room were
workmen are pricking the steel with points, and painting the marks and decorations on the blade. These
are then burned into the metal till they add a magic
glamour to its silver sheen. We then go into the
department where the grindstones are flying around
and the sparks are flying off. They are turned by
water-power. A part of the Tagus runs through the
factory, and works as it runs. While looking at the
men polishing the blades upon whirling walnut-wood
wheels, amidst the hum of the room, a clear-eyed boy
of twelve, catching some words of English I had
dropped, came up and said, politely : 'I am so glad to
see some one, sir, who speaks English. Excuse me for
asking if you are not English.' He had been three
years away from home, learning to make Toledo blades
and to speak Spanish. He was a Birmingham boy,
and as bright in his eyes as the blades he polished,
though as grimed as Vulcan in his face. I have since
learned that Birmingham manufactures great numbers
of Toledo blades ; but this is irrelevant. We purchased
some of the daggers, as mementoes of the spot. My
blades had to be paid for ; were they not genuine arms
from Toledo ? Was not Toledo, for a thousand years,
the fabricator of those swords most petted by the proud
Spaniards ? Were not these swords to them almost
sentient beings ; and, like Burke's rhetorical sabre,
ready to leap from the scabbard by their own inherent,
chivalric temper ? We paid well for our souvenirs ;
but, as the conductor remarked, 'Señor, one must pay
for his caprices !'

I have made out my case. It is this blade which,
more than anything else, gives Toledo its fame. Toledo
still preserves the art. The Fabrica is the only sign of
life in or about the city. Three hundred and fifty
workmen here earn their living making the *armas*

blancas for the Spanish cavalry. Until the nations
learn the art of war no more, and the sword is beaten
into the ploughshare (a most valuable improvement
for Spain, where the old wooden ploughs are to be
seen yet), this grand old capital of Spain will be known
by the glistening letters painted in umber and burned
with fire into its blades. These blades have a moral—
if not in their thrust, in their elegant elasticity. They
show that the finest metal, the most exquisite temper,
and the most elegant polish, are consistent with that
gracious bending which is the proof and essence of true
gentility. The stiff, unyielding, untempered iron is
easily shattered; the pliant, graceful, tempered steel,
like true courtesy, is for all time. We hear of heroes
who have backbone. Backbones, to be useful, must
be vertebrated and bend like the blade. The Toledan
blade of choicest quality hath its renown because in its
polish and elements it reflected the knightly gentleman
who bore it! I make my parting salute to Toledo—
with her own courtly blades! Swish! Allons!

CHAPTER XX.

ALCOLEA—CORDOVA AND ARANJUEZ.

'The temples and the towers of Cordova
Shining majestic in the light of eve.
The traveller who with a heart at ease
Had seen the goodly vision, would have loved
To linger——'

SOUTHEY's 'Roderick, last King of the Goths.'

WITHOUT telling you now how I reach Cordova, I spring—like the Toledo blade—elastic, at once thither. Having been over the road, through the waste places of La Mancha, celebrated every inch by the eccentricities of Don Quixote, am I not licensed, as was the hero of Irving's story, to seize upon enchanted means—say, a magic velocipede—for the journey? From the Tagus to the Guadalquivir we fly by ruined towers and ancient castles; now and then by imperial palaces and theatres of Roman eras; then by olive trees and vineyards watered by the Moorish norias; then without seeing it, by the dull, poor, but royal, city of Ciudad Real (how much royalty in dilapidation Spain has!); then by the Venta de Cardenas, where Don Quixote paid penance and liberated the galley slaves, through tunnels and over bridges, amidst scenes of Moorish and Spanish wars, over a battle-field where the chronicles report 200,000 Moors killed and only 25 Christians, through the mining shafts of Linares, leaving the impurpled peaks of the Sierra Morenas to the right, we come upon a locality where we hesitate and halt. Two objects are here memorable. Not the Guadalquivir. Its stream still runs red, as if the blood of the 200,000 Moors still ensanguined its flood. Here at Alcolea, on its banks—if I may drop into practical horse-sense—was once the breeding ground of the splendid Cordovese barbs. There was the great stable called La Regalada. Here were begotten the splendid steeds of ample breast and majestic head, descendants of the great Arabian Godolphin, on whose backs Spain's proud chivalry rode into the battle! The best horses, however, were taken away by the French and English in the Peninsular war. But the people have had good stock cheaply born as the result of these breeding establishments.

The other object worth seeing at Alcolea was pointed out to us—the marble bridge built by Charles III. It spans the river with grace. It has twenty arches. The marble is dark and rich. The bridge and site are now historic; the site because, as the name signifies, it was an outpost fortress of the Moors; a pivot and strategic point in campaigning. Many a battle has been fought here. A passenger in the train pointed out, with eager particularity, the scene of the battle on the 28th of September, 1868. It decided the fate of Queen Isabella, and perhaps of Spain. Serrano, now Regent, commanded the revolutionary force. Prim was then at Cartagena. Serrano drove the Queen's forces before him from the bridge, until they disbanded and surrendered. Fifteen hundred were killed, and the bridge was slippery and running with blood. The general in command of Isabella's forces had part of his face shot off. Serrano's popularity is not alone owing to the persuasive plausibility of his manners; he has substantial qualities as a general. Besides, as the passenger said, it was a touching sight to see him embrace, with tears, his old and vanquished comrade, the maimed general of the Queen, after the battle. He granted the most clement terms. In fact, the gentleness of his character is its salient and significant point. By it he has, for the present at least, reconciled conflicting elements, and wields the executive or kingly power.

But we are nearing Cordova. The battle bridge is but seven miles from the city. Along the rest of the route, the influence of the protecting mountains is seen in the vegetation. We are in Andalusia again! No more battle-fields, no Toledo sabres, no Isabellas to be fought for or against; but here are vegetable glories—plumed palms, bloody roses—

'Springing *blades*, green troops, in innocent wars,
Crowd every shady spot of teeming earth,
Making invisible motion visible birth!'

Here the generous soil—warmed by the sun, fed by the water, and clothed by the trees—was once, what Southey describes in the verses prefixed to this chapter, and is yet, the green setting of what the ancients called the 'gem of the earth'—Cordova. Flowers, fruits, and trees still surround her. They are the offerings which kind Nature lays upon the stony sarcophagus of dead glory. Cordova—once the pet of Pompey, the victim of Cæsar, the home of poor patricians from Rome, the birthplace of Seneca, and at last trodden under foot of Goth—had a splendid resurrection under the Moor! It was called the Athens of the West. It became the capital of Moorish Spain. Its importance, learning, and wealth read like fables. It had its sultans, seventeen in number. But they, at last, had their rivals. Then came treachery, then divisions, then the old story of men and empires. With divisions the Spanish conquerors came! When

Saxon England groaned under the Plantagenets—rude, boisterous, and enslaved; from the time of Charlemagne to Philip Augustus—Cordova had three hundred mosques, nine hundred baths, and six hundred inns! Now, after seven hundred years, how few are the Moorish relics! The Moorish walls and cisterns remain; the Moorish mills grind slowly and 'exceeding small' grists; the Alcazar and Mosque are partly preserved; but, as I said, Cordova is a stony coffin. Beggars importune where kings commanded. Donkeys trudge over paths where pranced the stately Moslem chivalry! Toledo I have described as asleep. Her books, her vaults, her pictures, and her sculptures give to Toledo an air of drowsy seclusion, which is, however, not quite the atmosphere of death; but Cordova is a corpse, or rather the coffin of a corpse. Forty thousand people wander and work about its narrow streets where once a million lived! The very years since its death are notched by the aloes, which disclose the time when the Moors were here. A few palms, too, overtop the Convent wall near one of the gates. These are souvenirs of the famous race who came from 'the fervid land which gave the plumy palm-tree birth.' For did not Abdurrahman, the mosque maker and caliph, sighing after his beloved Damascus, with his own hand plant the Oriental tree, that its graceful leafage and stately stem should remind him of his old home? Yet these palms have no—dates. I mean by that no facetiousness. They are all *male* palms in the convent garden, and do but flower. Fruit comes not to the lonely old bachelors of the—convent! But still bachelors and masculine palms have their uses. The fruitless palm is used in holy observances, even as the celibate monk has his honourable place within. Therefore, repine not, O Benedicts! Ye are vessels of clay—sometimes fine

porcelain, and sometimes poor pottery: but ye may
be made sacred by hallowing influences from heaven.

Our stay in Cordova was but for two days. We
exhausted its attractions in one; then rested on the
other in the loneliness of our Fonda. 'Oft in the
stilly night' the watchmen of the city, like those at
magnificent Murcia and splendid Seville, chanted the
hours. They gave us the weather, too, being incarnate
barometers as well as vocal chronometers. Thus
have they sung time away here since the Moor cried
muezzin from the minarets of the mosque.

Yet we were anxious to see the city for the sake of
what it once was. A Moorish tower appears at the
extremity of the bridge over the river. We cross the
bridge. On one side of the tower is a waggon factory; on the other a school for little girls. To ascend
the tower we must go into and through the workshop.
After the waggon-maker had expanded his tire of iron
in the ring of fire and fixed it upon the wheel—the
old process—and taken off his leather apron, he
gallants us to the top. Through dark, damp, and
devious ways, all out of repair, we reach the summit.
It is a good rule to find an eminence for your first
look at a strange city. The tower and its top were
not unlike in structure and form to those of the
Alhambra. Grass and moss hung about its tiles and
parapets. It furnished a fine view of the mountains
on the sky's rim, the falls of the river, the Roman
bridge over which we had come, and the city, which
is as compact as a fortress.

Coming down from the tower, we perceive the
school-room. The teacher is an old lady, who looked
at us kindly over her huge, round spectacles. We
asked permission to enter. The hum of the room
stopped. All the little girls rose and saluted us. One
little girl asleep near the door was awakened by my

dropping a coin into her lap. The old lady tolerated the twittering titter of the little ones which followed. The child awoke; saw the coin; thought, perhaps, I was an enchanter, lingering around the old Moorish tower; but, shaking off the spell, she followed me to the door to return the coin. The teacher said, 'It is yours.' She made the enchanter a *gracia*, and he departed. That was a feature of Spanish life worth noting—an offer to return a coin! *Credat!* The teacher receives only thirty cents a day for her teaching from the city government. Such is the wages of the ancient 'Athens of the West.'

There is to be seen in Cordova one thing which I have reserved. Call it still a mosque if you are Moslem; or a cathedral if you are a Christian; or, if a scholar, *La Mozquita*, Arabic for worship. It has never, under any rule, been changed from a temple of religion. It is enclosed in high walls, and within these walls are very old orange-trees. In fact, we saw the oldest orange-tree in Spain, as is said. It is called the Caliph's tree—planted by him. Parapets and spandrils, gates and cisterns, fountains and courts—all bespeak the Orient, though the Moors have not been here for nearly four hundred years! The tower of the mosque is yellow. Many additions and subtractions have been made. It is in good repair, although it has been shattered by hurricane and by war. You may have a good view of the city and its environs from its dizzy height. Very little of the buzz of life comes up from below; for there is no buzz when there is no business. You are curious to see the twenty roofs of the mosque, and these are duplicated on the other side of the court. You then descend; for your curiosity as to the mosque is to be gratified only by going inside. Let us enter!

One feature of this mosque—or cathedral—is at

once apparent. It is supported by 1096 monolithic columns! These are of every variety—the red Granada marble, the twisted Byzantine porphyry, the yellow French marble, &c. It once had 1200 columns, but

Mosque of Cordova.

there remain enough to confuse the vision, were there not great symmetry in size and position. Viewed from some angles, the maze of pillars looks like a forest of neatly-trimmed trees. This mosque was built out of the materials of an old Gothic temple, upon whose site it stands, though many of the columns, like those of the great mosque at Constantinople, came from other temples. The Pagan and Christian world was ransacked to prop and furnish this mosque. It was second only to Mecca and Jerusalem in the Mohammedan realm. It was copied from that of Damascus. It was not finished till the tenth century, though begun A.D. 786. At first glimpse the roof looks low; in fact, it is only 35 feet high. It is very unlike, therefore, in solemn grandeur, to the Gothic arch and vault. It looks lower than it is, as the area is so great. That is 356 by 394 feet. All the decorations of side-chapels, all the burning tapers, all the pictorial arrangements and displays, all the roominess of and under the central dome, with its choir and altar, all the filling up of the place with Christian signs, imagery, and symbols, cannot take from it the odour and presence of Mohammedanism. It is a mosque; it will ever so remain. Its dedication is one thing, as the sweet music informs me, but its form and structure are another.

As we leave it the sacred influences do not follow us as sweetly devotional as at Seville. There are pointed out to us, upon one of the marble columns near the door, the marks of the chain of a Christian captive, here ironed for a quarter of a century by the Moors. A cross marked in the marble is said to have been made by the captive with his finger-nail. Constant rubbing—like the water-drop—wears away stone.

One more enchanted, elastic spring. I land at Aranjuez. It is an hour's ride from Madrid. I cannot linger here. True, it is a seat of royalty—royalty as recent as Isabella's, and as faded. We stayed but two hours and a half. The city has 20,000 inhabitants, and it is supported, or was, by reason of the palace of the Queen and the villas of wealthy nobles. The

Queen spent her April, May, and June, here once.
She will spend them here no more. The palace is
called by the natives the 'Metropolis of Flowers.'
There are trees and gardens about it. It was a
shooting villa of Charles V. Ferdinand VII., Isabella's
father, finished the palace. Here he took his wives
and maidens, and enjoyed the cascades of the Jarama
and Tagus, which here unite to irrigate garden and
forest. They irrigate to some purpose. Long avenues
of Oriental plane-trees and English elms, filled with innumerable birds, are here. Everywhere is the sound of
running waters. I made a few notes, in my haste,
of this place. A hasty copy of them would answer all
the place demands of a tourist. My notes run thus :—
Saw labourer's cottage, fitted up with china, tapestry,
pictures; malachites, built by Charles IV., a fool, full
of whims; fine gardens, laid out by an Irishman,
Richard Wall, who became a Spanish Minister. N.B.
—Names of gardens changed; conspicuously painted,
' *Jardin Serrano*,' ' *Jardin Prim*,' ' *Jardin Topete*.'
Finest picture in palace—Boabdil leaving Alhambra.
Mem. for a companion picture—Isabella leaving Spain.
We rushed through palace, dashed through forests,
strawberries for breakfast, and a run of a mile, with
beggars following, to the depot; reached it an hour
before time! Depot awfully solemn. Erected probably by the Phœnicians. It seems like everything
about Aranjuez, deserted. A rumble, then an omnibus; another rumble, the train! Out of the depot we
fly; forests to the rear; poppy-fields,—a flash of red;
donkeys under hay and faggots, tails alone visible in
the receding landscape; huts thatched with straw. An
hour dashes by, and Madrid appears!

Having arrived at Madrid, I spent my first day with
Mr. Hale, our Minister, and his family. After returning with him from a walk about the city, I was

informed that the Governor of Madrid desired my
presence at the palace—the Casa del Ayuntamiento, in
the Plazuela de la Villa. One of his officials had called
for me in my absence. My courier had seen him.
In great trepidation the courier recounted his interview with the officer. The question arose, What did
it mean? what portend? What should I do? The
Minister rather thought that I was to be arrested for a
Republican speech made at Granada to the volunteers.
He was there at the time. This view I was inclined
to take also. But it was rather too much for me to
indulge in the hope of the political luxury of such
an imprisonment. At length we marched to the
Governor's office. What might not depend on the
interview! Would I retract? *Jamais.* I repeated
the *sum Romanus*, and felt the wing of our noble
bird fanning my patriotic brow! We passed up the
narrow stairway of the palace; the route which many a
criminal had trod before.

Mr. Hale has not lost his sense of humour, though
he had lost his place. I asked him, as we walked up
the worn steps, and with a tremulous treble, 'Mr.
Hale, have I a guilty look? and if so—of what?' He
seemed to apprehend that there was a serio-jocoseness
about the situation, and attributed it to irrepressible
elocution! He also agreed that it was rather pru-
dential and moderate elocution, considering the occasion, the speaker, the temptation, the wine, the enthusiasm, place, and theme! It had already cost me for
arrobas, bread, cigars, and 'federal republican' sau-
cissons, 492 reals, and was I to suffer more? Ye
gods! If so, let Spain look to her colonial possessions! This remark was not audible, but it had reference to an Ever Faithful Isle. Well, we were ushered
into the presence. A full-bodied and good-natured
man, smoking cigarettes with celerity, bowed. I

bowed grimly. I confess to a feeling very near akin
to a sense of guiltiness. I tried to look—I think I
did look—innocent of any intentional breach of international comity! The Governor inquires my name
and that of my companions; he then opens a volume
of despatches. He has one—a long one from the
Governor of Granada! So, it is Granada! Well,
hurrah for Cuba! The Minister whispers to me aside,
'There's a full despatch on you, Mr. C. Got a long
account against you, reported at length. It looks
formidable!' After much suspicious quiet smoking
and whispering side-conferences with an officer, the
Governor announces that—a draft of mine drawn at
Granada on Paris has not been paid, but protested!
'What a fall was there, my countrymen!' I need not
say that, while I healed my wounded credit with a
check, my patriotic pride and political self-devotion
also received a check of another quality.

Napoleon's Grotto (see p. 71).

CHAPTER XXI.

MADRID—THE CORTES—JUBILEE OF THE CONSTITUTION.

'A kingless people—for a nerveless State.'—BYRON.

SPAIN to-day turns over a new leaf, or rather, having perused a new political book, contented with the contents, she, to-day, snaps the clasp. Spain and all her colonial children have been for many years in a state of chronic disquietude. Her soil is the result and the political theatre of volcanic excitement. She has had more than her share of vicissitudes. From the time the Romans made her sons slaves in her silver mines, until the great era when she made the Americas tributary with the same chinking 'legal tender,' and from that era till now, she has been violently volcanic. But to-day she is republican in fact, though monarchical in form. She has no monarch, and yet she has in her constitution ample provision for his election and his government. She has no head, except the many-headed Cortes; yet she is preparing to make a provisional head, called a regent, till she can put up a toy dummy with a crown on it, or perhaps, put aside the idea altogether.

She has a new constitution. The little boys who sell the journals were crying it about the streets and plazas before the paper on which it was printed was dry. True, it has not been signed by all the members of the Cortes. I saw some of the lingering, reluctant members of the Cortes, called *moderate* republicans, on Thursday, 'step up to the captain's office' and write their names. But it is passed if not signed. The republican members, numbering some sixty, have not been in haste to sign a document which, in its second title and thirty-third article pronounces, '*La forma de gobierno de la nacion Española, es la monarquia.*' True, the preceding (thirty-second) article recognizes the sovereignty as essentially residing in the nation, from which are all powers. This latter was a little painted tub thrown to the republican whale. Still, the republicans have not been lively after it. Perhaps the republicans might have done better, in one sense, by signing. I have just seen in a funny journal, '*Gil Blas,*' a good caricature of the wives of two deputies: one is woebegone, thin and scraggy, from the furzy head on which her mantilla hangs in tatters, to her shoeless feet, and short ragged dress. The other señora has a towering, glossy, elegant head of hair. If not her own, is all snug to her lofty and classic brow. Her train sweeps in rich and regal style, and her Grecian bender is faultlessly high on the beauteous curve of her back! She addresses the unhappy sister, 'Why art thou so miserable?' The disconsolate responds, 'My spouse has not yet signed the constitution. We die of hunger.' 'Ah! if he had done as mine has,' rejoins the fine señora, 'your hair would be like mine—glossy with beauty.'

Perhaps, when the Cortes meets, the work of signing will be completed. It will then perhaps be definitely and legally determined who is to represent the provisional executive or be appointed regent, till the king is elected. It is not yet decided; but it is generally conceded that Serrano will be the man.

The ceremony of to-day, and for the next two days,

is the jubilee of the new constitution. First, we are to have it proclaimed from the temporary platform which I saw lately erecting in front of the House of Congress; after the manner of American inaugural ceremonies announced from the east front of the Capitol at Washington. Then we are to have in one of the plazas a new statue uncovered, commemorative of the event. Then wine is to flow, it is said, in one of the fountains, at which all will be at liberty to drink. Then the soldiers are to take the oath to the Constitution, in dramatic style. Then, the city is to be illuminated, and then comes the bull-fight, when the ceremonies will assume their most enthusiastic form. So much for the programme.

Already, at nine o'clock on this Sunday morning, the streets and plazas, and the public and private buildings, are decorated, as if for a holiday. Every balcony has a red strip of bunting, dashed with yellow. Red and yellow are the national colours. The streets are thronged. The Puerta del Sol (once a gate, where lazy people used to bask in the sun), near the Hotel where I stay, is a mazy show. The noise of vehicles and screams of newsboys and newswomen mingle with the excited rush of people to and fro. Yes, this is a *fête* day; all the provinces and towns near Madrid are represented. It is festive, but who knows ·whether it may not end as a funereal day.

Parties here are mixed. A stranger has no business to try to give an idea of their complications. For instance, take one division—monarchists. There are progressive monarchists—democratic monarchists— shading up to the 'strongest Bourbon;' the chivalric Carlist, and the moderate Montpensier. Each of these parties has its adherents. Somebody has determined that Madrid shall not forget that the Duke is a Bourbon. The walls are covered with placards. 'Bourbon—Orleans

—Montpensier.' But Spain recognizes democracy.
The preamble to the Constitution says: 'The Spanish
nation, by and through the Cortes, *elegidas por suffra-
gio universal,*' does so and so—following the routine
and almost the words of our American constitution.
But the resurrection of an obsolete election law cut off
from voting those who had not attained twenty-five
years of age. This defeated a full vote for Deputies.
The republicans, however, though numbering few in
the Cortes, are morally and intellectually potential.
They are under the lead of the discreet Figueras and
the eloquent Castellar. They are more nearly united
than any other party. It is suspected that General
Prim, who is to be second to Serrano—in a *rôle* in
which he thinks he ought to be a star—half inclines
to the republicans, and that in a confusion he might
go over to them, and consummate what is to-day the
desire of the people. But Prim disavows; and grows
more reserved. In a comic journal there is a commentary on this complication and division of parties.
It has especially an eye upon Prim. The scene is:—
Door of the Minister of War (Prim). 'Haloo!'
'Well?' 'Are you a Carlist?' 'No, Señor.' 'Republican?' 'Neither.' 'Moderado?' 'Que te que-
mas,' (now you are near burning yourself, *i.e.,* near the
mark!) 'De la Conciliation?' 'There you hit it!'
'Go in!'—and he went.

It is a great relief to the people of Spain that the
long discussion in the Cortes has closed. This ex-
citable nation has been unusually vivacious and nervous
about the result. Trades-people, especially, were beginning to think that contentment, even under Isabella,
with all the shame, would do better than uncertainty
by their own choice, and suspense under a perennial fountain of senatorial eloquence. The difficulties of making the constitution—growing out

of the church and monarchical questions, then out of
the Regency question, added to the financial measures
—have made the session long, and excusably so ; but
the people here soon become impatient. They have,
therefore, some cause to rejoice over the fact that an
end is made of the debates. While the military are
parading beneath my balcony, and the trumpets of
the cavalry are filling the air and prematurely crowding
the adjacent windows; while the crowds are gathering
round the squares, and in front of the Congress Hall
—and before I go out to see—I will give you a brief
summary of the new constitution.

There are eleven titles, and 112 articles. The
preamble is modelled after that of the United States.
It recites the desire of the Spanish nation for justice,
liberty, security, and well-being, and for the accom-
plishment thereof provides the following 'constitution.'
The first title is about citizenship and rights—a bill of
rights, in fact. It has in it habeas corpus, free speech,
fair trial, domicile inviolability, protection to private
property, and religious freedom, but not the dis-
establishment of Church by State. The second title
divides the public powers; the third treats of the
legislative, both Cortes and Senate; the fourth of the
King; the fifth of the succession to the crown and of a
regency; sixth, of a responsible ministry; seventh, of
the judiciary; eighth, of the provinces and corpora-
tions; ninth, of finances; tenth, of colonies; and
eleventh, of the amendments. One thing occurs to
me as a special clause. The right of domicile is to
be regarded—every man's house his castle—except in
case of fire or inundation, or something analogous.
That exception springs from peculiar circumstances,
and is very general and ominously uncertain. Another
thing to be noted as a part of the progress which
Spain has caught from England or America. Spain

is not to lose her Cortes. Parliament is omnipotent. The Regent, King, Ministry, all are at the will of the Potential Legislature. Its power is not to be abrogated. This is wise as well as conservative, provided the Cortes keeps the constitution. To read this constitution one would say it is liberal beyond all expectation. Will it be executed? *Ça dépend* on the people themselves. While I am here in Madrid, wine flows, flags fly, drums beat, trumpets resound, men march, people hurrah, and bulls die, for the constitution, in which there is monarchy; to-day in Saragossa, the ancient capital of Arragon, the people, with great demonstration, are burying the crown under their historic earth.

As I occasionally drop my pen to gaze from the balcony, I perceive that Madrid grows more and more military, or, call it festive. The cavalry have gone by, so also the infantry—volunteer and regular—and just now the artillery. The guns of the latter are drawn by mules, half shaven of their hair. The horses of the officers are splendid barbs; their manes are frizzed or crimped, and tails cut and cued. They caricole nimbly, as if dressed for a lady's chamber. The music swells all about us. Offenbach monopolizes the tunes. Now a familiar touch of General Boum—quite original and congenial, I think here; then, Barbe Bleue—a reminiscence of the Moors; then, the 'Belle Helène,' also Spanish; for the charms of a heroine in Spain, Count Julian's daughter, made civil war and set kings on fire. You should see the mounted men, not gendarmerie though they too are out in their Quaker coats, white pants, big boots, and cocked hats, but the mounted fusileers, each man galloping along with his finger on the trigger. I ask what that means? 'It means, Señor, fear of Republicans.' There are some fifty thousand troops hereabouts. I think that they are all under arms. This is a very festive day. It is the Jubilee of Liberty! It is the Jubilee of the 'Constitution'!

Now what is the Constitution? As one of the old advertisements for quack medicines used to say, 'It is certainly that which constitutes.' Good! But Spain has always had a Constitution, even when she had nothing which constituted. Theoretically, her Constitution may be traced, as the English, to the days of the Conquests, or to the German people before Conquests. England finds her Commons in the Witenagemote. Spain believes that her early invaders, the Suevi and Visigoths, replaced the power of Rome by an elective monarchy. This was the German way. Spain now proposes, after twelve centuries, to restore it. Thus you see that constitutions came out of the old German forests, the home of liberty and right. This fact was noted as long ago as Tacitus. The contest between privilege and prerogative,—between Commons and King, people and power—can be traced as clearly in Spain as it has been by philosophical historians in England. Spain has had her Magna Charta. Her '*fuero juzgo*,' drawn up by order of the Gothic Kings in the seventh century, was promulgated by the Alonzos of Leon. It proclaimed *Electione igitur non autem jure sanguineo olim Hispaniæ reges assumebantur*. The Cortes of Aragon, where they are now burying crowns, as early as 1094 obtained *fueros*, privileges and rights as precious as Habeas Corpus, and gave in return guarantees to their kings. Codes were agreed on for the transmission of the crown. These codes excluded female heirs until 1830. Then Ferdinand VII.—the big-lipped, brutal-faced, stupid-looking Bourbon—father of the present ex-Queen Isabella II.—repealed by a pronunciamiento the ancient code. He thus enabled his serene daughter Isabella to take the crown. This excluded her uncle Carlos.

Hence the Carlist war of 1833. Hence the Carlists of to-day. I find photographs of 'D. Carlos de Bourbon' selling here by the side of Prim, Serrano, Topete, and Montpensier. He is pictured as a Spanish soldier—a coming man—on a horse, saluting something with his sword. I suppose he salutes the future. He is the direct male heir of the Spanish line of kings, and has some supporters here, who think or fancy that he may win. They are very active and noisy.

Spain had a Constitution, even after the French departed in 1812. In 1837 she had another. Isabella II. was then guaranteed her throne, France, England, and Portugal going security. The bond is now forfeited. Another constitution was made in 1845. It was, ostensibly, in force till five days ago. That constitution defined the monarchy within the limits of liberty. It conserved the ancient rights of the component kingdoms of Spain. Isabella and Ferdinand, Charles V., and Philip II., made Spain, or rather themselves, absolute, or even worse. How horribly they treated the Moors and Jews, to the dishonour and detriment of Spain, and the misery and disgrace of humanity! This, history records. Buckle has philosophised over it, and nothing remains to be said. But all the time there was a current of democratic liberty running beneath the tyrannies of those times, like the pure streams which are found even beneath the grand palaces, in conduits from the free mountains to the thirsting cities! Spain has had her Sydneys and Russells. I have read their names painted on the Cortes panels. We have had in America so much Protestant or Puritan politics and philosophy, that we scarcely know that ultra-Catholic Spain has a long and illuminated roll of worthies who worshipped liberty even as they worshipped God. True, they were sometimes nurtured in the cloister; oftentimes in the hut or the court;

sometimes in the field; and very often in the universities; but ultra-Catholic Spain had the very eye, ear, head, muscle, heart, mind, genius, and soul of freedom. Why it was repressed, I do not ask now. Buckle answers. That it is now abroad, all know. When I write what I have seen in the provinces and in the capital, among the peasants and in the Cortes, you will re-read Buckle, with new annotations, and believe that Spain means much and means nobly by her new aspirations and struggle.

Spain is a unity, made by coalition, marriages, commerce, interest, physical boundaries, and olden pride. She is united from many diversities. We think Virginia was proud and imperious; but think of Castile! We reprehended that rash and undutiful daughter Carolina; but think of fiery Andalusia! We talk of cold, isolated, imperious, intellectual New England; but forget that Aragon, Leon, Navarre, Catalonia, Valencia, Biscay, and Mancha have a thousand years of fueros, independencies, kingdoms, parliaments, codified laws, and immemorial customs more peculiar than all the several virtues of New England. We, as provinces in America, or as thirteen colonies, had scarcely a century and a half of chartered rights, and as the United States hardly a century of State rights. What a splendid field for the illustration of a federal republic is Spain! Let me particularise: Biscay had peculiar laws. They were published as early as 1526. She was always free from conscription and taxes, except of her own levy. The Queen was only Lady of Biscay, not ruler. The families of Biscay were and are autonomous rulers. Barcelona had a commercial code, which became that of Spain. It was printed the year after Columbus reported, in Barcelona, to Queen Isabella (he made his report April, 1493) that he had found gems and gold, and that there was a new world!

For four centuries all Europe regarded the Barcelona
Consulado del Mar as the law of the sea.   The laws
as to real estate in Spain are as different as the costumes of the different provinces.   Catalonia has one
law and the different provinces of Castile others; the
Balearic Islands one, and the Canaries another; in fine,
as to all the subjects of human, national, and provincial
rights, Spain has more pride and pertinacity than she
has had success in their exercise.   If there is one
nation where provincial independence has been eager
to be recognized and harmonized with nationality, it is
Spain.   Hence, how easy, if fairly tried, to inaugurate
a system of federation here, and how hard it has been
to make any centralized system durable!   With the
forty-nine provinces, each having its Civil Governor and
Provincial Council and Assembly, we can understand
why, when revolutions may overturn, they do not altogether destroy, and why Spain retains her old franchises,
even when the world believes she is being torn to
shreds by ambitious chieftains and factions.

True, Spain has had revolutions.   They have not
been unassociated with bloodshed.   Recently, when
walking about this capital with the American Minister,
he pointed out to me where the balls had blistered the
walls and stones, where the cannon had pierced the
buildings, and where the gutters had run with blood.
This happened as lately as 1866.   Isabella II. then
was marked.   Her time had nearly come; but by
some misadventure O'Donnell was enabled, by his
artillery, to crush the revolt before the infantry could
combine with the artillery.   The latter had twenty-
four pieces, and were within two streets of the
palace.   Waiting for signals from their confederates in
the insurgency, they failed.   Not so Prim, Serrano,
and Topete.   Their work is to-day to be consummated;
I cannot yet say exactly, crowned!   They hold Spain,

under no crown and with no sceptre, because they have conciliated by liberality the republican elements. They pretend to make the monarchy an empty bauble and the republic a living fact. Hence there is a strange, and to some an unaccountable, indifference to the erection of a throne! But there is no indifference to the establishment of a constitution. If there were, what does all this trumpeting, mustering, and viva-ing mean which salute the ears?

Having an eye for displays, being fresh from Granada, Cordova, Seville, Aranjuez, and the Escurial; having seen the parliamentary halls at Valencia, with almost as much interest as those at London and Paris; and having, while here three weeks ago, heard Prim and Serrano and the republican deputies speak in the Parliament House, I have become so interested in Spanish politics as to rush off with the crowd to the place where the 'jubilee of the constitution' is to be most conspicuously celebrated!

I am now abiding at an hotel near a famous gate. That gate is now a plaza. My balcony looks out on the plaza and down the street of St. Geronimo. Through that street the crowds pass, and through that street the regiments march. Having secured a roof near the Cortes edifice for the ladies, I move in that direction to take possession. On the way we mingle with the throngs. For a half mile—about the distance from my hotel to the Cortes—the whole route is full of moving people. We move with the crowds. Directly we come to the neighbourhood of the Cortes. A statue in bronze appears. I have seen it before. It is Cervantes. He stands before the Parliament House, and before the palace of the Duke of Medina, and before the people. He is before them all. He is immortal.

Having placed my company on the top of a four-story house, I seek the crowd. The scene is splendid.

From a thousand windows and roofs ten thousand fans
of as many females are fluttering, like butterflies in a
flower garden.   The sun is hot.   The air simmers and
vibrates with caloric.   The red, maroon, yellow, gold,
and white canopies, hangings, and fringes from every
balcony make it a festive scene.   But that scene
would lack much if the maidenly modesty of many
hundred living Murillos were not *painted*, and framed,
too, and peeping from the windows of palace and
mansion.   But our eye seeks next the front of the Congress.   There it finds a scene worthy of a Dutch painter
for particularity of detail and individuality of person.
It is like the canopy of an olden tournament expanded to the proportions of a national exhibition.
There is a crowding of well-dressed people thither.   A
telegraph of ten strands of wire goes out of the Cortes.
There are safety and civilization in that idea !   There
above, the tops of the houses are covered with people.
The palaces of the grandees are also covered.   The
women above, like those beneath, are bonnetless ; but
they have mantillas.   I try to get near, in order to see
the dignitaries who are under the legislative canopy.
I am told there are one hundred generals and some
three hundred majors and as many deputies under
those yellow and red awnings.   So I press on.   I get
to a point where I can see the Church of the Ascen-
sion ; then the great palace of the Duke of Medina.
Everywhere are hangings and flags of red and gold and
white and yellow.   Now I can glance under the large
canopy.   It extends the whole front of the legislative
building.   Under it there are seats for the dignitaries,
and I creep down so near that I can smell the quality
of the cigarettes which the well-dressed young men in
the crowd and the gay young soldiers on the platform
puff into the faces of the señoras and señoritas who are
freely mixed in the mass around me.   A hurrah is

made; rather a viva! Four vivas! One for the Cortes! one for Serrano! one for Prim! one for the Constitution! The Constitution is last; but the vivas are executed spiritedly enough for a Spanish crowd. The guns fire! I change my position, having an eye on a gas post, in case of retreat. Near are several of the cavalry patrol. Their horses are restive; so are the people. The latter all have canes; and when a horse pushes near them several canes punch the horse! This makes the horse and the crowd mutually vivacious. A very restive horse makes the people very restive. The calendar of saints is run over in a hurried and profane way by excited Espagnols. A rush of horses and men is made. I strike for my lamp-post. I give a small boy, who is up the post, a peseta for his place, and obtain the privilege of climbing up! I climb. There is a wild dash of cavalry. It is to clear the way. There is an intense excitement. The crowd think that there is an *émeute*. How they run! up the main street and down the back streets. Everybody rushes. I am still and solid as iron; being anchored on my post. There was nobody injured, though London and Paris telegraphs reported great loss of life!

There is a vacant place cleared before the awnings, where the Cortes, with Serrano, Prim, and the Generals now are! I look all around. A shot of a pistol would make trouble. The house-tops are crowded. Another rush of cavalry! Then a sound, as of—'Hi! Hi!' We look up to a four-story house; girls are getting out of the attics on the roof; boys are chaffing and hallooing at them! Then, another rush and push; and from my perch on the lamp-post, I see on the platform the American Minister. I feel under his protecting shadow.

The scene in front of the Cortes was one which,

once seen, will not be forgotten. The building is of
white marble.   Imagine a platform gorgeously decorated with the national colours, crimson and gold,
with the shields and arms of the different provinces of
Spain emblazoned in endless variety.   This platform
is occupied by the two hundred and sixty members of
the Cortes, by the representatives of foreign nations,
and by the most distinguished civic and military
functionaries of Spain.   These grey-headed, scarred
militaires are exhibited to convince the people that the
army supports the Cortes and the new Constitution.   A
space of perhaps eighty feet in width in front of the
building is kept clear for the manœuvres of troops, a
double line of whom make a wall of bayonets to keep
back the crowd.   This crowd might well be numbered
among the wonders of the earth.   Of all nations, conditions, ages, and colours—from the Arab of the desert
to the yellow Peruvian and the dapper citizen of Paris.
England and America had few representatives.   The
fear of revolution has kept tourists away from Spain
this summer.   The representative men of Spain upon
the platform appeared well.   The Cortes, a fine-looking
body of men, occupied the centre, with the grandees
and notables upon their right, and the foreign envoys
upon the left.   At the table in the centre sat Serrano,
Prim, and Topete, the three most prominent members,
of whom caricatures were widely circulated, representing them as auctioneers offering the crown for
public competition.

Some fortnight ago I had the pleasure of visiting
the Cortes several times, and there I saw and heard
Prim and Serrano, but not much from Topete.   It is
not difficult to find the House of Congress.   It has a
gilded sign, '*El Congreso de los Diputados.*'   The
edifice is handsome, though small for a national
assembly.   It is new, having been completed in 1850.

In the centre of the principal façade is a triangular front. You hardly notice it to-day, amidst the decoration of flags above, under, and about it. Its figures are supposed to represent Spain receiving Law, accompanied by Power and Justice. At the steps two horrid bronze lions represent—something wild. The lions were formerly of stone, but a cannon ball nipped off one of the leonine heads in 1854, and bronze was substituted.

I propose, while the military are marching, trumpeting, drumming, and thundering through the city, and the dignitaries are waiting for the troops to pass in review before them and salute the Cortes and its new Constitution, to go inside. Not to-day, but *nunc pro tunc*, when I heard the debates more than a fortnight ago. It was somewhat difficult to obtain permission ; however, I sent my card to President Rivero, who sent a caballero to conduct me to the Diplomatic Gallery. I soon began to understand what was going on. There is a universal language. Parliaments, and parliamentary halls and debates have a common object. As I am taken through byways and corridors, up and round, I fancy I tread the maze of the Capitol at Washington, or am lost in the lobbies of the English Parliament. The Speaker's room at Washington has its counterpart in the *Sala de la Presidencia*. But the latter is not so much like a furniture establishment as the former ; for it is decorated by an elegant and significant painting, executed with skill, and eloquent with meaning. It far exceeds the flimsy frescoes of the rest of the House. Grisbert is the painter, and the subject is 'The Comuneros.' It is a representation of three popular leaders—Bravo, Padella, and Maldonada, who were executed as martyrs of liberty in the time of Charles V. It is a sad scene : similar scenes will one day be painted for the English Parliament when the people have their empire in England. The same will be done for America when Art assumes her sceptre there. Out of all the paintings in Europe—not sacred—which we have seen, how few speak of aught else than of the kings, grandees, and aristocracy! The heroes of the people—the Pyms, Hampdens, and Mirabeaus—are unpainted. Inside of the Cortes hall there are paintings representing royalty on the throne, and the deputies of the people at the footstool. The fresco of the dome is allegorical. The seats for the members are red, circular, and raised as they recede from the chair. The hall in size and appearance resembles the *old* Hall of Representatives at Washington; it has not much more gallery-room; yet here the accommodation is considered extensive. It is more so than in the Chamber at Paris, or the ridiculously little Parliament Hall in London.

The interior is very lofty and well ventilated. There is a marble pavement in the centre of the hall. That is neutral ground. Behind the President are red hangings, and above them, or above the canopy, is a symbolic castle. On either side of the chair are two remarkably dressed persons, who leave every once in a while, as if for a drink. They are replaced by others. They have on theatrical, gay caps, and long white feathers in them. They hold in their hands a gilt mace, or —something. Their dress is a long robe of red and gold. Making allowance for their standing still so long, their chief employment seems to be to uphold the dignity of the ancient Cortes, by gaping; as if they had just wearily waked up, like the seven sleepers, to the realities of a new era! The Secretaries, who are Deputies, sit on either side of the President. The hall has pendent from its centre a splendid chandelier. Around and below the galleries are the escutcheons of the provinces; each once a kingdom.

Speaking is going on! Opposite, on the left of the President, sit the Republicans. Below me, on the right, on one bench, called *banco azul*, sit the ministers. The members speak from their seats. They have little desks, which let down and shut up. They are but seldom used. The debating is very spirited. It is, to-day at least, humorous. A republican is uttering a diatribe upon the financial situation. He is cheered with *vivas!* The House is much more quiet than our Congress or than Parliament. Even the poorer speakers are not disturbed. Disorder is checked by a '*hist!*' or '*tisst!*' from the members. The President calls to order by a bell any member who is out of order. He had occasion to make several rings upon an eloquent, cool, determined, happy, good-natured, pungent speaker, who held the crowds in the gallery and the members very intent! It was Castelar! Bald-headed rather, like most of the members; black moustache; elegant contour of face and figure, and with a graceful *ore rotundo* voice, he is considered to be as pre-eminently the orator of the Cortes as he is undoubtedly the Liberal leader.

His speech for the republic and religious liberty, accompanied by his portrait, is hanging at all the book stalls and stores. Directly after he has concluded, an attack is made on Figuerola, the Finance free-trade Minister. He is a rather tall, thin man, with very little auburn hair; and, as I believe, a gray eye. He wore a long surtout, and spoke a little awkwardly, with one hand in his coat pocket. He spoke, however, with energy. For a student and professor of political economy, as he has been, he seems to take the rough handling of the politicians admirably. I heard, on another occasion, General Prim. He is the character of the Cortes. He is a short, compact man. He is a little bald-headed, but not old, say

forty-five; his hair is black, his whiskers and moustache
are black.   He is a reserved man.   He imitates Louis
Napoleon, it is said.   He is a self-made man.   His
mother was a Catalonian and a washerwoman.   He is
a capital, energetic, ardent, Spanish speaker.   He speaks
at first humorously, then pithily, then patriotically,
and then, with a lifting of his hand and eye, to God;
a touch of his finger on the region of the "waist-
coat," and then with Spain, *Spain*, SPAIN on his lips,
as a ringing climax, he sits down, with a loud viva as
his echo!   I did not fully understand him as he spoke,
but I read his speech afterwards.   It was a dedication
of his heart, soul, and life to Spain; repelling all
rumours and thought of failure in his steadfast duty
to her liberties!   Then more fine speaking followed
from a republican, interlineated with much bell ringing
by the indignant President.   Then one of the Ministry
speaks with great grace, and with a rhetorical shrug of
the eloquent shoulder.   Then Serrano rises!   There
is a hush.   Then a replacing of canes and eye-glasses,
and a settling into deep attention.   Castelar takes
notes.   Then the reporters, five, sharpen their pencils,
and one eagerly stands up to be sure of his hearing!
The *banco azul* grows more aristocratic; for is it not a
duke, a soldier, and *the* leader who is to speak?   He
looks like a polite, plausible man.   His light hair is
tinged with gray, or, rather, his moustache; for he,
too, is rather bald, like the rest.   He and Prim have
been passing some bon-bons, sent down from the
President either to prepare the larynx for smooth
utterance, or to cultivate the sweet amenities.   Serrano
fills my ideal of him as a persuasive and popular man.
What he says is received with vivas, and, what is
better, with respectful decorum.   Topete, the sailor,
also said something, but I could not catch the pur-
port of it.   He is a plain, blunt man, and affects the

sailor style. Such are the leaders of the September revolution.

Full in the midst of one of these harangues, a deputy from the provinces rushes in and cries out: ‘ I claim the *word !*’—not ‘ the floor.’ But as the orator who *has the word* has not concluded, the eager and rustic member is laughed at and laughed down, just as in the English Parliament or the American Congress.

From a general survey of the Cortes, I cannot but accord to them a more than ordinary style of oratory and an extraordinary measure of intellect. They speak out boldly and plainly their thoughts. They never dribble them through written essays to empty benches, like American Parliamentarians. No man is heard very long who does not say something; for the lobbies are near, smoking is common and attractive, and the exit is speedy. There is a striking similarity in the appearance of most of the deputies, at least from the gallery. They are far above the average for intellect, according to my observation. I will not say that they outshine the American Congress; but they are certainly equal to the English Parliament. I have had opportunities to make the comparison.

This is the body which has just finished for Spain an organic law. To-day it is proclaimed with the imposing ceremonies which I have described.

There is no space in my unpretending volume to describe the swelling scene in the Cortes and on the portico in its front. Certainly not if I had to copy the titles of the grandees as they are recorded in Spanish heraldry. In the Cortes—such as I have pictured it—at 2 o’clock, President Rivero assumes the chair. He declares the session begun. ‘ Conforming to the order,’ he says, ‘ the Constitution voted by you will be proclaimed, and all the “ Señores diputados” will, on leaving their seats here, be shown seats outside !' He leads the way with the secretaries, and, together with the Constitutional Commission and the ministers, he is seated on the *banco azul !* The Supreme Judicial Tribunal, the provincial deputations, and those from the cities and towns, and the scientific bodies, surround them. All the civilians are together, and their appearance is celebrated by more salvos of artillery and cheers from the people. Two secretaries relieve each other in the reading of the Constitution. More salvos and vivas announce its conclusion. The deed is done. The members return to their chamber. There they are sworn to support the instrument.

The processions move on, on, for two hours or more. I leave my place on the lamp-post, and retire to my leased roof. I am on a four-story house. Below me are Portuguese and French in the crowd; also, Galician, Castilian, Basque, Valencian—men, women, and children of every Spanish province, each known by the handkerchief on the head or the pantaloons and sash; and all eager to see the 50,000 soldiers who are filing by to salute the Cortes and its work. For two and a half hours the volunteers and regulars march by. In blue, red, black, green— infantry, cavalry, artillery, the mounted police or gensd'armerie, the shining bayonets, the glittering helmets and cuirasses, the flags of crimson and gold —all marching, and marching as no other soldiers do, to quick step. One tune is dominant. It is the Marseillaise of Spain—the hymn of Riego. It is named after a Liberal General, who was a pet of the people, and died for them. It is music, splendid and stirring. A few months ago, people were shot for singing or whistling it here. For nearly a dozen streets, past the palaces of the grandees and down the wide avenue in front of the gaily-decked platform before the Cortes,—these fifty thousand march! The officers

salute, the men touch caps, the cheers go up. It
does seem as if Spain had hope! This is all for a
new order—a better day. God help them! How
splendidly and dashingly the men march! How the
long lines and areas of human heads, decked with
colours or hid under gay fans and silken sun-shades,
lean forward to see each new regiment! How handsomely the cavalry ride their superb Arab steeds!
Never did I see such horses! Each one is fit for
the model of the steed of Aurelius done in bronze
in the Roman Capitol. The Spaniard is enough of
a Moor to ride like him; and his horse is enough
of a barb to make horse and rider a centaur! George
Eliot, in her new poem, touches my fancy with historic
and physiological verity and artistic and facile pencil,
when she says :—

> 'Spain was the favourite home of knightly grace,
> Where generous man rode steeds of generous race,
> Both Spanish, yet half Arab; both inspired
> By mutual spirit, that each motion fired
> With beauteous response, like minstrelsy—
> A fresh fulfilling—fresh expectancy.'

Perhaps she had her hint from the Cid and his horse,
as Lockhart translates the ballad :—

> ' And all that saw them, praised them,—they lauded man and horse,
> As matched well and rivalless, for gallantry and force.
> Ne'er had they looked on horseman, might to this knight come near,
> Nor another charger worthy of such a cavalier.'

The appearance of the cavalry, all on black horses,
clad in shining mail and helmets plumed with white,
was like a blaze of burnished mirrors. The eye could
not bear the Oriental blazon. It was a relief when
they had passed by, and when the cavaliers, called those
of the Prince—(the name is to be changed to-day)
—dashed after them. The music, all mounted, preceding the cavalry and artillery, is like a thousand

organs—slow, solemn, thunderous, and grand. In
some of the bands there were eighty pieces. Some
of these pieces were large enough for an ordinary
American 'string band'—to hide in. They sounded
with a diapason that made the air rumble, and between
their notes there was occasionally a solemn hush like
that after the summons to judgment from the trumpet
of Gabriel!

Immediately below me (for I am yet on a roof) is
the Commanding-General of Madrid—General Milan
del Bosque. I know him by his grey moustache and
beard. He is a great friend of Serrano, a member
of the Cortes, and a dashing soldier. How his sword
flashes as it salutes the flag and the officers as they
go by! Here come some cavalry, all a blaze of
scarlet. Surely, Spain has the most theatric of people!
All this seems like a tournament. Has the bull-fight
given a taste for this bloody colour? The saddle-
cloths are burning with red and gold; the facings of
the soldiers and their plumes are red or yellow. This
is the guard around the Cortes. They move. It is
a sign that all is over. Our companions on the roof
—Spanish fashion—though unknown to us, salute as
they depart. We look up and down the streets. The
stand in front of the Cortes is deserted of its functionaries, grandees, and generals; but the people push
up to see where their superiors have just been.

The Constitution is affirmed. The people and the
military have done it. We leave for home and dinner,
only to go out at night and see Madrid in another
blaze. Every house is illuminated. Gas is indispensable to a 'Democratic Monarchy'! We observe
mottoes in gas jets, like this: 'Viva la Milicia!' Ah!
The people, the volunteers, the militia, are to be
conciliated. Voilà—encore! 'Viva Constitucion De-
mocratica!' Good! When a constitution making

monarchy means democracy, and when all these ex-
penditures and galas are to placate the people; and
whereas they (the people) used to be shot down here
by dukes, kings, and grandees, by Narvaez, O'Donnell
& Co.; and whereas they are now to be pleased with
pyrotechnics and gas, therefore, hurrah for the Democratic, Republican, Monarchical Sovereignty! No
sooner are the words uttered than another illumination in gas appears to my eye: ' *Viva el Gobierno
Nacional !'* This is blazing on the Department of
the Interior.

We press through guards and people—whither?
From what we learn, there are a hundred thousand
strangers from the provinces here, to herald and
confirm—by presence, by hurrah, by gas and bull-
fighting—the new constitution! Therefore, the press
of people is on every side. We go to the main
avenue, the Prado, thence to the Salon del Prado.
Salon means a place to rest, though out of doors.
Here are seats, and all crowded. Thence, through
splendid gates, to the Gardens of Buen Retiro. We
are then in the 'Central Park' of Madrid. Here are
beautiful groves, walks, rides, and lakes! Here, yet,
we find wildness, if not wilderness. Scarcely any
woods remain about Madrid; though when Columbus
discovered America, or even as late as 1582, Madrid
was a royal residence, because the country was a good
cover for boars and bears. The arms of Madrid are
a green tree, with fruit, gules, and a bear climbing a
tree. Madrid was a cool spot, and it is yet. Charles
V. was anxious about making this place grand, and so
he made his court here at an elevation of 2500 feet
above the sea. He afterwards went into a convent to
meditate on *dust.* This elevation on a windy plateau
gave the name to Madrid; for *Majerit* in Arabic
means a current of air—a *Buenos Ayres* of dust.

In trying to reach one of the eminences which art and wealth have here decorated, apart from the dust of Madrid, we pressed on, till we found the beautiful artificial lake in the gardens of Buen Retiro. We found there temples all aflame, the shores all alight, the lake all covered with illuminated boats and caravels! We found a hundred thousand people surrounding the water, moving amidst the paths and groves, and wondering at the exquisite duplicates of the pyrotechnic temples and boats mirrored in the lake! And this was the last act in the drama of the new Constitution !

Travelling Harem (see p. 179).

CHAPTER XXII.

MORE OF MADRID AND POLITICS—THE ESCU-RIAL—MURILLO AND HIS ART.

THE scene which I have pictured in the preceding chapter may not, in one sense, have much significance at the close of the year 1869. While the reader peruses my words, there may be new revelations of Spanish politics, if not revolutions. But my anticipations are in favour of the permanency of the present established order. The Carlists will make trouble; but Prim has the army, and he will ruthlessly suppress their attempt. Isabella has but little prospect of restoration, and her son still less. Montpensier is nothing, unless the Triumvirate —Serrano, Prim, and Topete — decree his elevation. So long as Serrano is nominally at the head, and Prim really the ruler—with a Cortes whose fiercest extreme left has been already, and may be more, conciliated by delay in the election of a monarch—so long the body of the people will be content. The priests are much berated for their incivism; but I believe that the great body of them, especially those who minister in charity and kindness to the masses of the people, and who therefore have their confidence, are acquiescent in the present situation.

If these political pictures of Spain have the tint of optimism, I must plead my earnest interest in the cause of national self-government, and beg the reader to make the necessary allowances.

The fiery processes are still going on in Spain. We hope for the best, yet fear the worst. The constitution has been accepted. It is for the most part as unexceptionable for Spain as that of the United States is for us. What is most needed is a just, firm, and honest administration. On General Serrano, the hero of Alcolea, this depends. The recent discussions in the Cortes show that there is much effervescing, if not writhing, among the Republicans. They fear the appearance of Montpensier, as if royalty were already established in his person. But there must be a fiery, tempestuous ordeal before that is accomplished. If the present regency could be continued in perpetuity—without a king—it would be better. The reality of a republic would then remain, though the form were monarchical. The Progressists, Union Liberals, and Democrats, when a republic impended, said: 'Give us rather an interregnum for all time than a republic for a day.' It is a curious condition of things—is it not? I am not sure but that General Prim means the continuance of a regency; though, perhaps, Serrano may not. Last week, the discussion in the Cortes was continued with fresh excitement and rare eloquence. It was opened by a Carlist orator, Ochoa. He desired Charles VII. to be elected by a plebiscite, and with him as king he predicted a liberal rule. Here was a bow to the democratic tendencies of the time from an absolutist advocate. He was answered by a 'Democrat,' Beccara, who sustains with a peculiar solecism of nomenclature the monarchical form and the Serrano republican, revolutionary regency. He thought that Serrano only could give full force to liberty by consummating the revolution of which he was the hero. After him, Castellar, the gifted republican orator, 'took the word,' and, with his usual brilliance and fertility of historical resources, electrified the

Cortes. He analyzed and enlarged on the difficulties in the establishment of a monarchy, paid a handsome tribute to Serrano, and hoped that the revolution would be accomplished and liberty subserved without a monarchy. Then the vote was taken—194 to 45 for the Regency and Serrano.

Spain has, therefore, an Executive, under the fifth title of her organic law, with all the functions of a king, except that it cannot dissolve the Cortes or suspend its sessions. The Cortes, like other parliamentary bodies in history, takes care to perpetuate itself.

Then General Prim had *his* grand performance. The troops of Madrid, 20,000 in number, must take the oath to the constitution. They are drawn up in line in the Prado and in the promenade of the Atocha. There Prim presents himself before their regimental banners. He cries out in a high key, and he has a sonorous voice: 'You swear to guard faithfully and loyally the constitution of the Spanish monarchy, decreed and sanctioned by the Cortes of 1869?' All the officers and men enthusiastically answer, 'We swear!' Prim then rejoins: 'If this you do, God and the nation will recompense you; if not, they will call you to account!' He then fixes some ribbons on the banners to commemorate the ceremony. Then the troops are reviewed by him amidst the immense concourse. Cannon roar till long after nightfall.

The day is done! The Spanish capital is pleased with these frequent appeals to their eye and ear. The crowds kept it up till long after honest people should have been abed. Strange to say, Madrid seems never to sleep, unless by day, like the owl. The newspapers are issued about ten at night. The streets then begin to be thronged. The shrill voices of women news-venders begin. *La Correspondencia!* is sung with a Castilian ring, and seems to have the noisest vendors and the most frequent vendees. It is a little paper, badly printed, with plenty of news in short paragraphs, and the latest telegrams. It has a sort of semi-official authority, or, at least, authentic sources of information. Some twenty other papers, all small, a good many of them comically intended, called *The Cat, Don Quixote, The Padre, The Mosquito*, &c., are screeched, buzzed, hawked, and circulated about till morning. While at Madrid, at my hotel near the 'Gate of the Sun,' where men most do congregate after eleven at night, the tumult of the city concentrates and continues till daylight! It is as if night were turned into day for social, peripatetic, political, and journalistic purposes.

Two months' sojourn in Spain and much observation of her politics have made me somewhat suspicious of the permanency of the present arrangement, in some measure because parties are so indistinctly defined and combinations may be made so readily and disastrously. There are many inflammable questions. For instance, some of the bravest and most honoured of the generals are Republicans. General Prim had to say, in answer to a question in the Cortes, that if they did not swear to the constitution off their names would go from the army rolls! This was not balm to the wounds of the Republicans. Another, and the most perilous, question is that of finance. It is estimated that there will be a deficiency of many millions of dollars this year. The present government inherited from that of Isabella a large deficit. It was compelled to make new loans under difficulties. The finance minister must look the further fact in the face that his revenues are failing. I do not count this fact as at all ominous for Spain. Better have a revenue too small for the expenditure than *vice versa*. What Spain wants is economy. She never had it, nor could have it, under her previous governments. They were the results of court and military intrigues. They were fomented and fed by—*spoils*. Narvaez and O'Donnell—O'Donnell and Narvaez—these, for years past, were the expensive and disastrous oscillations of the political pendulum. The people have found out recently, by the free discussions of forum and press, what it costs, and ought to cost, to carry on a government in the Peninsula of sixteen millions of people. The disbursements have been in excess of receipts by several millions. It cost an enormous outlay to collect and disburse these sums. Figuerola opposes such extravagances. But in spite of all these financial portents, Spain is advancing. The last decade shows an increase of population of over a million. The children are better educated. It is said, that whereas at the beginning of the century only one in 340 was educated, now there is one in every fifteen. I know it is the custom of travellers to depreciate Spain. They laugh at her pretensions and ridicule her performance. They sneer at her religion, and, in their hurried transit and superficial observation in the peninsula, see nothing but poverty, laziness, and beggars; but Spain is growing. She cannot grow worse. The leaven of progress has entered the lump. The country is peaceful and orderly. I was struck with the perfect decorum everywhere, even amidst the wild excitements incident to the past few months. Her volunteer militia are a stable security to keep the peace. But this orderly condition is owing to the provincial and municipal governments. Make a note of that, ye rulers of discontented states and people! Suppose the treasury is empty, and its credit down; suppose bankruptcy stares the nation in the face; suppose Isabella, or Don Carlos, do make disaffection here and there—yet Spain, like other nations I wot of, is improving in spite of certain well-defined drawbacks. These drawbacks are the slinking, cowardly self-exile of her nobles and rich families. They have run from her troubles to spend their time and money in inglorious ease at St. Sebastian, Paris, or Biarritz. Trade, therefore, languishes in many of the large places, as at Murcia and Saragossa. Again, it is said that Spain has had to import 50,000,000 dollars worth of grain to feed her people the past year. This fact would seem incredible to one who has seen the breadth of land, yellow for the sickle, the past two months from Andalusia to Biscay. Notwithstanding the departure of her rich fugitives, her outlay for foreign bread — and other impediments — still Spain grows. Her *fête* days are still as numerous and as well attended; her public displays are still as gorgeous and imposing; her bull fights are still patronised as numerously and as noisily; her hospitals and churches are still supported abundantly; and all she wants is that fixed tranquillity which a substantial civil government, reposing on popular liberty and private right, can insure. Then she may begin a new career and grow with more blossom and fruit. Under such a rule there would be no foreign wheat imported; the waters which the Moor managed to direct into fruitful highways and byways would soon fill the land with plenty; for labour is not wanting, nor is it reluctant among the population of Spain. When she has this new order I will not reprehend the guitar and the dance. They need not be abolished, because Spain becomes more industrious and free. The tauromachian heroes and gentry might and would be a little less patronized under a better system, which would insure to industry its prompt and proper reward. The gaudily caparisoned mules—which, every Sabbath, drag out the dead horses by the score, and dead bulls by the dozen— from the arena, amidst the shouts of the populace—

might do something better than such unproductive Sabbath work. Instead of fructifying a few feet of the soil by their carcases, the horses and bulls might be utilized for more remunerative agricultural purposes.

To the end of stimulating the industry and enterprise of Spain, it has been suggested to hold a grand exhibition. The Escurial is named as the place. It would be a magnificent site, and the idea is excellent. Besides, it satisfies the Spanish desire for congregating and enjoying themselves *en bloc*.

Upon the Saturday before we left Madrid, we made a visit to the Escurial. The journey is two hours from Madrid, on the Northern Railway. It passes through a bleak and uninhabited country. The road rises until at the Escurial it is at least 2700 feet above the sea. Nothing, hardly a flock of sheep, attracts attention on the way, until at length we see a mighty mass of granite. It is an edifice with glassless windows, and at first sight seems empty of all that is alluring and comfortable. This is the Escurial, the mausoleum of Spain's dead royalty. It is monastery and cloister— palace and minster—all in one; and that one utterly and sublimely dreary! Here are entombed the mortal remains of Charles V. and Philip II., the proudest and greatest of Spanish kings. Here, under the savage shadows of the Sierra de Guadarama, stands this shadowy shell of magnificence. It is of itself a gloomy shadow of the gloomiest of Spanish potentates. So far removed from human life and its ebb and flow, it awes one by its very isolation. How lonely! How sad! I never visited a place where everything so contributed to the heaviness of the associations; everything—but the happy company with whom we made the visit. The building is a rectangular parallelogram, nearly 600 by 800 feet. It boasts of 1111 windows outside and 1562 inside, 1200 doors, 16 courts, 15 cloisters, 86 staircases, 89 fountains, 3000 feet of frescoe painting, and some 90 miles of—promenading! It was erected by the second Philip, whose sour and gloomy visage, and gray, cold eye chill you from the canvas as does that of Charles V., opposite to him, in the palace which the former built to honour and entomb the latter.

Alighting at the depôt, we find, as it is Saturday, that there is a crowd. They at once take possession of the omnibuses and carriages. We, therefore, must walk up the hill, a mile nearly, to the palace. All is rock and desolation above and around, save this walk. Happily, as the sun is out, there are on the way grateful shade trees and stone seats. The cathedral we first visit. It is enormous in size, and produces something of the awe-inspiring effect of St. Peter's. It is without the ornamentation of the latter, and it fails to satisfy the taste. The dome is fine. The paintings of value have been removed to Madrid. Indeed, why not remove all beautiful associations hence? The storms of wind and rain, the storms of war—French and civil war—the storms of bankruptcy—civil and ecclesiastical —all the storms, have burst upon this grand, gray " eighth wonder of the world;" and nothing of serene beauty or gentle repose is left, or ought to be left. The very images of the Hebrew kings in the great court; the figures in the great picture of the Judgment (great in the size of its canvas) in the church; the very coffins and urns in the vault, where are the bones of Charles V., Philip II., and their descendants, and where there is one niche still for Isabella—which she does not come forward to claim—all seem to take the prevailing dreariness, as if from the heart, soul, mind, and features of the severe and gloomy founder of the Escurial, Philip II. Paradoxically the tomb is the most cheerful of all the objects. This Pantheon, as it is called, is also the most interesting place in the Escurial. It is something to have seen the cinerary urns of great potentates. There is an *amour propre* about each person which arrogates to itself a borrowed dignity from the departed royalty. In this crypt there are six rows of niches, and six urns in each row. The death-chamber is lofty. The top is lined with black and red and other coloured marbles. It has not a very sepulchral look. It is gaudy with bronze. It is immediately under the high altar. The celebrant, when he elevates the host, raises it immediately over the dead. In descending the wide, yellow, jasper staircase, one must take care. The marble is polished and is slippery with wax from the tapers. The pathway that leads to the tomb is ever slippery; but to this tomb exceedingly so. The great chandelier, the gilt ornamentation, and the poor paintings do not impress one so funereally. Indeed, another Philip did this part of the business. The second Philip planned a tomb of the humblest dimensions and the gloomiest character. Immediately above, in the church, or in the oratories, on each side of the great altar, some of the kings and queens who are buried below appear in full effigy, kneeling before the King of Kings. The effect is more impressively serious there than in the decorated tombs below in the vault.

Three mortal hours afoot we trudged through these deserted halls. In the palace there was some relief. The rooms are hung with tapestry, rather bright; but old tapestry is never at best very cheerful. Then, there are three little gems of rooms in 'marqueterie' and gold and silver trimmings. The ladies were here in raptures. Old cabinets, old clocks, embroidered silks—all these are the ushers to the little gloomy side-rooms. Here Philip II. sat to hear mass when sick. A little door opened here into the church. It was here that he died that lingering death which has been represented as so remorseful and terrible. They show us his chair in which he sat in summer; then the one in which he sat in winter; then the chair whereon he rested his lame foot; and then the stain of the oil from its bandages. All these are shown, and we retire satisfied.

I have no room or time to tell of the many corridors and courts; of the pretty box-rows in the flower-gardens, which you may see from the windows of the palace portion of the Escurial; nor of the fountains, the pretty temple, the statues of the Apostles, some, alas! partly gone to decay, and others going entirely; or of the little palace in the garden, for the 'spoiled child' of royalty, and all its exquisite paintings and decorations. Even this did not relieve the heaviness of the granitic scene and its sombre surroundings. A library, which is to me ever a cheerful spot, where 'hourly I converse with kings and emperors'—calling their victories, if unjustly got, to a strict account, like the 'Elder Brother' of Beaumont and Fletcher—was here very repellent. All the books, dressed in their toilette of pig-skin, turn their backs from us, and refuse to show us even their titles, as it were to relieve us from the sorrowing sight of so many dead titles and names, and useless works.

Far better would it be to change this monster mass of granite and mortar from its dull mortuary purposes into something else. Where the half monkish King Philip ruled with so severe a sceptre, let the gay Spanish people—for the present by their own grace sovereign—come to a jubilee of industry within the courts of their dead oppressors. Let the wine and oil, the grain and marbles, the pictures and statues, all art, science, labour, skill, and wisdom which Spain hath, here be gathered! Aye, even then and there let the guitar twang, the castanet click, and the tambourine resound to the steps of dancing feet! No better dedication of the Escurial can be made. And all the world, of Europe at least, which regards Spain as apart from its routes and its interests, may be tempted to come, see, admire, and—invest!

As we leave the Escurial a crowd of beggars approach. One of the beggar boys addresses me in a confidential, ironical way, pointing to a tatterdemalion in a rusty cloak, hiding a skirtless body. '*That, Señor, that is, or rather is not, the son of the blind Cornelio!*' Well, as I had not known that Cornelio was thus optically afflicted, or that he lived hereabouts, or had a son at all, or that, in fact, there was any such person as Cornelio, blind or otherwise, now or ever, I asked: 'Who in the—name—of curiosity *is* Cornelio?' '*He is the false guide, Señor, against whom all are to beware!*' 'Ah! what does he do to travellers? Shut them up in dark crypts or haunted cloisters; starve them in lonely rooms, under spring locks, or under granite basements; or drop them neatly into unseen cisterns, from which cometh no sound or bubble? Or in unfrequented paths doth he plant the perfidious poniard beneath the unsuspecting rib?' '*None of these, Señor. He imposes on tourists by representing himself to be what he is not; for the veteran Cornelio, señor, the royal guide of guides of the Escurial, died in* 1863, *unmarried and childless!*' This was the climax of the Escurial desolation. I dropped a figurative tear over the childless blind Cornelio, and a peseta into the outstretched itching palm of my garrulous informant.

How pleasantly we spent the hours on returning to Madrid; how the evening wore away at the Minister's hospitable home in cheerful chat with the cordial Nuncio of the Pope, whom I had the honour there to

meet; how the next day we visited the royal palace—
but that must be told; and yet what is there new to
tell of palaces? As one parlour is like another at home;
as a face is the counterpart of that which it reflects in
the mirror, so is one palace in Europe very like to all
the rest. Tapestry, pictures, malachite, mosaic tables,
statues and candelabra; old clocks, escritoires, and
bedsteads, &c.; but there *is* something magnificent in
this palace of Madrid, as I recall it. Not the like-
ness of Narvaez, nor of the recent royal family who
lived here; not the peculiar white stone of Colmenar,
of which it is built; nor the stone statues of the royal
line, adorning the Plaza del Oriente opposite; not
because Joseph Bonaparte (of Bordentown, New
Jersey, formerly) here lived a brief time as king; not
because Wellington drove him hence after the battle
of Salamanca, and lodged here in this palace; not the
rich and precious marbles (in which, indeed, all parts
of Spain are so opulent) in floor and doorway; not
the crystal chandeliers, colossal looking-glasses; not
the mediocre frescoes, illustrating the dead majesty of
Spain; not the apotheosis of kings on the ceiling, and
the rare china work, as brittle as the apotheosis with
which the walls are decorated; not the views from the
windows—for there has been no Moor at Madrid to
decorate the hills with verdurous loveliness; not the
royal library of 100,000 volumes, to which now no
bookworm except the moth hath access; none of
these interested me, for I have been palled with their
iteration. What I looked at in this palatial museum
with wonder and interest were the splendid paintings
upon wall and ceiling, especially those in fresco, of
Spanish America. The new world—new Spain—
every part of her once proud and rich empire—here,
in a prouder day, was drawn and coloured by the hand
of genius. The commentary is to be found at this

day, in her struggle to hold the last of her trans-
atlantic possessions.

There is much beside to describe at Madrid. The
Armoury, where the mediæval age of arms is illustrated; where one may see Columbus, Cortes, and
Pizarro in their own armour, on horseback or foot, as
you may please. Then there is the convent of the
Atocha. That merits a better notice. Why? Not
because it is a convent, for, of late, convents generally
have become uninteresting in Spain. In fact, they
have been under the law mostly suppressed. But this
is a place held very sacred, for many reasons. It was
built by the confessor to Charles V. It was rich—
was once, but is not now—in gold and silver. It has
had the fate of war and pillage. To the chapel of
Atocha Ferdinand VII. came to worship. So did his
daughter, the deposed queen. She used to drive here
in great style, drawn by eight mules, with husband and
children. It was here that the attempt was made on
her life by a crazed monk with a dagger. In wandering
about its now empty cloisters—once full of Dominican
friars, and of holy zeal and life, one cannot but remark
the effect of the revolution in unsettling the reverence
for royalty. The custodians joke and smile at the
exiled queen; they laugh as they tell us the government has an inventory of this and every other religious
place in Spain. But the reverence for this place is
not entirely gone. General Prim chose it as the place
for the oath to the constitution which he administered
to his soldiers. It is regarded as an especial honour to
be buried here.

While wandering about the cloisters and walls of the
Convent of Atocha, I found the tomb of Marshal
Narvaez, Duke of Valencia, who died in April, 1868,
and whose gross, sensual, imperious features I saw the
day before yesterday in full-length frame in the Queen's

apartments at the palace.   He was a red-faced, strong-
bodied man.   He sleeps within the same walls as his
great rival, O'Donnell!   With no effigy, no monument, no display, the great minister—of Irish descent
as his name indicates—lies here interred.   Only this
inscription marks his resting-place: ' *Enteramiento del
Exmo. (Excellentissimo) Señor D. Leopoldo O'Donnell,
Duque de Tetuan (Morocco), fallecio in Biarritz el
dia 5 de Noviembre de* 1867.'   This was all.   Change
the name of the Hiberno-Iberian statesman, one
letter only, and you have O'Connell; change the *locus
in quo* of sepulture from Madrid to Ireland; and, lo!
what a contrast between the great minister of the now
exiled Queen in his almost nameless tomb, and that of
the Great Commoner of the Irish isle, honoured by a
nation of fervid hearts.   Only yesterday did I read that
after a quarter of a century, an imposing ceremonial
was performed in Glasnevin Cemetery, when the
remains of Daniel O'Connell were removed to a crypt
under the Round Tower!   A requiem mass—a fervid
oration—the presence of cardinal and bishops and
thousands of Irishmen.   These were the witnesses of
the great virtues still perpetuated in remembrance as
characterising O'Connell.   The little inscription in the
Atocha convent is all that marks the memory of
the now hated and almost despised minister of royal
spites and despotic power!

   There are many excursions to be made around
Madrid, besides the one to the Escurial.   Indeed,
after the month of June, excursions are indispensable.
There are baths in the mountains near the confluence
of the Tagus with the Cifuentes, called the Baths of
Trillo; there are baths also at Sacedon, famous from
Roman days, and now much resorted to.   You may,
in a few hours, reach La Granja, where the thermometer is about 68 degrees in mid-summer when it is

83 degrees in Madrid. This was the Queen's favourite resort. It is nearly 3000 feet above the sea, while there is still above it a mountain, La Penalara, three times that height. The scenery is Alpine. Here Philip V. built his farm-house—La Grange. Here he would live; here he abdicated and afterwards resumed the throne; here he died; and here he chose to be buried. His French affiliation made him dislike the Austrian associations of Spanish royalty, and so he would not allow his *corpus defunctum* to sleep in the company which we have seen holding their Court of Death in the Pantheon of the Escurial. Here, at La Grange, Godoy, the 'black-eyed boy,' immortalized by 'Childe Harold,' signed the treaty which made Spain, for a time, a fief of France. Here the father of the exiled Queen, in September, 1832, promulgated a decree abolishing the Salic law, and made Isabella heiress to the throne. From this little source in the mountain has thus sprung a devastating torrent of civil affliction.

I should be flagrantly unjust if I did not, before quitting Madrid, at least refer to its museum. It is, beyond doubt, the finest collection of paintings in the world. It had such patrons, such wealth, and such artists that every school of early and modern art received here warm welcome. Not for its architecture or founders did I care to visit it again and again. Architecture had better samples to show in Spain, and its founders have other more fitting monuments. But at the new dawn of art, when the finest flush of talent overspread the European horizon; when Titian, Velasquez, and Rubens were honoured in the palaces of Spain; when Murillo, Van Dyck, Claude, Canos, Paul Veronese, Teniers, and Albert Durer gave new splendour to churches and palaces; when Andalusia gave her Oriental imagination to Murillo and Velasquez—and

Raphael and Titian gave their genius to the Continent; then there was in Spain a generous enthusiasm in the encouragement of painting which has not since been known, and has never been surpassed. This Museum is for Spain what the Medici Chapel, or the Patti Palace, is for Italy. It marks the era in the world of Art—and, I may add, of letters and commerce, which the biographer of Lorenzo, the magnificent, thus analyses :—'The close of the fifteenth and the beginning of the sixteenth century, comprehend one of those periods of history which are entitled to our minutest study and inquiry. Almost all the great events from which Europe derives its present advantages are to be traced up to those times. The invention of the art of printing, the discovery of the great Western Continent, the schism from the Church of Rome, which ended in the reformation of many of its abuses and established the precedent of reform, the degree of perfection attained in the fine arts, and the final introduction of the principles of criticism and taste, compose such an illustrious assemblage of luminous points as cannot fail of attracting for ages the curiosity and admiration of mankind.'

It was a splendid ordination of Providence that Murillo and his compeers should have been contemporaneous with this dawn of Art.

When Murillo arose—the Chaucer, or rather the Spenser, of Spanish Painting—there were many to applaud and there was much to encourage, but 'the soul of Adonais, like a star,' waited long for a throne. At length Madrid erected one. It is in the Museum. There are to be found forty-six Murillos, sixty-four Velasquez, ten Raphaels, and forty-three Titians; what a company of Olympians! Are they not all 'Grand Masters'?

In another museum of San Fernando there are

several Murillos, and among them the ' Artist's Dream,' so celebrated as the representation of sleep. It is the ideal realized in the artist's own family, for the features and forms are copied from his Andalusian wife and relatives ; but the dream is beyond conception unreal and tranquil. It is the story of the Roman patrician, who dreamed of the building of the Santa Maria Church at Rome. The Virgin appears to point out the spot for its erection. There is a companion picture, painted in Murillo's *vaporoso* style, in which the lines are not so well defined, but the sweet hues and charming forms are blended as if under some spell of enchantment. Many of these pictures have had their heroic experiences. They have been prisoners of war, have been exchanged, and returned home.

But I must linger no longer at Madrid, not even with my pen and memory. I must travel, where the portly Isabella trod before, en route to France. Leaving my Murillos to be packed (copies I mean), and the charming originals in museum and chapel ; leaving my heart with those wonderful originals and with our kind friends at Madrid ; leaving the Spanish capital in a state of excitement, which scarcely ever subsides, we arrange for a night ride toward the Pyrenees, with Saragossa for our objective point.

What we passed by—at Alcala, once a proud seat of learning, fostered by Ximenes, now its light extinguished ; what at Guadalajara, so full of memories of Moors and of Mendoza—is in the dark. To sleep, perchance to dream, through those enchanted realms, every mile of which has a history or a romance, is more of a necessity than a pleasure, as most of the trains in Spain are nocturnal—to avoid the heat. We awake at Calatayud. We are on genuine Aragonese soil and in one of the genuine towns. This was the birthplace of the Roman poet Martial, but we did not

think of him.   At the depôt, and along the route, we
began to see the peasants in their native costumes.
Knee-breeches take the place of pantaloons.   Broad
brims take the place of head handkerchief, velvet
sombrero, and Phrygian cap.   Broad silken, gaudy
sashes all the men wear ; and the red kirtle and blue
boddice are worn by the Aragonese women.   They are
as picturesque as any painter could wish for, as sisters
or 'sitters' for the 'Maid of Saragossa.'

CHAPTER XXIII.

*SARAGOSSA—THE MAID—OVER THE BORDER—
OUT OF SPAIN.*

E reached Saragossa early in the morning. As we alight at the depôt and drive to the hotel, my eye glances about for the ' Maid.' It was a strictly historical optic. I saw her. My first glance was at her bronze figure at the public fountain, where, in graceful posture, she is for ever emptying water from an upturned classic pitcher. Besides, I saw her in photograph, as we rode by, in the narrow streets. I bought pictures of her firing off a cannon, while the dead lover lay near, weltering in his blood. I knew that she was an artillery-man, but I was not prepared for the anachronism of her photograph. Perhaps it was a spiritualist one. I saw her (or her descendants) carrying babies about; for the ' Maid of Saragossa' is a mother now. I saw her bearing on her head a basket of clothes to the brink of the Ebro, for a day's washing. I saw her with her face tied up as with the tooth-ache or mumps. Finally, I saw her at work in one of the cool, stony houses on the first floor, of a narrow street, with one of Wheeler and Wilson's American sewing-machines! The heroine of Saragossa, plying her plump little satined foot, and using the heroiç glance of her death-defying eye upon a Yankee sewing-machine! Well! well! all I can do, is to present her, as she was and is.

What of the maid! Is she a myth? Does she

vanish when you approach her home? Saragossa,
from Roman days—from earliest Christian days—from
the early wars with Goth and Frank and Moor—and
later in the French wars of 1707—and later still in the
Napoleonic wars, was brave to the very death and
starvation point! Free, beyond all other Spanish

The Maid of Saragossa.

provinces—having a *fuero*—or magna charta, or decla-
ration of independence of her own; having, in fact,
republican liberty, with a congress of four branches,
each a check on the others, and all jealous of monarchical prerogative and encroachment. Aragon, whose
capital Saragossa was, became renowned for the bravery,
persistence, and chivalry of her sons, in which patriotic attributes her daughters shared.

What of the 'Maid'? Byron writes that the
enemies of Spain were 'foiled by a woman's hand
before a battered wall;' and adds, in a note, that,
'When the author was at Seville, she daily walked on

the Prado, decorated with medals and orders, by command of the Junta.' We did not see her—perhaps; but we saw many of her sisters, or her progeny. We went to the battered wall, near the north-west gate. There you will see the place where she fought by her lover's side. There, when he fell, she took the flaming match. There she worked the thunderous gun! Prose fails to tell what I would. I therefore quote the poet :—

> ' Is it for this the Spanish maid, arous'd,
>   Hangs on the willow her unstrung guitar—
> And, all unsex'd, the Anlace hath espoused,
>   Sung the loud song, and dared the deed of war?
>
> Ye who shall marvel when you hear her tale,
>   Oh! had you known her in her softer hour,
> Mark'd her black eye, that mocks her coal-black veil;
>   Heard her light, lively tones in lady's bower.
> Seen her long locks, that foil the painter's power,
>   Her fairy form, with more than female grace,
> Scarce would you deem that Saragoza's tower
>   Beheld her smile in danger's Gorgon face.
>
> Her lover sinks—she sheds no ill-timed tear;
>   Her chief is slain—she fills his fatal post.
> Her fellows flee—she checks their base career;
>   The foe retires—she leads the sallying host:
> Who can appease, like her, a lover's ghost?
> Who can avenge so well a leader's fall?'

This is all very well; but it is due to truth to say that her legitimate business was to sell cooling drinks. However, she has gained, by fighting Frenchmen, an immortality for bravery in Aragon which has been shared by a few others of her sex. In the famous sieges of the city there are recounted many stories of heroic devotion like hers, and among the photographs we bring home are those of two other heroines who fought in the early wars for the safety of their homes and city.

Saragossa is a republican city. As I said on a

former occasion, on the Sunday before we came, there were 10,000 people in the Plaza del Toros to bury the crown. That was on the day when they were proclaiming the monarchy at Madrid. Along the walls, as we rode to the hotel, we saw posted in large letters, 'FUNERAL! ALL INVITED TO THE OBSEQUIES!' We noticed that the soldiers here were very plentiful. and early and late were displaying themselves in drill and otherwise. We drove round the walls; thence out into the country to the falls and canals, where the Ebro River, uniting with the waters of the Great Canal of Aragon, furnished this part of Spain with creative irrigating water power. People were gathering their grain, or directing the water into its channels of irrigation. The great plain was all green with verdure, or yellow with the ripe harvest. There are beggars in plenty; but the people, especially the peasants, are industrious and well off. In their red plaid shawls, knee breeches, wooden shoes, and independent air, we find that we are nearing the mountains: for the mountaineers of Aragon are not unknown to history for their defiant love of independence and their ability to maintain it. The 'Maid' has left many descendants worthy of her pluck.

It was hard to leave Saragossa. We were in fact driven thence by the heat, and we press ardently for the French frontiers. As we pass through Navarre and then into Biscay, the glorious Pyrenees appear in their misty mantles, and begin to fling their cool shadows from their snowy tops upon us. The fields of Navarre and Biscay are picturesque with women at work. They wear their broad hats tied close under their chin, and in violently red gowns they blaze like big animated poppies over the fields. We are yet to hear the Basque people talk! It is conceded that Adam talked Basque. The primeval guttural of the natives of Biscay is proverbial. In the other provinces, the Spaniards say, ' The Basque folks write " *Solomon* " —and pronounce it " *Nebuchadnezzar !* " ' The houses on our route still remind us of Africa; for the Moors have been here also. Where have they not been? These houses have square windows, and look forbidding beside the deep, green, cultivated valleys in which they are placed. We pass many spurs of the Pyrenees, as the tunnels indicate. Swiss cottages appear in Biscay. The country begins to lose its calcined, desolate appearance. It becomes sylvan, in its green groves and running waters, and flocks of sheep and goats. We look for Pan and his pipe. All that we have heard of these beautiful intervales is more than realized. We watch the panorama of cultivated loveliness till the evening comes on to melt all outlines into one harmoniously beautiful scene.

Then appears in a dusky light St. Sebastian, where Isabella ate her last Spanish breakfast. In fancy, if not in reality, she had bid hurried adieux to her palaces at La Granja, Aranjuez, and Madrid; and here, upon the frontier of a foreign land and a new home, she who ruled by defrauding her cousin Charles of his (so-called) right, and by the forced repeal of the ancient law of descent; she who ruled, too, in defiance of decency, of the people, and of God—here bid a long and last farewell to all her greatness. 'These are bad times for us,' she said to a friend who bid her good bye at the railway station. Compressing her lips and hiding her eyes, to conceal her depression and her tears, the same railway which bore plebeian and peasant, carried the last Spanish Bourbon over the border.

As we enter St. Sebastian, the lofty green mountain, surmounted by a splendid fort—the scene of many a battle—rises on the left. We are in the town, and amidst the noises, nurses, and children of the beautiful

promenades of this finest of watering-places. Groves
are all about us, and promenades in plenty. The dusty,
hot ride of the day is forgotten in the beautiful
prospects. Soon we pass through rocky, mountainous
defiles, and are at Hendaye, just on the border. The
Pyrenees are pierced! Our trunks are searched. A
grim Spanish Custom-house officer says to me: 'Any
cigars, caballero?' 'No.' 'Pass!—no, stop! What
is in that long box?' Now that was my box of
Murillos. It looked formidable and suspicious. It
had been taken once for a coffin. I related succinctly
that it held pictures. He doubted. He was about to
open it, when a superior officer came and stopped him.
The inferior said that he thought it might contain—
arms! Arms—to be carried out of Spain and into
France! What for? I was not able to solve this
strategico-political problem before the cars hurried us
to that beautiful spot on the Bay of Biscay—Biarritz!
I have not solved it since.

Out of Spain! Snugly ensconced in the uttermost
south-western corner of France, with the proud peaks
of the Pyrenees along the coast to the west and to the
south, with the Bay of Biscay making a right angle
here with the coast line—if one angle there is right,
where all is so crooked and rugged—and with the
sweetest refreshment of air from the sea, we begin,
here and now, to realize that we are out of Spain!
The Sunbeams of the Peninsula began to warm and
warn. Thereupon we came to the north and to the sea-
side. Looking out upon the west, with the waves slightly
curling and capped with white, the mind turns to the
home-land beyond—beyond—beyond! There are all
our hopes and loves. We seem nearer to them now than
at any time in our travels. Upon the shore are a score
or more of unshapely masses of isolated rock, under,
over, and on which the sea pounds and froths, and

under which, even on calm days, it sulks and roars. These rocks furnish a breakwater for the harbour and smooth the sanded bathing beach. These caves are the counterfeit presentment of those of the Jersey Isle, where Victor Hugo's sea-toiling hero found the devil-fish, or the devil-fish found him. Along the shore are the villas of the French nobles and millionnaires; and there, on that hill by the shore, is the newly-fashioned palace, or lodge, where the Emperor and Empress of France summer away the solstice in (perhaps) measure-less content. Beyond it still, on the route to Bayonne, is the tall, white light-house, nearly 300 feet high, from which you may perceive the slight line of yellow sand, running far out into the sea, and helping to make the harbour of Bayonne 'in the Bay of Biscay, O!' and the line of the Pyrenees, running from the sea far to the inland in one grand continuous line of loftiness!

Wandering amidst the palace pleasure-grounds by the lakes, and (by permission) through the palace itself—just fitting up, and soon to be occupied—or driving along the beautiful road to Bayonne, shaded with pines, and to the lakes in the rear of Biarritz, and through the (so-called) Bois de Boulogne; mingling with the fishermen on the shore and going with them in their little shells of boats far out into the open sea, trolling as we go, and in lieu of luck, watching the porpoises flounder, flop, and wallow like hogs—*cochons de mer*, as they are well named; strolling and clamber-ing over the straggling rocks, which are anchored to the sounding shore by the neatest of fairy-arched bridges; seeing the new faces, mostly Spanish (absent from the home perils of revolution), with nurses and babies; observing the throngs sauntering and sunning themselves along the beach, or the bathers, young and old, of both sexes, rejoicing under the sweet sun and in the glad waves—thus have the days gone by since we pierced the Pyrenees. Most delightfully; for Biarritz is by nature and art the very pearl of a summer resort. From the verandah or terrace in front of the palace not only is there the finest sea view, but the freshest air which ever salt sea gave to the famishing lung.

And yet one cannot always linger here, even amidst the Sweet Sunbeams, nor dwell on what surrounds

Pierced Rock at Biarritz.

Biarritz. Indeed, my mind has not yet been relieved of its memories of Spain. I love to be on the sand of the shore or on the grass by the lakes, and paint my castles in Spain with fancy's pencil. The political excitements which come from Paris, and which quicken the newsmongers here in this far corner of the empire, scarcely interest me—at least not yet; for the Spanish

interest is yet unabated. I should not be surprised to
hear at any time of fresh troubles and new complications, even of a sanguinary kind, in Spain. Not in
Madrid; but at Malaga, Seville, and Cadiz, where
there is much rampant dissatisfaction with the present
arrangement. No one seems to fear that the dynasty
of Louis Napoleon will be disturbed by the *émeutes* at
Paris. They are the froth of the elections. France,
more than Spain, has been solidified and compacted in
her polity and policy. Spain is yet in the nonage and
experiment of parliamentary and public liberty. An
illustration which I saw the other day in an English
paper, applied to France, has a much more apt application to Spain. The writer, after praising England
for the substantial foundations of her government,
indicates that France, owing to repeated shakings, is
still unsettled. Its component particles have never yet
found their level. There is no stratification, scarcely
even have the various elements had time to crystallize.
He does not find in France, and it is still harder to
find in Spain, a strong granite basis, the result of many
fiery processes it may be, but formed and welded
together into an indissoluble whole. Upon that kind
of foundation alone the superimposed strata lie easily
and firmly. We shall see!

From Biarritz, securely aloof from all the cares,
tremors, and ills of Spanish travel, I can cast an eye
backward, and reverently upward—to the Source of
that Light of which I have been in search—to thank
God that he has permitted me to see so much of what
is rare in nature, beautiful in art, kind in courtesy,
and sacred in worship. Only one thing doth Spain
need. Not royal pageantry; not historic memories;
not pictured or statuesque forms; not heroic qualities;

'——not the spirit of religious chivalry
In fine harmonic exaltation'—

20

not grand cathedrals, or the still Grander Presence; not mountains swelling splendidly from alluvial plains; nor seas almost encircling its Peninsular borders with a blue girdle of beauty. Let the following fable, which Ford records, illustrate the one great need of Spain, and suggest the problem which she is now attempting to solve: When San Ferdinand captured Seville from the Moor, and bore the conquest to heaven, the Virgin desired her champion to ask from the Supernal Power any favour for Spain. The King asked for a fine climate and sweet sun. They were conceded. For brave men and beautiful women. Conceded. For oil, wine, and all the fruits and goods of this teeming earth. This request was granted. 'Then, will it please the beauteous Queen of Heaven to grant unto Spain *a good government?*' 'Nay, nay! That can never be. The angels would then desert Heaven for Spain!'

That angelic advent will never happen until something nobler shall absorb the Spanish mind than the perpetual parade of nearly a quarter of a million of expensive soldiers, even though they march to the soul-inspiring, liberty-born Hymn of Riego!

THE END.

RECENT PUBLICATIONS.

I.

THE PATHOLOGY OF MIND. Being the Third Edition of the Second Part of the "Physiology and Pathology of Mind," recast, enlarged, and rewritten. By HENRY MAUDSLEY, M. D., author of "Physiology of the Mind," "Responsibility in Mental Disease," etc. One vol., 12mo, 580 pages. Price, $2.00.

SUMMARY OF CONTENTS: Sleep and Dreaming; Hypnotism, Somnambulism, and Allied States; The Causation and Prevention of Insanity: (A) Etiological; The Causation and Prevention of Insanity: (B) Pathological; The Insanity of Early Life; The Symptomatology of Insanity; Clinical Groups of Mental Disease; The Morbid Anatomy of Mental Derangement; The Treatment of Mental Disorders.

II.

THE CHEMISTRY OF COMMON LIFE. By the late JAMES F. W. JOHNSTON, F. R. S., etc., Professor of Chemistry in the University of Durham; author of "Lectures on Agricultural Chemistry and Geology"; "Catechism of Agricultural Chemistry and Geology," etc. A new edition, revised, enlarged, and brought down to the Present Time, by ARTHUR HERBERT CHURCH, M. A., Oxon., author of "Food: its Sources, Constituents, and Uses"; "The Laboratory Guide for Agricultural Students"; "Plain Words about Water," etc. Illustrated with Maps and numerous Engravings on Wood. In one vol., 12mo, 592 pages. Price, $2.00.

SUMMARY OF CONTENTS: The Air we Breathe; The Water we Drink; The Soil we Cultivate; The Plant we Rear; The Bread we Eat; The Beef we Cook; The Beverages we Infuse; The Sweets we Extract; The Liquors we Ferment; The Narcotics we Indulge in; The Poisons we Select; The Odors we Enjoy; The Smells we Dislike; The Colors we Admire; What we Breathe and Breathe for; What, How, and Why we Digest; The Body we Cherish; The Circulation of Matter.

III.

PROGRESS AND POVERTY. An Inquiry into the Cause of Industrial Depressions and of Increase of Want with Increase of Wealth: The Remedy. By HENRY GEORGE. One vol., 12mo. 512 pages. Cloth. Price, $2.00.

"I propose to seek the law which associates poverty with progress, and increases want with advancing wealth; and I believe that in the explanation of this paradox we shall find the explanation of those recurring seasons of industrial and commercial paralysis which, viewed independently of their relations to more general phenomena, seem so inexplicable."—*Extract from Introduction.*

IV.

GREAT LIGHTS IN SCULPTURE AND PAINTING. A Manual for Young Students. By S. D. DOREMUS. One vol., 12mo. Cloth. Price, $1.00.

"This little volume has grown out of a want felt by a writer who desired to take a class through the history of the great sculptors and painters, as a preliminary step to an intelligent journey through Europe."—*From Preface.*

For sale by all booksellers; or sent, post-paid, to any address in the United States, on receipt of price.

D. APPLETON & CO., Publishers, New York.

Recent Publications.

I.

A Class-book History of England.

Illustrated with numerous Woodcuts and Historical Maps. Compiled for Pupils preparing for the Oxford and Cambridge Local Examinations, and for the Higher Classes of Elementary Schools. By the Rev. DAVID MORRIS, Classical Master of Liverpool College. First American from fifteenth English edition. 1 vol., 12mo. Cloth, price, $1.25.

II.

The English Language,

AND ITS EARLY LITERATURE. By J. H. GILMORE, A. M., Professor of Logic, Rhetoric, and English, in the University of Rochester. 1 vol., 12mo. Cloth, price, 60 cents.

III.

Gems of Thought:

Being a Collection of more than a Thousand Choice Selections of Aphorisms, etc. Compiled by CHARLES NORTHEND, A. M. 1 vol., 12mo. Cloth, price, 75 cents.

IV.

Macaulay's Essays.

ESSAYS, CRITICAL AND MISCELLANEOUS. By Lord MACAULAY. In two vols., 8vo. Cloth, price, $2.50.

This is a remarkably cheap edition of Macaulay's Essays. It is printed in good style, and handsomely bound.

V.

The Spectator.

A New Edition, carefully revised. With Prefaces, Historical and Biographical, by ALEXANDER CHAMBERS, A. M. This is an *édition de luxe* of "The Spectator," being printed in large type, on choice paper, in perfect style, and bound in vellum cloth with gilt top. In six volumes 8vo. Cloth, price, $12.00.

VI.

Poems by Henry Abbey.

12mo. Cloth, gilt top. Price, $1.25.

"The stories on which the several poems are based are, many of them, well-known stories of history. They are told with much beauty of diction and with rich poetic feeling."—*Leeds Mercury.*

For sale by all booksellers; or sent by mail, post-paid, to any address in the United States, on receipt of price.

D. APPLETON & CO., Publishers, New York.

Completion of "Picturesque Europe."

NOW READY, complete in Three Magnificent Volumes.

Royal quarto, price, in half morocco, $48.00; full morocco, $54.00; morocco, extra gilt, $57.00.

PICTURESQUE EUROPE.

With 63 EXQUISITE STEEL PLATES, *and nearly* 1,000 ILLUSTRATIONS ON WOOD, *from Drawings made expressly for this Work, by Birket Foster, Harry Fenn, J. D. Woodward, and other eminent Artists.*

LIST OF ARTICLES.

Windsor.
Eton.
North Wales.
Warwick and Stratford.
The South Coast, from Margate to Portsmouth.
The Forest Scenery of Great Britain.
The Dales of Derbyshire.
Edinburgh and the South Lowlands.
Ireland.
Scenery of the Thames.
The South Coast, from Portsmouth to the Lizard.
English Abbeys and Churches.
The Land's End.
Old English Homes.
The West Coast of Ireland.
Border Castles and Counties.
Cathedral Cities.
The Grampians.
Oxford.
The West Coast of Wales.
Scotland, from Loch Ness to Loch Eil.
The West Coast of Wales.
The Lake Country.
Cambridge.
The South Coast of Devonshire.
South Wales.
North Devon.
The Isle of Wight.
Normandy and Brittany.
The Italian Lakes.
The Passes of the Alps.
The Cornice Road.
The Forest of Fontainebleau.
The Rhine.
Venice.

The Channel Islands.
The Pyrenees.
Rome and its Environs.
The Bernese Oberland.
The Rhine, from Boppart to Drachenfels.
Spain (North and Old Castile).
Auvergne and Dauphiné.
Old German Towns.
Naples.
Norway.
Spain (New Castile and Estremadura).
The Lake of Geneva.
The Frontiers of France (East and South).
North Italy.
Norway (The Sogne Fjord, Nord Fjord, Romsdal).
Spain (Cordova, Seville, and Cadiz).
The Frontiers of France (West and North).
Calabria and Sicily.
The Black Forest.
Sweden.
The Tyrol.
Gibraltar and Ronda.
Dresden and the Saxon Switzerland.
Eastern Switzerland.
Constantinople.
Belgium.
The High Alps.
Granada and the East Coast of Spain.
Russia.
The Jura.
Athens and its Environs.
Holland.
The Danube.

D. APPLETON & CO., PUBLISHERS, NEW YORK.

THE
TURNER GALLERY.

A SERIES OF

One Hundred and Twenty Engravings on Steel,

FROM THE WORKS OF
J. M. W. TURNER, R. A.

Each plate is accompanied by historical and critical remarks, compiled from authentic sources, so that the whole affords a most instructive guide to the study of Turner's unrivaled pictures.

Two folio volumes. Price, half morocco, $32.00; full morocco, $36.00.

TURNER, the world-renowned English painter, is not only acknowledged to be the greatest landscape-painter England has produced, but he is, by general consent, placed next to, if not by the side of, Claude Lorraine, the most distinguished of the great Continental masters in landscape-art. Turner's paintings, being remarkable for breadth of effect and of shadow, and brilliant representation of light, are peculiarly adapted for engraving. It is, indeed, remarkable that, although the most vivid colorist of modern times, no painter's works are so susceptible of reproduction by the graver.

THE
POET AND PAINTER:
Or, Gems of Art and Song.

An imperial 8vo volume, containing Choice Selections from the English Poets.

Superbly illustrated with Ninety-nine Steel Engravings, printed in the best manner on the page with the text.

New edition: cloth, extra, $12.00; morocco, antique, or extra, $20.00.

D. APPLETON & CO., PUBLISHERS, NEW YORK.

CLASSICAL WRITERS.

Edited by JOHN RICHARD GREEN.

16mo. Flexible cloth. - - - *Price, 60 cents.*

UNDER the above title, Messrs. D. APPLETON & Co. are issuing a series of small volumes upon some of the principal classical and English writers, whose works form subjects of study in our colleges, or which are read by the general public concerned in classical and English literature for its own sake. As the object of the series is educational, care is taken to impart information in a systematic and thorough way, while an intelligent interest in the writers and their works is sought to be aroused by a clear and attractive style of treatment. Classical authors especially have too long been regarded as mere instruments for teaching pupils the principles of grammar and language, while the personality of the men themselves and the circumstances under which they wrote have been kept in the background. Against such an irrational and one-sided method of education the present series is a protest.

It is a principle of the series that, by careful selection of authors, the best scholars in each department shall have the opportunity of speaking directly to students and readers, each on the subject which he has made his own.

The following volumes are in preparation:

ENGLISH.

MILTON.....................................Rev. Stopford Brooke. [*Ready.*
BACON......................................Rev. Dr. Abbott.
SPENSER...................................Professor J. W. Hales.
CHAUCER..................................F. J. Furnivall.

GREEK.

HERODOTUS..............................Professor Bryce.
SOPHOCLES...............................Professor Lewis Campbell.
DEMOSTHENES..........................S. H. Butcher, M. A.
EURIPIDES.................................Professor Mahaffy.   [*Ready.*

LATIN.

VIRGIL......................................Professor Nettleship.
HORACE....................................T. H. Ward, M. A.
CICERO.....................................Professor A. S. Wilkins.
LIVY...W. W. Capes, M. A.

Other volumes to follow.

D. APPLETON & CO., NEW YORK.

SCHOOLS AND MASTERS OF PAINTING.

With an Appendix on the Principal Galleries of Europe. By A. G. RADCLIFFE. 1 vol., small 8vo. Cloth, $3.00.

"The volume is one of great practical utility, and may be used to advantage as an artistic guide-book by persons visiting the collections of Italy, France, and Germany, for the first time. The twelve great pictures of the world, which are familiar by copies and engravings to all who have the slightest tincture of taste for art, are described in a special chapter, which affords a convenient stepping-stone to a just appreciation of the most celebrated masterpieces of painting. An important feature of the work, and one which may save the traveler much time and expense, is the sketch presented in the Appendix, of the galleries of Florence, Rome, Venice, Paris, Dresden, and other European collections."—*N. Y. Tribune.*

STUDIO, FIELD, AND GALLERY.

A Manual of Painting, for the Student and Amateur. With Information for the General Reader. By HORACE J. ROLLIN. 1 vol., 12mo. Cloth, $1.50.

"The work is a small one, but it is comprehensive in its scope; it is written as tersely as possible, with no waste sentences, and scarcely any waste words, and to amateur artists and art-students it will be invaluable as a hand-book of varied information for ready reference."—*N. Y. Evening Post.*

"A want which has long been felt is now filled by the issue of a manual under the title, 'Studio, Field, and Gallery.' It is a clear, practical hand-book of art, by the aid of which the student may post himself upon the various subjects relating thereto, without wading through long and intricate works on each topic. It is a most useful and practical work, one of real merit, and which will take its position as such."—*Boston Globe.*

GATHERINGS FROM AN ARTIST'S PORTFOLIO.

By JAMES E. FREEMAN. 1 vol., 16mo. Cloth, $1.25.

"The gifted American artist, Mr. James E. Freeman, who has for many years been a resident of Rome, has brought together in this tasteful little volume a number of sketches of the noted men of letters, painters, sculptors, models, and other interesting personages whom he has had an opportunity to study during the practice of his profession abroad. Anecdotes and reminiscences of Thackeray, Hans Christian Andersen, John Gibson, Vernet, Delaroche, Ivanoff, Gordon, the Princess Borghese, Crawford, Thorwaldsen, and a crowd of equally famous characters, are mingled with romantic and amusing passages from the history of representatives of the upper classes of Italian society, or of the humble ranks from which artists secure the models for their statues and pictures."—*New York Tribune.*

"'An Artist's Portfolio' is a charming book. The writer has gathered incidents and reminiscences of some of the master writers, painters, and sculptors, and woven them into a golden thread of story upon which to string beautiful descriptions and delightful conversations. He talks about Leslie, John Gibson, Thackeray, and that inimitable writer, Father Prout (Mahony), in an irresistible manner."—*New York Independent.*

D. APPLETON & CO., Publishers,

1, 3, & 5 Bond Street, New York.

EARLY CHRISTIAN
LITERATURE PRIMERS.

EDITED BY

Professor GEORGE PARK FISHER, D.D.

The "Early Christian Literature Primers" will embody, in a few small and inexpensive volumes, the substance of the characteristic works of the great Fathers of the Church. The plan recognizes four groups of works:

1. *The Apostolic Fathers, and the Apologists*, A. D. 95–180.
2. *The Fathers of the Third Century*, A. D. 180–325.
3. *The Post-Nicene Greek Fathers*, A. D. 325–750.
4. *The Post-Nicene Latin Fathers*, A. D. 325–590.

These groups are to be embraced in four books. In the first book are given exact translations of the principal works of the Apostolic Fathers and the Apologists, preceded by introductions upon the writings of the period, and by sketches of the several authors. Nearly every known author of the period is mentioned, and his place pointed out. Only genuine works, as translated from the latest critical texts, have been admitted, and of these a very large part have been brought in.

BY REV. GEORGE A. JACKSON.

THE APOSTOLIC FATHERS, AND THE APOLOGISTS.

A. D. 95–180.

CONTENTS: Introduction—The Earlier Patristic Writings—THE APOSTOLIC FATHERS—Clement of Rome—Sketch, Epistle to Corinthians, and Clementine Literature; Ignatius—Sketch, and Epistle to Romans, Ephesians, and Polycarp; Polycarp—Sketch, and Epistle to Philippians; Barnabas—Sketch, and Epistle. Associated Authors. Hermas—Sketch, and the Shepherd; Papias—Sketch, and Fragments.

THE APOLOGISTS.—Introductory Sketch—Notice, and Epistle to Diognetus; Justin—Sketch, First Apology, and Synopsis of Dialogue with Trypho; Author of Muratorian Fragment, and the Fragment; Melito—Sketch, and Fragment; Athenagoras—Sketch, Chapters from Mission about Christians, and Final Argument on the Resurrection.

In 16mo. Cloth. Price, 60 cents.

[NOW READY.]

D. APPLETON & CO., PUBLISHERS, 1, 3, & 5 BOND STREET, N. Y.

EARLY CHRISTIAN

LITERATURE PRIMERS.

EDITED BY

Professor GEORGE PARK FISHER, D. D.

IN PREPARATION.

THE FATHERS OF THE THIRD CENTURY.

CONTENTS: Introduction (A. D. 180-325), on the Influence of Origen in the East and of Cyprian in the West—Irenæus—Hippolytus—Clement of Alexandria —Origen—Methodius—Tertullian—Cyprian.

THE POST-NICENE GREEK FATHERS.

CONTENTS: Introduction (A. D. 325-750), on the Schools of Alexandria and Antioch—Eusebius of Cæsarea—Athanasius—Basil—Gregory of Nyssa—Gregory Nazianzen—Epiphanius—John Chrysostom — Theodore of Mopsuestia — Theodoret—Cyril of Alexandria—The Historians of the Fifth and Sixth Centuries.

THE POST-NICENE LATIN FATHERS.

CONTENTS: Introduction (A. D. 325-590), on the Influence of the Roman Jurisprudence upon the Latin Church Writers — Lactantius ; Hilary; Ambrose; Jerome; Augustine; John Cassian; Leo the Great; Gregory the Great; the Historians Rufinus, Sulpicius, Severus, and Cassiodorus.

D. APPLETON & CO., PUBLISHERS, 1, 3, & 5 BOND STREET. N. Y.

Appletons' Periodicals.

Appletons' Journal:

A Magazine of General Literature. Subscription, $3.00 per annum; single copy. 25 cents. The volumes begin January and July of each year.

The Art Journal:

An International Gallery of Engravings by Distinguished Artists of Europe and America. With Illustrated Papers in the various branches of Art. Each volume contains the monthly numbers for one year. Subscription, $9.00.

The Popular Science Monthly:

Conducted by E. L. and W. J. YOUMANS. Containing instructive and interesting articles and abstracts of articles, original, selected, and illustrated, from the pens of the leading scientific men of different countries. Subscription, to begin at any time, $5.00 per annum; single copy, 50 cents. The volumes begin May and November of each year.

The North American Review:

Published Monthly. Containing articles of general public interest, it is a forum for their full and free discussion. It is cosmopolitan, and true to its ancient motto it is the organ of no sect, or party, or school. Subscription, $5.00 per annum; single copy, 50 cents.

The New York Medical Journal:

Edited by FRANK P. FOSTER, M. D. Subscription, $4.00 per annum; single copy, 40 cents.

CLUB RATES.

POSTAGE PAID.

APPLETONS' JOURNAL and THE POPULAR SCIENCE MONTHLY, together, $7.00 per annum (full price, $8.00); and NORTH AMERICAN REVIEW, $11.50 per annum (full price, $13.00). THE POPULAR SCIENCE MONTHLY and NEW YORK MEDICAL JOURNAL, together, $8.00 per annum (full price, $9.00); and NORTH AMERICAN REVIEW, $12.50 per annum (full price, $14.00). APPLETONS' JOURNAL and NEW YORK MEDICAL JOURNAL, together, $6.25 per annum (full price, $7.00); and NORTH AMERICAN REVIEW $10.50 per annum (full price, $12.00). THE POPULAR SCIENCE MONTHLY and NORTH AMERICAN REVIEW, together, $9.00 per annum (full price, $10.00). APPLETONS' JOURNAL and NORTH AMERICAN REVIEW, together, $7.00 per annum (full price, $8.00). NEW YORK MEDICAL JOURNAL and NORTH AMERICAN REVIEW, together, $8.00 per annum (full price, $9.00).

D. APPLETON & CO., Publishers, New York.